Chromium Toxicity

Nitish Kumar
Editor

Chromium Toxicity

Challenges and Solutions

Springer

Editor
Nitish Kumar
Department of Biotechnology
Central University of South Bihar
Gaya, Bihar, India

ISBN 978-3-032-14322-8 ISBN 978-3-032-14323-5 (eBook)
https://doi.org/10.1007/978-3-032-14323-5

This Springer imprint is published by the registered company Springer Nature Switzerland AG
The registered company address is: Gewerbestrasse 11, 6330 Cham, Switzerland

Preface

Environmental chromium (Cr) pollution has drawn a lot of attention in recent decades. Only trivalent chromium (Cr (III)) and hexavalent chromium (Cr (VI)) are commonly encountered in the natural environment, despite the fact that chromium (Cr) is a polyvalent element that can exist in a wide range of oxidation states. The biogeochemical processes that control Cr (VI) mobilization, accumulation, bioavailability, and toxicity in soils are covered in this book. Additionally, potential threats to the ecosystem are addressed. Reduced growth and development, as well as decreased agricultural productivity and quality, are observed in plants growing in soils contaminated with Cr (VI). Furthermore, the generation of free radicals brought on by Cr (VI) exposure results in oxidative stress, which alters plant morpho-physiological and biochemical processes at the tissue and cellular levels. To ensure their survival, plants may, nevertheless, respond to Cr (VI) toxicity by developing robust cellular and physiological defense mechanisms. In order to combat Cr (VI) toxicity, plants either prevent themselves from absorbing Cr (VI) from the soil or activate their detoxifying system, which includes generating both enzymatic and non-enzymatic antioxidants to scavenge reactive oxygen species (ROS). This book also highlights new developments in remediation techniques, such as bioremediation/phytoremediation, remediation by microbes, exogenous use of organic amendments (biochar, manure, and compost), and nano-remediation supplements, which reduce potential health and environmental risks and significantly clean up Cr (VI)-contaminated soil and water. Knowledge gaps and future research needs are also discussed. By having a thorough understanding of the mechanistic basis of Cr (VI) toxicity, signaling pathways, and tolerance mechanisms, the book's observations should help develop innovative and practical strategies for limiting Cr (VI) bioavailability, toxicity, and sustainably managing Cr (VI)-polluted soils and water, thereby lowering its environmental hazards.

Gaya, India | Nitish Kumar

Acknowledgments

First and foremost, I would like to praise and thank God, the almighty, who has granted countless blessing, knowledge, and opportunity to accomplish the book project work. Thanks to all the authors of the various chapters for their contributions. It had been a bit of a long process from the initial outlines to developing the full chapters and then revising them in light of reviewer's comments. I sincerely acknowledge the author's willingness to go through this process. I also acknowledge the work and knowledge of the members of our review panels, many of which had to be done at short notice. Thanks to all the people at Springer Nature, especially Dr. Zachary Romano and Ms. Tina Silva with whom I corresponded for their advice and facilitation in the production of this book.

Gaya, India Nitish Kumar

Contents

Sustainable Mitigation Strategies and Solutions

About the Editor

Dr. Nitish Kumar is an associate professor at the Department of Biotechnology, Central University of South Bihar, Gaya, Bihar, India. Dr. Kumar completed his doctoral research at the Council of Scientific and Industrial Research–Central Salt and Marine Chemicals Research Institute, Bhavnagar, Gujarat, India. He has published more than 50 articles in leading international and national journals, more than 25 book chapters, and 20 books. He has a wide area of research experience in the field of sustainable agriculture and development. Dr. Kumar is a recipient of the Young Scientist Award from the Science and Engineering Research Board (SERB) in 2014. He has received many awards/fellowships/projects from various prestigious government organizations like CSIR, DBT, ICAR and SERB-DST, BRNS-BARC, among others. He is a reviewer for various international journals.

Chromium Toxicity: Source, Distribution, Challenges, and Remediation

Md. Kamruzzaman, Rysul Hasan Hredoy, Farjana Jannat Shimo, Md. Shihab Uddine Khan, Md. Sadiqul Amin, Sadia Afroz Ritu, and Shihab Uddin

Abstract Chromium (Cr), a transition metal of significant industrial and environmental relevance, primarily exists in two oxidation states: the relatively stable Cr(III) and the highly toxic, mobile, and carcinogenic Cr(VI). This chapter provides an interdisciplinary synthesis of Cr's environmental behavior, toxicity mechanisms, primary sources, and remediation strategies. Naturally found in minerals such as chromite and fuchsite, Cr is released into the environment through both geogenic and anthropogenic activities, most notably from leather tanning, electroplating, pigment manufacturing, textile dyeing, and stainless steel production. Complex redox reactions, pH-Eh conditions, organic complexation, colloidal transport, and microbial processes control its mobility and speciation. Cr(VI), due to its high solubility and bioavailability, poses acute ecological and public health risks, including oxidative stress, DNA damage, and apoptosis mediated by reactive oxygen species. Human exposure can occur through inhalation, ingestion, dermal contact, and even transplacental routes, affecting multiple organ systems, including the respiratory, hepatic, renal, immune, and reproductive systems. Ecologically, Cr disrupts plant physiology, microbial nutrient cycling, and aquatic food webs via bioaccumulation and trophic transfer. Case studies from Bangladesh, India, China, Peru, and Brazil underscore the socio-environmental dimensions of Cr pollution, including groundwater contamination, occupational health risks, and environmental justice concerns. The chapter

---

Md. Kamruzzaman · F. J. Shimo · Md. S. Amin
Soil, Water and Environment Discipline, Khulna University, Khulna, Bangladesh

Md. Kamruzzaman · F. J. Shimo · S. Uddin (✉)
Department of Soil Science, Faculty of Agriculture, Bangladesh Agricultural University, Mymensingh, Bangladesh
e-mail: shihab.ss@bau.edu.bd

R. H. Hredoy
Department of Environmental Science, Bangladesh University of Professionals, Dhaka, Bangladesh

Md. S. U. Khan
Agricultural Research Station, Bangladesh Agricultural Research Institute, Satkhira, Bangladesh

S. A. Ritu
Department of Crop Botany, Bangladesh Agricultural University, Mymensingh, Bangladesh

N. Kumar (ed.), *Chromium Toxicity*, https://doi.org/10.1007/978-3-032-14323-5_1

evaluates both conventional and emerging remediation approaches, such as chemical reduction, precipitation, adsorption using biochar and activated carbon, microbial transformation, and phytoremediation, while emphasizing that their efficacy depends on site-specific geochemical and speciation factors. By integrating geochemical, toxicological, and ecological perspectives, this chapter offers a systems-level understanding of Cr toxicity and sustainable management strategies.

Keywords Chromium contamination · Textile dyeing · Chromium dynamics · Human exposure · Phytoremediation

1 Introduction

Chromium (Cr) is a naturally occurring element in the Earth's crust, commonly found in two environmental oxidation states: trivalent chromium (Cr(III); Cr^{3+}) and hexavalent chromium (Cr(VI); Cr^{6+}) (Genchi et al., 2021; Uddin et al., 2023). Cr(III) is an essential trace element involved in glucose, lipid, and protein metabolism (Genchi et al., 2021; Monga et al., 2022; Shadreck & Mugadza, 2013). It contributes to insulin function and may regulate glycemic levels, triglycerides, and cholesterol (Ray & Jankar, 2022). In contrast, Cr(VI) is highly toxic, carcinogenic, mutagenic, and teratogenic, with no known biological function (Monga et al., 2022). It is more mobile and readily penetrates cell membranes compared to Cr(III), leading to severe health effects, including liver and kidney damage, skin lesions, and various cancers (Mishra & Bharagava, 2016; Ray & Jankar, 2022; Shadreck & Mugadza, 2013). Cr(VI) contamination affects soil and water ecosystems, reduces plant growth and yield, alters metabolic activity, and enters the food chain (Dotaniya et al., 2014; Kapoor et al., 2022). It also disrupts key microbial processes such as carbon mineralization and nitrogen cycling (Das & Mishra, 2008; Dotaniya et al., 2014).

Even at low concentrations, Cr can bioaccumulate in human and animal tissues, causing long-term health impacts (Kapoor et al., 2022; Prasad et al., 2021). Exposure primarily occurs through ingestion, dermal contact, and inhalation of contaminated soil, water, or air (Pokhrel & Pokhre, 2022). Various remediation methods exist, but selecting an effective, sustainable, and economically viable solution remains a challenge. This situation demands a comprehensive examination of Cr sources, distribution patterns, associated risks, and remediation strategies (Kapoor et al., 2022; Prasad et al., 2021).

Anthropogenic activities such as leather tanning, electroplating, stainless steel production, textile dyeing, and chemical manufacturing release high concentrations of Cr into the environment (Das & Mishra, 2008). Improper waste management has led to the widespread Cr(VI) contamination of terrestrial and aquatic systems (Mishra & Bharagava, 2016). Its high solubility and oxidative potential facilitate rapid environmental transport and biological uptake. Climate change, particularly flooding, further intensifies this risk by remobilizing previously stable contaminants (Bolan et al., 2024; Ponting et al., 2021). Flooding alters soil redox potential, pH, dissolved organic matter (OM), and microbial activity, thereby affecting the availability and mobility of toxic elements, which makes Cr's environmental behavior increasingly unpredictable.

Environmental pollution disproportionately affects vulnerable populations in low- and middle-income countries (LMICs), raising critical concerns about environmental justice (Levy et al., 1991). Industrial activities in LMICs expose hundreds of millions of people to health risks from toxic pollutants, with Cr being a major contributor (Friedrich, 2016). The environmental justice movement emphasizes equitable protection from environmental hazards and highlights the need for cross-sectoral collaboration across science, health, policy, and society (Claudio, 2007).

This chapter explores the journey of Cr from its source to environmental sinks and its interactions with biological systems. It examines the mechanistic pathways of Cr toxicity and exposure, linking scientific understanding to practical remediation strategies to ensure environmental sustainability and protect public health.

2 Chemical Properties of Chromium

Chromium is a hard, lustrous, steel-gray metal with the atomic number 24. It belongs to Group 6 of the periodic table and is classified as a transition metal (Focardi et al., 2013; Teklay, 2016). Chromium is both industrially valuable and environmentally significant due to its distinct physicochemical properties. It has a high melting point (1907 °C) (Gasik et al., 2020) and a body-centered cubic lattice structure, which provides corrosion resistance and high hardness (Li et al., 2017). One of the most notable characteristics is its ability to form stable and colorful compounds across a wide range of oxidation states, from − 2 to + 6 (Leimbach et al., 2019), which contributes to its prevalence in both natural systems and industrial processes.

The redox behavior and environmental mobility of Cr are regulated by its chemical speciation, which is highly dependent on pH and redox potential (Eh) (Avudainayagam et al., 2003). Under oxidizing conditions, Cr^{3+} can be converted to the more toxic and mobile Cr^{6+} form:

$$2\,Cr^{3+} + 3\,H_2O_2 + 10\,OH^- \rightarrow 2\,CrO_4^{2-} + 8\,H_2O$$

Conversely, under reducing conditions, such as in the presence of Fe^{2+} or OM, Cr^{6+} is reduced back to the more stable Cr^{3+}:

$$Cr_2O_7^{2-} + 6\,Fe^{2+} + 14\,H^+ \rightarrow 2\,Cr^{3+} + 6\,Fe^{3+} + 7\,H_2O$$

Cr(VI) is 100 times more toxic than Cr(III) due to its higher water solubility and mobility (Saha et al., 2011) and is estimated to be 10–100 times more toxic via oral exposure and 100–1000 times more cytotoxic (Katz & Salem, 1993). Despite its

Table 1 Comparative overview of chromium forms across environmental compartments

Parameter	Cr(III)	Cr(VI)
Common forms	Cr^{3+}, $Cr(OH)_3(s)$, Cr-organic complexes	$CrO_4{}^{2-}$, $Cr_2O_7{}^{2-}$
Solubility	Low (precipitates as $Cr(OH)_3$)	High (soluble anions)
Mobility	Low to moderate	High to very high
Biological uptake	Requires chelation/complexation; slow uptake	Rapid uptake via anion transporters
Toxicity	Low; essential at trace levels	High; mutagenic, carcinogenic

toxicity, Cr is widely used in various industries, including stainless steel manufacturing and chrome plating, due to the stability of Cr^{3+} oxides under typical environmental conditions (Barnhart, 1997). Furthermore, Cr(III) can form soluble organo-complexes, which may enhance its mobility under specific environmental scenarios (Puzon et al., 2005). The Cr behavior in the environment is presented in Table 1.

Due to its exceptional corrosion resistance and ability to enhance surface hardness (HV0.11700 after heat treatment) (Li et al., 2017), Cr is extensively used in metal plating, particularly in chrome plating, to improve both the durability and aesthetic appeal of metal products (Lindsay Jr, 1999). Beyond electroplating, Cr has wide industrial applications, including leather tanning, textile dyeing, pigment production, wood preservation, catalysis, and chemical manufacturing (Dotaniya et al., 2014; Mishra & Bharagava, 2016). Cr(III), in the form of Cr picolinate [$Cr(C_6H_4NO_2)_3$], is also marketed as a dietary supplement, although its efficacy and safety remain subjects of scientific debate. In electrodeposition, Cr's stereochemistry, electrochemistry, and redox reactions are critical for process optimization. Its versatile chemistry also underpins its value in catalysis, particularly in reactions such as ethylene polymerization.

3 Environmental Source, Occurrence, and Behavior of Chromium

Chromium enters the environment through both natural and anthropogenic processes (Gutterres & Mella, 2014). Naturally, it is released from the weathering of Cr-bearing rocks and minerals, as well as volcanic eruptions. However, anthropogenic activities such as leather tanning, electroplating, stainless steel production, textile dyeing, and the improper disposal of industrial effluents significantly contribute to the Cr contamination.

3.1 Natural (Geological) Sources of Chromium

Chromium concentrations in the Earth's crust typically range from 100 to 300 μg g^{-1} (Focardi et al., 2013). It commonly occurs in association with iron and other metal oxides (Barnhart, 1997). The most commercially important Cr ore is chromite ($FeCr_2O_4$), which can contain up to 55 wt% Cr_2O_3 (Matrosova et al., 2020).

3.1.1 Major Chromium-Bearing Minerals

Chromite ($FeCr_2O_4$)

Chromite is a primary source of Cr and forms in layered mafic intrusions during the crystallization of high-temperature magma, as observed in large igneous complexes such as the Bushveld (South Africa), Great Dyke (Zimbabwe), and Stillwater (USA) Complexes (Barnes et al., 2020, 2023). It crystallizes in cotectic proportions with silicate minerals, such as olivine and pyroxene (Barnes et al., 2023). Although chromite deposits are often irregular in shape, they are economically significant, particularly in regions such as Albania, Turkey, and Oman (Ali & Khan, 2013; Steblez, 1994). According to Barnes et al. (2023), chromite and silicates initially co-precipitate but later separate due to the differences in viscous flow, resulting in the formation of distinct chromite layers. Owing to its high specific gravity and chemical durability, chromite is resistant to weathering and often accumulates in heavy mineral sands (Gujar et al., 2010). However, mining and processing activities can lead to the release of Cr(VI), making the management of chromite-bearing waste a significant environmental concern.

Crocoite ($PbCrO_4$)

Crocoite is not typically abundant enough to pose a direct environmental concern; however, it serves as a valuable geochemical indicator of Cr(VI) formation. Its presence suggests that Cr has undergone oxidative mobilization, a process associated with environmental Cr contamination (Fendorf et al., 2000). Crocoite is a striking mineral, known for its brilliant orange-red color and monoclinic crystal structure (Cooper et al., 2011; Crane et al., 2001). It is a rare secondary mineral that forms under specific geochemical conditions, where chromate-rich fluids interact with lead-bearing minerals such as galena (PbS) (Bao et al., 2021; Keim & Markl, 2015). These conditions are uncommon, as they require (i) a source of CrO_4^{2-}, which is limited in most geological settings due to prevailing reducing conditions (Guo et al., 2024; Hausladen et al., 2018), and (ii) sufficient lead in an oxidizable form capable of releasing both components. Notable occurrences of crocoite include the Dundas region in Tasmania (Cook, 2007) and the Adelaide mine in Australia (Jang & Topal, 2020), with smaller deposits reported in the Ural Mountains (Russia), Brazil, and the United States (Yur'Evich, 2022).

Fuchsite [K(Al,Cr)$_2$(AlSi$_3$O$_{10}$)(OH)$_2$]

Fuchsite is a green, Cr-rich variety of muscovite mica, with its distinctive coloration resulting from the substitution of Cr^{3+} for Al^{3+} in the octahedral sites of the muscovite crystal lattice (Randive et al., 2015). It typically forms under low- to medium-grade metamorphic conditions in Cr-bearing mafic and ultramafic rocks subjected to regional metamorphism (Argast, 1995; Raase et al., 1983). Commonly associated with phyllites, schists, and quartzites in greenschist and amphibolite facies, fuchsite is also found in talc-carbonate rocks and serpentinites (Austrheim et al., 2021). Unlike the toxic Cr(VI), the Cr(III) incorporated into fuchsite is environmentally benign and structurally stable. As such, fuchsite serves not only as a mineralogical indicator of Cr-rich metamorphic conditions but also provides insight into the redox stability and geochemical cycling of Cr in natural systems. Additionally, it holds aesthetic and commercial value, particularly in ornamental stones, such as in "ruby-in-fuchsite" assemblages where it coexists with corundum (Martyn & Johnson, 1986). Overall, fuchsite represents a key pathway for the natural immobilization of Cr within silicate mineral structures.

Vauquelinite [Pb$_2$Cu(CrO$_4$)(PO$_4$)(OH)]

Vauquelinite is a rare but chemically complex secondary mineral composed of chromate, phosphate (PO_4^{3-}), and often arsenate (AsO_4^{3-}) groups (Frost, 2004). It typically forms in lead-bearing hydrothermal ore deposits under oxidizing conditions (Southwood & Viljoen, 1986). Its generalized formula is $Pb_2MeII(Z)(XO_4)(YO_4)$, where Me^{2+} represents a bivalent cation such as Cu^{2+} or Mn^{2+}, Z is a hydroxyl group or water molecule, and X and Y can include S, Cr, V, P, or As (Fanfani & Zanazzi, 1967). Vauquelinite commonly occurs in association with other secondary lead minerals, including crocoite, pyromorphite, mimetite, and vanadinite (Inegbenebor et al., 1989). The presence of chromate (CrO_4^{3-}) groups in its structure indicates the oxidative mobilization of Cr^{3+} to Cr^{6+}, which typically occurs in oxygen-rich environments where oxygenated groundwater (GW) systems are present (Kazakis et al., 2015; Oze et al., 2007). As such, vauquelinite serves as a mineralogical indicator of oxidizing conditions and the geochemical pathways involved in Cr transformation.

3.2 *Anthropogenic Sources of Chromium*

Chrome Plating

Chrome plating is a metal finishing process in which a thin layer of Cr is electrochemically deposited onto metal or plastic substrates to enhance corrosion resistance, surface hardness, and aesthetic appeal. There are two main types: decorative chrome plating, which involves the deposition of an ultra-thin Cr layer (0.02–0.05 mil) over a nickel base to provide a bright, reflective surface, commonly used in automotive trim, household fixtures, tools, and furniture (Beacom et al., 1964; Lindsay, 1999); and hard chrome plating (or industrial chrome), which applies significantly thicker coatings ($\geq$ 0.9 μm) for enhanced water resistance and reduced friction, making it suitable for hydraulic cylinders, engine components, molds, and industrial tools (Protsenko et al., 2011). In addition to its mechanical advantages, Cr plating resists tarnishing, maintains surface integrity under thermal stress, and performs effectively in both aesthetic and high-performance contexts. A summary of Cr's functional benefits alongside its associated toxicological risks is presented in Table 2.

As Alloying Agent on Stainless Steel

Chromium is an essential alloying element in stainless steel, typically constituting at least 10.5% of the alloy by weight (Pariser et al., 2018). Its primary function is to form a passive, self-healing oxide layer (Cr_2O_3) on the steel surface upon exposure to oxygen, which imparts excellent corrosion and oxidation resistance, even under harsh conditions such as marine or industrial environments (Rao et al., 2003). This protective layer is thin, adherent, and stable, effectively preventing rust and surface

Table 2 Dual-edged utility: industrial benefits and toxicological concerns of Cr usage

Anthropogenic sources	Cr form used	Purpose/benefits	Toxicity concerns
Chrome plating	Cr^{6+} compounds (e.g., chromates)	Provides corrosion resistance, smooth finish, and wear resistance in tools, fixtures, and auto parts	Generates toxic Cr^{6+} waste; occupational exposure risks
Stainless steel alloying	Metallic Cr (oxidizes to Cr_2O_3)	Improves corrosion resistance, strength, and heat tolerance in steel	Minimal direct concerns; steelmaking processes can have CO_2 emissions
Leather tanning	Cr(III) sulfate	Stabilizes collagen, producing soft, durable, water- and heat-resistant leather	Improper waste disposal contaminates water and soil
Textile dyeing	Cr^{6+} compounds (e.g., dichromates)	Fixes dyes to fabric, improves colorfastness, and brightness	Cr^{6+} toxicity; increasing demands for safer, eco-friendly mordants
Pigment production	Cr_2O_3, $PbCrO_4$	Produces stable, UV-resistant green/yellow pigments for paints, plastics, ceramics	Lead and Cr^{6+} toxicity
Wood preservation	Cr^{6+} in chromated copper arsenate	Protects wood from rot, insects, and fungi	Leaching of Cr, Cu, and As into the environment
Catalysis	Cr_2O_3 and related oxides	Catalyzes dehydrogenation, polymerization (polyethylene, polypropylene production)	Risk of toxic Cr residues

degradation. Cr's presence also enhances the mechanical strength and aesthetic appearance of stainless steel, contributing to its widespread use in kitchenware, medical instruments, architectural structures, construction, transportation (automobile and aerospace), and critical components in nuclear reactors. Additionally, the clean, lustrous finish of Cr-alloyed steel makes it highly desirable for decorative and architectural applications.

Leather Tanning

Chromium salts, particularly Cr(III) sulfate [$Cr_2(SO_4)_3$], are widely used in leather tanning to stabilize collagen fibers and enhance the durability of leather (Zhang et al., 2018). Through coordination chemistry, Cr(III) forms stable complexes with the carboxyl groups of collagens, increasing the leather's resistance to heat, water, and microbial degradation (Brown et al., 2016). In leather tanning, Cr(III) forms coordinate bonds with the carboxyl groups of collagens:

$$\left[\mathrm{Cr(H_2O)_6}\right]^{3+} + \mathrm{collagen} \rightarrow \left[\mathrm{Cr(collagen)}\right] + 6\,\mathrm{H_2O}$$

Due to its efficiency and ability to produce soft, flexible leather in a relatively short time, chrome tanning has largely supplanted traditional vegetable tanning methods. It now dominates the global leather processing industry (Abbott et al., 2015).

Textile Dyeing

In textile dyeing, Cr compounds are employed as mordants—substances that fix dyes to fabrics, such as wool and silk, to improve colorfastness and enhance the brightness of certain natural and synthetic dyes (Islam et al., 2016; Manian et al., 2016). During the dyeing process, the Cr^{6+} from potassium dichromate is reduced to Cr^{3+}, which subsequently forms coordination complexes with dye molecules and functional groups (e.g., amino or carboxyl) on the textile fibers. The underlying redox transformation is typically represented as

$$\mathrm{Cr_2O_7^{2-}} + 6\,\mathrm{e^-} + 14\,\mathrm{H^+} \rightarrow 2\,\mathrm{Cr^{3+}} + 7\,\mathrm{H_2O}$$

Wood Preservation

Chromium, in combination with copper and arsenic, is a key component of chromated copper arsenate (CCA), a wood preservative used to protect against weathering (Morais et al., 2021). In CCA, Cr^{6+} acts as an oxidizing agent that chemically bonds Cu^{2+} and As^{5+} to wood lignin and cellulose, thereby "fixing" them in place; a simplified reaction is $2CrO_4{}^{2-} + 3Cu^{2+} + 2AsO_4{}^{3-} + H_2O \rightarrow$ CCA complex (fixed in wood). Despite this fixation, CCA-treated wood can leach Cr, Cu, and As into surrounding soil and water bodies (Coles et al., 2014); consequently, its use has been restricted or banned in residential applications in many countries.

3.3 *Geological and Biogeochemical Processes of Chromium Occurrence*

Geological and Geochemical Processes

Geological processes, such as erosion, fracturing, rock alteration, and tectonic uplift, can expose Cr-bearing minerals to environmental conditions. Under these conditions, the natural weathering of peridotites and serpentinites can release substantial Cr into soils and sediments, often exceeding 4000 ppm (Chrysochoou et al., 2016; Oze et al., 2004). Tectonic deformation and faulting further enhance water–rock interactions, accelerating mineral breakdown and increasing Cr bioavailability over geologic timescales. Mobility is controlled by various factors, including pH, Eh, organic complexation, and colloidal transport. While Cr^{3+} typically binds strongly to soils, Cr^{3+}-organic complexes exhibit reduced sorption and higher mobility (Cao et al., 2011). The environmental pathway of Cr, from natural and anthropogenic sources to its sinks and biological impacts, is summarized in Fig. 1, illustrating key transformations, mobility processes, and exposure routes that shape its fate in ecosystems.

pH-Dependent Reactions (Hydrolysis and Adsorption/Desorption)

Alkaline pH conditions favor Cr^{3+} mobility by reducing its adsorption to quartz sand, whereas acidic conditions enhance adsorption to the soil particles through protonation (Bartlett & Kimble, 1976; Ju et al., 2022). Adsorption generally increases with pH due to the formation of insoluble hydroxides and cation exchange–adsorption of hydrolyzed Cr(III) species (Griffin et al., 1977; Xu et al., 2020). The hydrolysis reaction, $Cr^{3+} + 3H_2O \rightleftharpoons Cr(OH)_3(s) + 3H^+$, remains shifted toward dissolved Cr^{3+} at low pH, maintaining its mobility. In contrast, higher pH favors precipitation as $Cr(OH)_3$, thereby reducing its aqueous concentration. However, the presence of organic acids or humic substances can form soluble complexes with Cr^{3+}, preventing precipitation even under alkaline conditions and sustaining its mobility in solution.

Redox Reactions and Chromium Dynamics in the Environment

The oxidation of Cr^{3+} to Cr^{6+} is a dynamic and reversible process, occurring primarily under oxidizing conditions. In soil, MnO_2 is one of the most important natural oxidants of Cr^{3+} (Hai et al., 2020), abundant in well-aerated soils and sediments, following the reactions:

$$2\,Cr^{3+} + 3\,MnO_2(s) + 8\,H^+ \rightarrow 2\,CrO_4^{2-} + 3\,Mn^{2+} + 4\,H_2O$$

This reaction is strongly pH-dependent and favored under acidic conditions. Although oxidation can also proceed theoretically via molecular oxygen. The reaction follows:

$$4\,Cr^{3+} + O_2 + 10\,H_2O \rightarrow 4\,CrO_4^{2-} + 20\,H^+$$

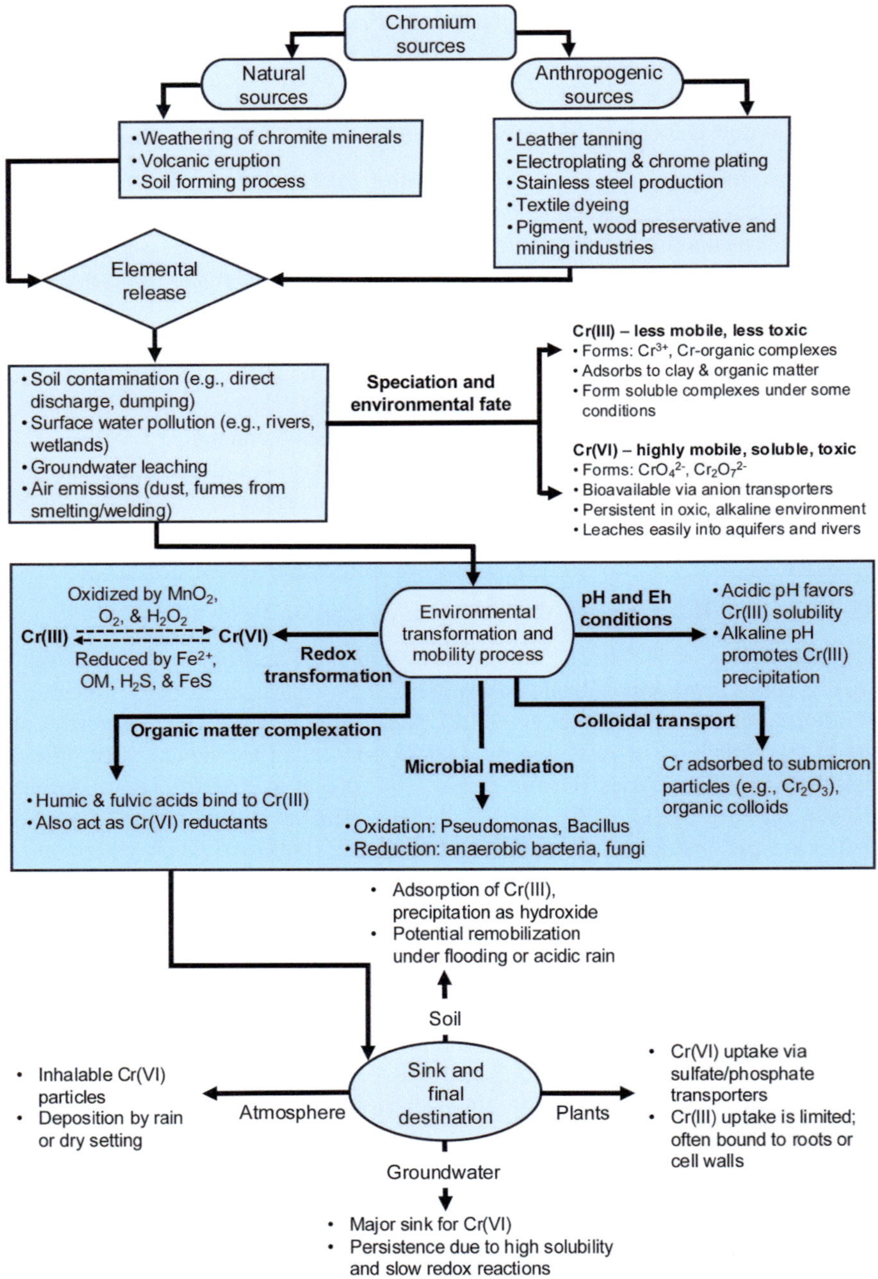

Fig. 1 Source-to-sink dynamics: chromium's environmental journey

However, this pathway is rarely significant under ambient conditions. Hydrogen peroxide, often present in contaminated or industrial environments, can also oxidize Cr^{3+} to Cr^{6+} (Rock et al., 2001). The reaction in alkaline media is expressed as

$$2\,Cr^{3+} + 3\,H_2O_2 + 10\,OH^- \rightarrow 2\,CrO_4^{2-} + 8\,H_2O$$

Furthermore, microbial oxidation of Cr^{3+} has been observed, especially under aerobic conditions. Certain bacteria, such as species of *Pseudomonas* (Wu et al., 2005) and *Bacillus* (Murray & Tebo, 2007), can enzymatically catalyze the oxidation of Cr^{3+} using oxygen or other electron acceptors. In general, the biological oxidation can be denoted as

$$Cr^{3+} + [O]_{bacteria} \rightarrow CrO_4^{2-}$$

Conversely, the reduction of Cr^{6+} to Cr^{3+} is a crucial process in remediation efforts (Chirwa & Molokwane, 2011). Organic matter plays a significant role in this process. Humic and fulvic acids contain phenolic and carboxylic functional groups that act as electron donors (de Melo et al., 2016; Ratasuk & Nanny, 2007). A general reduction reaction involving OM is

$$CrO_4^{2-} + 3\,e^- + 4\,H^+ \rightarrow Cr^{3+} + 4\,H_2O$$

Fe^{2+} ions in anoxic or waterlogged soils act as a powerful reductant of Cr^{6+} (Loyaux-Lawniczak et al., 2001), as do the sulfide species (H_2S, FeS). This pathway is especially relevant in submerged soils, wetland sediments, and contaminated aquifers.

$$CrO_4^{2-} + 3\,Fe^{2+} + 4\,H_2O \rightarrow Cr^{3+} + 3\,Fe^{3+} + 8\,OH^-$$

$$CrO_4^{2-} + 3\,H_2S + 4\,H^+ \rightarrow Cr^{3+} + 3\,S^0 + 4\,H_2O$$

$$CrO_4^{2-} + 3\,FeS + 8\,H^+ \rightarrow Cr^{3+} + 3\,Fe^{2+} + 3\,S^0 + 4\,H_2O$$

Organic Matter Complexation

Under natural environmental conditions, Cr(III) readily forms stable complexes with functional groups present in OM, including carboxyl (– COOH), hydroxyl (– OH), and phenolic groups (Gustafsson et al., 2014; Hao et al., 2022). These interactions lead to the formation of soluble Cr(III)-organic complexes, particularly with low molecular weight organic acids such as citric, oxalic, and malic acids (Chen et al., 2018). Such complexation enhances the mobility and bioavailability of Cr(III) in the environment, particularly in organic-rich soils and waters.

Humic acids, due to their abundant aromatic and aliphatic structures, exhibit a high affinity for metal cations. When Cr^{3+} interacts with humic substances, it forms stable coordination complexes that are more soluble than simple inorganic Cr^{3+} species (Gustafsson et al., 2014). This interaction not only increases the solubility of Cr(III) but also promotes the dispersion of Cr(III)-bearing colloids by preventing aggregation through electrostatic and steric stabilization mechanisms (Furukawa et al., 2009). Consequently, Cr(III) becomes more mobile in both terrestrial and aquatic ecosystems. In addition to complexation, humic substances also contribute to Cr redox dynamics. They can serve as natural reductants of Cr(VI), thereby playing a dual role in both detoxification and mobility enhancement. This insight highlights the critical role of OM not only in controlling the speciation and transport of Cr(III) but also in mitigating the environmental risks associated with Cr(VI) contamination.

Colloidal Transport

Colloids are highly reactive carriers that facilitate the long-distance transport of contaminants in the environment. Chromium can associate with mineral or organic colloids, and colloid-bound Cr(III) remains suspended, allowing it to move through both soil and GW systems. Submicron Cr_2O_3 particles have been detected in alkaline soils and GW (Zhu et al., 2023), indicating that even ostensibly insoluble forms of Cr(III) can exhibit notable mobility when incorporated into colloidal carriers.

3.4 *Environmental Occurrence*

Soil, Water, and Air

The environmental fate of Cr is primarily governed by its chemical speciation, which is influenced by redox conditions, pH, and its affinity for various environmental matrices. The different oxidation states of Cr exhibit markedly distinct solubility, mobility, and biological uptake behaviors. Cr(VI), commonly found as chromate (CrO_4^{2-}) or dichromate ($Cr_2O_7^{2-}$) under oxidizing and alkaline conditions, structurally mimics essential anions such as sulfate (SO_4^{2-}) (Bouhadi et al., 2023) and phosphate (PO_4^{3-}) (Gheju & Balcu, 2017). This mimicry facilitates Cr^{6+} uptake into plant root cells through anion transporters (Guo et al., 2024). In contrast, under neutral to alkaline pH conditions, Cr^{3+} predominantly precipitates as Cr hydroxide [$Cr(OH)_3$], thereby limiting its mobility and bioavailability (Papassiopi et al., 2014). Due to high solubility and negative charge, Cr^{6+} is highly mobile, readily leaching into aquatic systems as chromate or dichromate anions. Additionally, redox cycling of Cr in the environment is mediated by interactions with Fe(II)-bearing minerals, manganese oxides, and organic matter, which influence its speciation and environmental persistence.

GW flow is a key factor controlling the transport rate of Cr species through the aquifer system. Aquifer porosity, which defines the volume of void spaces, directly influences the storage and movement of water and dissolved contaminants; higher porosity facilitates greater fluid movement, whereas lower porosity slows migration and enhances retention. Hydrological events, such as heavy rainfall, flooding, irrigation return flows, and GW pumping, can further mobilize Cr, especially in urban-industrial landscapes. Elevated concentrations of Cr(VI), typically ranging from 5 to 70 $\mu g\ L^{-1}$, are commonly observed in shallow, porous aquifers with low GW velocities (Kazakis et al., 2015). In environments rich in organic complexing agents, Cr(III) may exist in colloidal or soluble forms, potentially increasing its mobility. Chromium can persist in sediments and GW for decades, undergoing dynamic redox transformations governed by pH, OM, and microbial activity (Losi et al., 1994). In wetland ecosystems, Cr speciation is controlled by complex processes, including diffusion, precipitation, complexation, and redox reactions (Masscheleyn et al., 1992). Under anaerobic conditions, Cr^{6+} can be reduced by OM or ferrous iron (Zha et al., 2024).

$$Cr_2O_7^{2-} + 6\,Fe^{2+} + 14\,H^+ \rightarrow 2\,Cr^{3+} + 6\,Fe^{3+} + 7\,H_2O$$

The resulting Cr^{3+} may precipitate as insoluble Cr hydroxide (He et al., 2004)

$$Cr^{3+} + 3\,OH^- \rightarrow Cr(OH)_3(s)$$

Conversely, in oxygenated zones such as wetland rhizosphere or surface waters, Cr^{3+} can be re-oxidized to Cr^{6+} by oxidants like hydrogen peroxide:

$$2\,Cr^{3+} + 3\,H_2O_2 + 10\,OH^- \rightarrow 2\,CrO_4^{2-} + 8\,H_2O$$

In the atmosphere, Cr primarily originates from industrial emissions, combustion of fossil fuels, metal refining, and welding operations, primarily as fine particulate matter. These airborne particles can remain suspended and be transported over distances ranging from hundreds to thousands of kilometers—factors such as wind speed, direction, atmospheric turbulence, and precipitation influence Cr's dispersion patterns. Wet deposition removes Cr-containing aerosols through scavenging by raindrops onto terrestrial or aquatic surfaces. In contrast, dry deposition occurs through gravitational settling, especially during dry periods, as part of its biogeochemical cycle. Chronic exposure to airborne Cr(VI) is linked to adverse health effects, including respiratory tract irritation, pulmonary inflammation, lung tissue damage, and an elevated risk of lung cancer, posing significant threats at both local and global levels. Table 3 summarizes the effects of environmental parameters on Cr speciation, mobility, and toxicity.

Biological System

Chromium assimilation and behavior in biological systems vary widely with species and environmental conditions. Cr(VI), mimicking essential anions like sulfate and phosphate, enters root cells via anion transporters and is often reduced to Cr(III) inside cells, where it binds to cell wall components or chelates with organic acids, thereby limiting translocation and toxicity. Excessive Cr accumulation disrupts photosynthesis, enzymatic activity, and water uptake, leading to oxidative stress and growth

Table 3 Redox-driven transformation of Cr: implications for environmental risk

Environmental condition	Dominant Cr species	Form	Redox process	Mobility	Toxicity	Fate
Oxic, Acidic	Cr^{6+}	CrO_4^{2-}/ $Cr_2O_7^{2-}$	Oxidation of Cr^{3+} by MnO_2 or O_2	High	High	Leaching into aquifers
Oxic, Alkaline	Cr^{6+}	CrO_4^{2-}	Potential oxidation of Cr^{3+}	Very high	Very high	Stable and mobile
Anoxic, Acidic	Cr^{3+}	Soluble complexes	Reduction of Cr^{6+} by OM, Fe^{2+}, H_2S	Moderate	Low	Partial mobility
Anoxic, Alkaline	Cr^{3+}	$Cr(OH)_3(s)$, Cr-organic complexes	Precipitation of $Cr(OH)_3$; reduction continues	Low	Very low	Precipitates; immobilized in soil

inhibition, with numerous studies reporting Cr(VI)-induced inhibition of germination in diverse plants species, including *Phaseolus vulgaris, Cucumis sativus, Lactuca sativa, Panicum miliaceum, Medicago sativa, Echinochloa colona, Triticum aestivum* (Datta et al., 2011; Lopez-Delacalle et al., 2020), *Glycine max, Vigna radiata and V. angularis, Apium graveolens, Beta vulgaris, Raphanus sativus, Daucus carota, Solanum melongena, and Lycopersicon esculentum* (Lakshmi & Sundaramoorthy, 2010), *Brassica oleracea* var. *acephala* (Ozdener et al., 2011), and *Casuarina equisetifolia* (Singh et al., 2013).

Microorganisms, including bacteria, fungi, and actinomycetes, play critical roles in Cr cycling through biosorption, bioaccumulation, and biotransformation, often sequestering Cr via extracellular polymeric substances, intracellular compartmentalization, or cell wall binding. In aquatic systems, species such as rotifers, daphnids, amphipods, polychaetes, and zebrafish show high sensitivity to acute and chronic toxicity; however, Cr(VI), reduced by citric acid in ice, exhibits markedly diminished toxicity in water (Saha et al., 2011). To provide a more straightforward overview of the effects of different Cr species and reduction pathways, Table 4 summarizes the toxicity endpoints and biochemical responses observed in multiple aquatic species, as reported in Saha et al. (2011). Toxicity is commonly assessed using the LC_{50} (Lethal Concentration 50%), which is the concentration causing 50% mortality within a defined period (e.g., 24, 48, or 96 h). Lower LC_{50} values indicate higher toxicity. In addition to the LC_{50}, sublethal effects such as survival rate, reproduction, growth, and biochemical responses (e.g., enzyme activity or oxidative stress) are monitored to evaluate both acute and chronic toxicity on aquatic organisms.

Some bacteria utilize Cr(VI) as a terminal electron acceptor during respiration, facilitating its redox transformation in soils. In contrast, fungi such as *Aspergillus* and *Penicillium* can immobilize Cr through similar mechanisms. These microbial processes are critical for bioremediation and are often integrated with hyperaccumulator plants in phytoremediation strategies. In animals and humans, exposure to

Table 4 Summary of Cr speciation and toxicity: species-specific responses in aquatic ecosystems

Species	Endpoint	Cr(VI) toxicity	Cr(III) toxicity	Cr(VI) reduced in water	Cr(VI) reduced in ice
B. calyciflorus	24 h LC_{50}	0.3 mg L^{-1}	11.9 mg L^{-1}	Moderate mortality	Low mortality
B. plicatilis	24 h LC_{50}	2 mg L^{-1}	18.2 mg L^{-1}	Moderate mortality	Low mortality
Daphnia magna	48 h LC_{50}/ reproduction	81 μg L^{-1}/↓ reproduction	946 μg L^{-1}/ ↓ at ≥ 100 μg L^{-1}	↓ Mortality and better offspring	Drastically better outcomes
Hyalella azteca	28 and 42 d survival	↓ at ≥ 6.25 μg L^{-1}	↓ at ≥ 25 μg L^{-1}	Mixed effects	↑ Survival at ≤ 25 μg L^{-1}
Polychaete	96 h LC_{50}/ burrowing delay	2.59 mg L^{-1}/ significant delay	Not tested	Mild improvement	Significant improvement
Danio rerio (Zebrafish)	Embryo and larval survival	Mortality at > 93 mg L^{-1}	Not tested	Slight improvement	Strong improvement

Cr occurs primarily through ingestion or inhalation; once inside the cell, Cr(VI) is reduced to reactive intermediates that cause extensive damage to DNA, proteins, and cell membranes. Unlike Cr(III), which functions as a micronutrient in glucose and lipid metabolisms, Cr(VI) is highly toxic even at low concentrations. Ecologically, Cr can enter and persist in terrestrial and aquatic food webs, with aquatic uptake by plankton, mollusks, and fish leading to bioaccumulation and potential biomagnification through trophic levels. Similarly, soil-bound Cr enters the food chain via plant uptake and microbial interactions, posing long-term ecological and health risks, particularly for top predators, including humans, who consume contaminated food.

4 Human Exposure to Chromium Toxicity

Chromium exposure occurs primarily through inhalation, ingestion, and dermal contact (Pokhrel & Pokhre, 2022; Shin et al., 2023), with each route involving distinct mechanisms, sources, and associated risks. The chemical form of Cr plays a critical role in determining its bioavailability and toxicity.

4.1 Inhalation Pathway

The health impacts of Cr inhalation are significant, particularly for chromates and dichromates, which are highly soluble and capable of penetrating the alveolar region of the lungs due to their small size (0.2–10 μm) (Friess, 1989). Once deposited in the epithelium, Cr(VI) is absorbed across cell membranes via nonspecific anion channels (Narayani & Shetty, 2013). Inside cells, Cr(VI) undergoes a stepwise reduction to Cr(III) through intermediate valence states, a process mediated primarily by intracellular reductants such as ascorbate and glutathione (Wong et al., 2012). This redox cycling generates reactive oxygen species (ROS), including hydroxyl radicals (·OH) and superoxide anions (O_2—·), which induce oxidative stress and DNA strand breaks (Nickens et al., 2010). These effects trigger inflammatory responses in the lung tissue, contributing to the development of chronic bronchitis, asthma, fibrosis, and tissue remodeling. Acute exposure may cause coughing, wheezing, irritation of the nasal mucosa, and shortness of breath, while persistent genotoxic stress from Cr(VI) metabolites can lead to mutations, chromosomal aberrations, and an elevated risk of lung cancer.

4.2 Ingestion Pathway

Ingestion is a significant exposure route for Cr, particularly Cr(VI), through contaminated water and food (Costa & Klein, 2006). Drinking water contamination is of concern near industrial zones, mining areas, or landfills, where chromate (CrO_4^{2}—) and dichromate ($Cr_2O_7^{2-}$) species can persist. Although the gastrointestinal tract can reduce part of Cr(VI) to the less toxic Cr(III), a fraction, estimated at 10–20% for low doses, can evade reduction and enter systemic circulation (Finley et al., 1997; Kerger et al., 1997; Zhitkovich, 2011). Unlike Cr(III), Cr(VI) enters intestinal cells via nonspecific anion transporters, mimicking sulfate and phosphate (Standeven & Wetterhahn, 1989; Wang et al., 2017; Witt et al., 2013), and undergoes stepwise intracellular reduction to Cr(III), generating ROS (Hu et al., 2017; Singh et al., 2022). This oxidation stress leads to lipid peroxidation, DNA damage, and protein oxidation (Jomova & Valko, 2011; Valko et al., 2005), contributing to cytotoxic and genotoxic effects in the gastrointestinal tract, liver, and kidneys (Goetz & Luch, 2008; Hossini et al., 2022; Iyer et al., 2023). Table 5 further illustrates multimodal human exposure to Cr, including routes, sources, and systemic outcomes. In vitro and in vivo studies indicate that 1–4% of applied Cr can penetrate the skin within 5–24 h, and hypersensitivity reactions may occur at Cr(VI) concentrations as low as 10 ppm (Bagdon & Hazen, 1991). Chromium can also infiltrate the food chain, as plants absorb it via sulfate transporters and accumulate in edible tissues (Xu et al., 2021), further contributing to dietary exposure.

Table 5 Summary of Cr exposure pathways and health effects

Exposure pathway	Primary sources	Cr species	Mechanism of exposure	Health effects
Inhalation	• Chrome plating-welding • Stainless steel manufacturing • Leather tanning (Srivastava, 2013)	Mainly Cr^{6+} (e.g., CrO_3, chromic acid mist)	• Inhalation of fine particulate fumes and mists (Serita et al., 1990; Vorpahl et al., 1976) • Poor ventilation enhances exposure	• Lung, nasal, and sinus cancers (Yatera et al., 2018) • Chronic bronchitis, asthma, emphysema, pulmonary fibrosis • Acute respiratory irritation (coughing, wheezing) (Saha et al., 2011) • Oxidative stress and DNA damage (Valavanidis et al., 2013)
Ingestion (Costa & Klein, 2006)	• Contaminated drinking water • Cr-laden food (bioaccumulated in crops) (Srivastava, 2013)	Primarily Cr^{6+} (e.g., $CrO_4{}^{2-}$, $Cr_2O_7{}^{2-}$) Trace Cr^{3+} in food	• Absorption through anion transporters (Standeven & Wetterhahn, 1989; Wang et al., 2017; Witt et al., 2013)	• Gastrointestinal irritation (nausea (Yatera et al., 2018)) • Hepatotoxicity and nephrotoxicity • Digestive and liver cancers • Oxidative damage, neurological and immune dysfunction (Krikorian et al., 2010; Wu, 2017)

(continued)

Table 5 (continued)

Exposure pathway	Primary sources	Cr species	Mechanism of exposure	Health effects
Dermal contact	• Chrome plating • Leather tanning • Pigment and dye production • Environmental contact (Srivastava, 2013)	Mostly Cr^{6+} (e.g., chromates, dichromates)	• Skin absorption • Environmental contact through contaminated soil/water	• Allergic and irritant contact dermatitis • Chronic eczema and skin ulcers (Hessel et al., 2021; Shelnutt et al., 2007)
Placental transfer	Transplacental movement of Cr from mother to fetus (Banu et al., 2017)	• Maternal exposure to Cr crosses the placental barrier (Agrawal, 2012; Banu et al., 2017)	• Pregnant women and fetuses (Banu et al., 2017) • In particular, male fetuses (Banu et al., 2018)	• Fetal malformations-Preterm birth, low birth weight

4.3 *Dermal Contact: Occupational and Environmental Exposure*

Dermal contact is a significant exposure route for individuals involved in industrial activities that involve Cr compounds (Srivastava, 2013). In chrome plating operations, workers are exposed to fine mists of Cr derivatives, while tannery workers encounter Cr salts during leather processing. Mechanistically, Cr(VI) penetrates the skin via paracellular (between cells) and transcellular (through cells) pathways. Within the epidermal and dermal layers, Cr(VI) is reduced to Cr(III) by intracellular antioxidants such as glutathione and ascorbate (Nickens et al., 2010; Standeven & Wetterhahn, 1989).

Dermal exposure to Cr(VI) in water up to 22 mg L^{-1} is generally insufficient to overwhelm the skin's or blood's detoxification capacity (Paustenbach et al., 2003). A review of eight patch test studies estimated that 10% of Cr-sensitized individuals react to ~ 150 ppm potassium dichromate (Paustenbach et al., 1992). Cr(VI) is a potent skin sensitizer, capable of causing allergic contact dermatitis and ulceration (Shelnutt et al., 2007). Sensitization prevalence ranges from 2% in the general population to 20% in patients with dermatitis (Stern et al., 1993). For soil contamination, concentrations of 350–500 ppm Cr(VI) are generally considered safe in relation to allergic dermatitis.

5 Global Chromium Reserves, Production Trends, and Emerging Challenges

Chromium reserves are unevenly distributed worldwide, with Kazakhstan and South Africa collectively holding over 75% of identified shipping-grade chromite reserves, followed by India. Turkey, despite its smaller reserves, remains a significant producer due to sustained mining and strong export potential (Gao et al., 2022). According to the U.S. Geological Survey (USGS), global Cr resources exceed 12 billion tons, far surpassing current reserves, indicating a sufficient long-term supply to meet industrial demand.

In 2023, South Africa was the leading producer, yielding ~ 8 million metric tons annually, despite Kazakhstan's larger reserve base. South Africa's dominance is primarily due to the Bushveld Igneous Complex, which hosts some of the wealthiest and most accessible chromite deposits globally. Kazakhstan, producing 1.8 million metric tons, ranked second, with its industry dominated by Eurasian Resources Group's Donskoy Mining and Processing Plant, one of the world's most extensive chromite operations. India remains a major Asian producer, with the Sukinda Valley in Odisha hosting most of its chromite reserves and mining activity; however, the sector faces serious environmental and health challenges from Cr(VI) contamination of soil and water. Turkey produced ~ 1.1 million metric tons in 2023, benefiting from low extraction costs and proximity to European markets (Fig. 2).

Collectively, these four countries accounted for the majority of global production in 2023. Between 2019 and 2023, global output fell by ~ 8%, from ~ 15.8 to 14.5 million metric tons. This decline reflects multiple factors, including COVID-19-related disruptions to mining operations, supply chains, and industrial demand, particularly in 2020 and 2021, as well as evolving market dynamics, stricter environmental regulations, and geopolitical tensions.

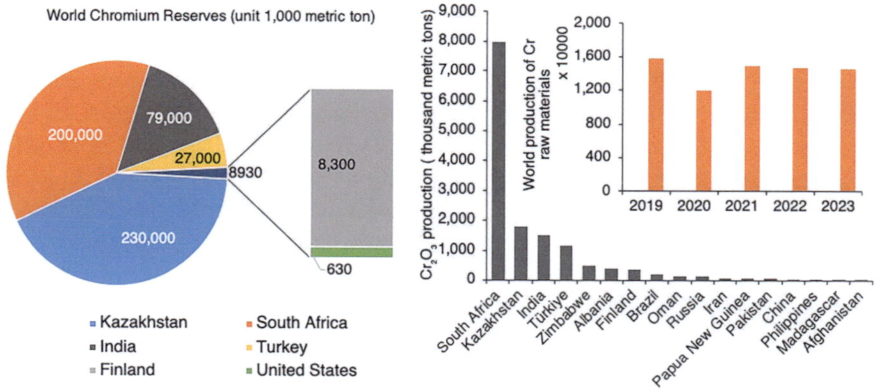

Fig. 2 Global chromium resources, production trends (2019–2023), and leading producers by Cr_2O_3 output. (*Source* compiled from World Mining Data (2024) and U.S. Geological Survey Mineral Commodity Summaries (2024)

6 Global Case Studies—Sources, Health Impacts, and Remediation Approaches

6.1 *Hazaribagh, Bangladesh*

Hazaribagh, located in old Dhaka, Bangladesh, is internationally recognized as a hotspot for Cr pollution, mainly due to its high concentration of leather tanning industries (Islam et al., 2015; Yoshinaga et al., 2018). Every day, the processing of raw hides releases large volumes of wastewater containing both Cr(III) and Cr(VI) (Yoshinaga et al., 2018). Environmental monitoring has confirmed that soluble Cr detected in the nearby Buriganga River primarily originates from Hazaribagh, with canal water containing concentrations of up to 0.7 μM Cr^{6+} and 1705 μM Cr^{3+} (Yoshinaga et al., 2018). Soil contamination is severe, with Cr concentrations reaching as high as 37,000 mg kg^{-1}, and the pollution extending to depths of 10–20 feet (Karim et al., 2013).

Environmental degradation extends beyond water and soil contamination, encompassing a broader range of issues. A critical concern is the improper use of skin-cut waste from tanning, which is often repurposed into poultry and fish feed, as well as fertilizers, despite containing Cr levels as high as 2.49% (Rafiqullah et al., 2008). This practice facilitates the entry of Cr into the food chain, posing a potent ecological and human health disaster. Although bioremediation techniques using native microbial isolates such as *Escherichia sp.*, *Staphylococcus aureus*, and *Pediococcus pentosaceus* have shown promising Cr^{6+} reduction efficiencies ranging from 82 to 100% (Rafiqullah et al., 2008), these methods remain vastly underutilized. This gap highlights the pressing need for an integrated pollution control strategy that combines remediation, regulation, and monitoring.

The public health implications in Hazaribagh are equally concerning. According to the Bangladesh Society for Environmental and Human Development, nearly half a million people residing in the area are chronically exposed to hazardous chemicals, as 6000 cubic meters of liquid effluent and 10 tons of solid waste are discharged daily (Maurice, 2001). Among tannery workers, many between the ages of 30 and 35, health surveys indicate high incidences of gastrointestinal disorders (58%), skin diseases (31%), hypertension (12%), and jaundice (19%) (Maurice, 2001). In addition, airborne pollutants such as sulfur dioxide released from tanning processes exacerbate respiratory risks by converting into corrosive sulfuric acid upon contact with atmospheric moisture. These threats underscore the urgent need for government-led interventions, including the relocation of tanning operations to environmentally regulated industrial zones. Without robust enforcement and sustainable management strategies, the environmental and health crises in Hazaribagh will continue to escalate.

6.2 Kanpur, India

Kanpur, a central industrial hub in northern India, has emerged as a significant hotspot for Cr contamination, primarily due to extensive leather tanning operations and the unregulated disposal of chromite ore processing residue (COPR) (Gowd et al., 2010; Gupta et al., 2024). The uncontrolled discharge of tannery effluents and the indiscriminate dumping of COPR have severely polluted the region's soil, water, and sediments, posing grave threats to both environmental and human health.

Field investigations and geochemical analyses have revealed alarmingly high levels of Cr contamination in surface and GW sources. Chromium concentrations in some water samples exceed 40 mg L^{-1}, far surpassing permissible safety limits. COPR residues contained Cr concentrations up to 9.6 wt%, while nearby soils showed Cr levels as high as 3.83 wt% (Gupta et al., 2024). A broader survey covering the industrial areas of Jajmau and Unnao reported soil Cr concentrations ranging from 161.8 to 227.8 mg kg^{-1}, with an average value of 2652.3 mg kg^{-1} (Gowd et al., 2010). In addition to Cr, other heavy metals such as Ba, Cu, Pb, Sr, V, and Zn were also detected at elevated levels, indicating widespread multi-metal contamination. These results were obtained through X-ray fluorescence spectrometry conducted on soil samples collected from the top 15 cm layer (Gowd et al., 2010).

The health implications for the local population are severe. A study involving 197 tannery workers and 117 individuals from the general population revealed a significantly higher morbidity rate among the workers (40.1%) compared to the control group (19.6%) (Rastogi et al., 2008). Respiratory ailments were the most common among tannery workers (16.7%), likely due to chronic inhalation of toxic chemicals and particulates. In contrast, gastrointestinal disorders were more prevalent in the control group.

In response to this environmental and public health crisis, researchers have explored phytoremediation as a sustainable and cost-effective solution. Field-scale studies have identified several plant species, such as *Cynodon dactylon, Chrysopogon zizanioides, Cyperus sp.*, and *Typha angustifolia*, as effective agents for remediation. These plants exhibit strong associations with Cr-reducing bacterial strains, including *Pseudomonas sp., Clostridium sp.*, and *Bacillus sp.*, which enhance Cr(VI) reduction and aid in the stabilization of Cr in contaminated soils (Gupta et al., 2024). Such biological systems offer promising potential for large-scale remediation of Cr-polluted environments in Kanpur and similar regions.

6.3 China

China faces serious environmental challenges from widespread Cr contamination, primarily driven by industrial activities such as ferroalloy production, leather tanning, electroplating, pigment and dye manufacturing, and mining operations. The scale and severity of Cr pollution are particularly alarming in provinces such as Hubei

(Suizhou) and Yunnan (Qujing), where the unregulated disposal of COPR and Cr-containing slag has led to extensive ecological degradation.

One of the most notorious incidents occurred in Qujing, Yunnan Province, in August 2011, when over 5000 tons of untreated Cr slag were illegally dumped into the mountainous area and the Nanpan River. As a result, Cr^{6+} concentrations in surface water exceeded permissible limits by up to 2000 times (Gao & Xia, 2011). Additionally, approximately 288,400 tons of untreated Cr slag were stored merely one meter from the river without any detoxification. The consequences were devastating, including livestock deaths, increased cancer cases in the nearby Xinrong village, and a severe public health crisis. The incident eventually led to factory closures and a government crackdown (Gao & Xia, 2011). In Suizhou, Hubei Province, the disposal of Cr slag similarly led to significant contamination of soils and food crops. Soil samples from both industrial and agricultural lands revealed total Cr concentrations as high as 1910 mg kg^{-1}, far exceeding safety thresholds (Wang et al., 2011). The implications for food safety and human health are particularly concerning, given the potential for Cr accumulation in edible crops.

Electroplating industries also contribute substantially to Cr pollution in China. At a typical chrome plating site, total soil Cr levels ranged from 7 to 7735 mg kg^{-1}, with Cr^{6+} alone averaging 139 mg kg^{-1} (Sun et al., 2022). Vertical soil profile studies showed that Cr^{6+} can leach through porous soil layers. However, iron-rich loam has the potential to reduce and immobilize the toxicant by converting it into less mobile forms, such as $Cr(OH)_3$ and Cr_2O_3, providing a degree of natural attenuation.

In response to these threats, China has introduced stringent environmental regulations that mandate the detoxification of all Cr-containing wastes prior to disposal, along with compulsory registration and tracking of Cr waste. Nonetheless, enforcement often lags policy, as evidenced by the Qujing incident. Strengthening compliance mechanisms and monitoring systems is essential to mitigating the long-term risks associated with Cr pollution in China.

6.4 Peru

Chromium pollution in Peru is primarily associated with industrial discharges from the leather tanning industry, as well as urban and mining activities. Several major rivers, including the Coata (Yana et al., 2019), Rímac (Juárez, 2006), Chili (Pinto et al., 2017), and Lurín (Rivera et al., 2007), have shown elevated Cr concentrations, posing threats to aquatic ecosystems and human health.

The tanning sector is a major contributor, consuming 60–80% of Cr in traditional leather processing (Córdova Bravo et al., 2014). Improper handling and disposal of untreated effluents have resulted in Cr^{6+} levels reaching up to 7.7 mg L^{-1}, well above the Peruvian Maximum Permissible Limit (MPL) of 2.5 mg L^{-1} (Pinto et al., 2017).

A significant concern arises from the use of contaminated water for crop irrigation, often without the farmer's knowledge. Many rely on visual indicators like sewage or garbage to assess pollution, underestimating the risks posed by invisible chemical contaminants like Cr. Sediment studies from the Rímac, Lurín, and Chillón rivers have confirmed the presence of multiple toxic metals, including Cr (Rivera et al., 2007), with downstream dispersion intensifying due to sediment transport processes (Juárez, 2006; Rivera et al., 2007). Inadequate waste management and weak monitoring systems exacerbate the chronic exposure risk to both ecological and human populations.

To address Cr pollution, Peru has introduced various technological and regulatory interventions. Electrocoagulation using aluminum electrodes has demonstrated over 99% Cr removal efficiency, representing a promising treatment method (Aguilar-Ascón et al., 2019; Gilpavas et al., 2011). In parallel, bioremediation using native filamentous fungi such as *Trichoderma viride* and *Penicillium citrinum* has demonstrated potential for Cr^{6+} detoxification (Zapana-Huarache et al., 2020). These fungal strains exhibit notable tolerance and enzymatic response under Cr stress. However, these technologies require broader implementation and integration into Peru's wastewater infrastructure. On the policy front, Peru has established environmental regulations, including Environmental Quality Standards (EQS) and Maximum Permissible Limits (MPL). Nevertheless, enforcement remains inconsistent, limiting its effectiveness. Strengthening policy enforcement, promoting cleaner production practices in the tanning industry, enhancing inter-agency coordination, and expanding environmental monitoring are key strategies needed to reduce Cr pollution and safeguard public and environmental health in Peru.

6.5 *Brazil*

Brazil, one of the world's largest leather producers, faces significant environmental challenges related to Cr contamination, primarily from the tanning industry. Chromium sulfate is extensively used in tanning processes, and in many regions, leather shavings and untreated tannery effluents are directly discharged into local water bodies. A notable case is the Monte Alegre Settlement in Motuca, where over 40 tons of Cr-laden leather waste were indiscriminately dumped, causing long-term soil and subsurface contamination (Helene et al., 2016). In addition to anthropogenic sources, natural (geogenic) Cr contamination is also reported. In the Adamantina aquifer, weathering of pyroxene-rich sandstones has released Cr^{6+} into GW, with parent rock concentrations reaching up to 6000 mg kg^{-1} and averaging 221 mg kg^{-1} (Bourotte et al., 2009). This geogenic contribution complicates the identification of contamination sources.

Environmental monitoring in Franca, São Paulo, revealed Cr concentrations in urban stream sediments ranging from 0.83 to 26.25 mg kg^{-1} (Matamet & Bonotto,

2013). Although below the average global shale standard, these levels pose ecological risks due to the potential remobilization. Groundwater samples from Urânia, São Paulo, showed Cr concentrations between 0.12 and 0.13 mg L^{-1}, exceeding the Brazilian drinking water limit of 0.05 mg L^{-1} (Bourotte et al., 2009). This contamination is believed to originate from the weathering of chrome-diopside in aquifer sandstones. The mobility and persistence of Cr in Brazilian ecosystems are influenced by environmental factors such as pH, redox conditions, and interactions with OM. Sediment core analysis from the Ribeirão dos Bagres River, using ^{210}Pb dating, indicates increasing Cr accumulation over the last two centuries, with sedimentation rates ranging from 0.6 to 3.2 cm $year^{-1}$ (Matamet & Bonotto, 2013). Geophysical surveys at the Motuca site also reported high chargeability values (>0.67 mV V^{-1}), indicating strong Cr retention in subsurface layers and confirming the persistence of contamination (Helene et al., 2016). Ecotoxicological studies further demonstrate the biological impact of Cr pollution. In *Oreochromis niloticus*, exposure to Cr-contaminated water resulted in DNA strand breaks and the formation of micronuclei (Matsumoto et al., 2006). Similarly, *Allium cepa* root-tip assays revealed chromosomal abnormalities such as stickiness, bridges, and laggards, indicating genotoxic effects (Matsumoto et al., 2006).

In conclusion, both industrial and natural sources contribute to widespread Cr pollution worldwide, with profound implications for environmental and public health. A comparative summary of anthropogenic Cr footprints across different countries is presented in Table 6.

7 Remediation and Mitigation Strategies

The remediation of Cr-contaminated environments requires a multifaceted approach that addresses both Cr^{6+} and Cr^{3+} species due to their distinct chemical properties, mobility, and toxicity. Among physicochemical methods, the reduction of Cr^{6+} to the less toxic and less mobile Cr^{3+} is the most widely implemented strategy. Common reductants include ferrous sulfate ($FeSO_4$), sodium metabisulfite ($Na_2S_2O_5$), zero-valent iron (ZVI), and sulfide compounds (Yang et al., 2021), which promote Cr^{6+} reduction under acidic conditions. A typical redox reaction using ferrous iron is

$$Cr_2O_7^{2-} + 6\,Fe^{2+} + 14\,H^+ \rightarrow 2\,Cr^{3+} + 6\,Fe^{3+}7\,H_2O$$

7.1 Chemical Precipitation Techniques

Chemical precipitation is a widely used and cost-effective technique for removing Cr from industrial effluents, particularly tannery wastewater. This process involves

Table 6 Transboundary Cr contamination: quantitative insights from industrial clusters

Location	Main Cr sources	Cr concentration (water)	Cr concentration (soil/sediment)
Hazaribagh, Bangladesh	Leather tanning industries	0.7 μM Cr^{6+}, 1705 μM Cr^{3+} (Yoshinaga et al., 2018)	Up to 37,000 mg kg^{-1} (Karim et al., 2013)
Kanpur, India	Leather tanneries, COPR disposal (Gowd et al., 2010; Gupta et al., 2024)	> 40 mg L^{-1} (Gupta et al., 2024)	162–6228 mg kg^{-1}; up to 3.83 wt% (Gowd et al., 2010)
China	Ferroalloy production, tanning, electroplating, pigment/dye manufacturing	Cr^{6+} 2000 × above limit in Qujing incident (Gao & Xia, 2011)	Up to 7735 mg kg^{-1} (soil) (Sun et al., 2022)
Peru	Tannery effluents, urban and mining activities	Up to 7.67 mg L^{-1} Cr^{6+} (Pinto et al., 2017)	Detected in Rimac, Lurin, and Chillon sediments
Brazil	Leather tanning industry, geogenic sources (chromite-bearing rocks)	Cr(VI): 0.12–0.13 mg L^{-1} (Bourotte et al., 2009)	15–11,066 mg kg^{-1} (sediments), up to 6000 ppm in rocks (Jordão et al., 1999)

converting soluble Cr^{3+} into insoluble chromium hydroxide [$Cr(OH)_3$] by raising the pH with alkaline reagents such as sodium hydroxide (NaOH), calcium hydroxide [$Ca(OH)_2$], or magnesium oxide (MgO) (Minas et al., 2017). Owing to the extremely low solubility product of $Cr(OH)_3$ ($Ksp = 6.7 \times 10^{-31}$), this method ensures efficient precipitation under optimal pH conditions.

$$Cr^{3+} + 3\,OH^- \leftarrow \left\{ \begin{array}{c} NaOH \rightarrow Na^+ + OH^- \\ Ca(OH)_2 \rightarrow Ca^{2+} + 2\,OH^- \\ MgO + H_2O \rightarrow Mg(OH)_2 \rightleftharpoons Mg^{2+} + 2\,OH^- \end{array} \right. \rightarrow Cr(OH)_3 \downarrow$$

NaOH achieves approximately 99.97% Cr removal efficiency within the pH range of 8.6–9.1, while $Ca(OH)_2$ performs similarly at pH 9.1–10.3. In contrast, MgO offers the highest efficiency (99.98%) within the pH range of 9.8–10.3, forming denser sludge due to the formation of granular precipitate (Minas et al., 2017). Process efficiency depends on several factors, including rapid mixing followed by gentle agitation, settling time, sludge volume, and the solubility of the reagent. Although MgO has slower dissolution kinetics ($Ksp = 1.5 \times 10^{-11}$), it generates less gelatinous sludge, facilitating more straightforward disposal and Cr ‘recovery (Minas et al., 2017).

In tannery wastewater treatment, where Cr^{6+} is predominant, chemical precipitation is typically preceded by reduction to Cr^{3+} using ferrous sulfate ($FeSO_4$) under acidic conditions (optimal at pH ≤ 2). In this process, Fe^{2+} reduces Cr^{6+} to Cr^{3+} while being oxidized to Fe^{3+}, and the resulting Cr^{3+} is precipitated as $Cr(OH)_3$ by adjusting the pH to around 8 (Sheeja, 2016).

$$Cr_2O_7^{2-} + 6\,Fe^{2+} + 14\,H^+ \rightarrow 2\,Cr^{3+} + 6\,Fe^{3+}7\,H_2O$$

This integrated reduction–precipitation strategy achieves a removal efficiency of up to 99.85% (Sheeja, 2016). However, the presence of complexing agents, such as ammonium chloride, tartrate, or citrate, may slightly hinder precipitation by stabilizing Cr ions in solution (Sheeja, 2016).

7.2 Adsorption-Based Treatments

Adsorption is widely used for Cr removal due to its operational simplicity and cost-effectiveness. Various adsorbents, including activated carbon (AC), zeolites, and biochar, have been explored, with bio-adsorbents derived from natural or waste materials gaining particular attention as sustainable alternatives. These lignocellulosic materials are rich in functional groups such as hydroxyl (–OH), carboxyl (–COOH), amino ($–NH_2$), and phenolic (–PhOH), which facilitate Cr(VI) binding through ion exchange, complexation, chelation, and physical adsorption. Despite their simplicity, these bio-adsorbents offer high efficiency, biodegradability, and low cost, making them promising substitutes for conventional materials like AC or other synthetic resins.

Biochar

Biochar, a C-rich material produced through pyrolysis of organic biomass, is a practical, sustainable, and low-cost sorbent for Cr remediation in water. Cr^{6+} removal occurs via a dual mechanism: (i) adsorption onto oxygen-containing functional groups (e.g., –OH, –COOH, and phenolic) and (ii) in situ reduction of Cr^{6+} to Cr^{3+} at the biochar surface (Qian et al., 2016). Under acidic conditions (pH 2–5), where Cr^{6+} primarily exists as $HCrO_{4-}$ and $Cr_2O_7{}^{2-}$ (Dong et al., 2011; Rajapaksha et al., 2018; Sinha et al., 2022), these species are electrostatically attracted to the positively charged surface of biochar. Phenolic and other electron-donating groups donate electrons to Cr^{6+}, reducing it to Cr^{3+}, via the simplified redox reaction:

$$Cr_2O_7^{2-} + 14\,H^+ + 6\,e^- \rightarrow 2\,Cr^{3+} + 7\,H_2O$$

The reduced Cr^{3+} then either precipitates or forms stable surface complexes with biochar. Synchrotron-based X-ray absorption spectroscopy confirms Cr^{3+} formation on the biochar surface (Rajapaksha et al., 2018). Biochars produced at higher pyrolysis temperatures (550–650 °C), especially from invasive plants such as *Ulex europaeus* (Gomez et al., 2021) and agricultural residues like soybean and bur cucumber (Rajapaksha et al., 2018), exhibit improved Cr removal due to enhanced surface area, porosity, and aromaticity. In some cases, these biochars outperform commercial AC by more than twofold in Cr removal efficiency (Gomez et al., 2021), underscoring their potential as eco-friendly and practical materials for aqueous Cr remediation.

Activated Carbon

Activated carbon is a highly porous material with a large internal surface area (300–3000 $m^2\ g^{-1}$) that enables excellent adsorption of contaminants, gases, and dissolved pollutants (Sharma et al., 2022). It is produced by carbonizing organic matter (e.g., wood, coconut shells, or coal) under limited oxygen, followed by physical (steam or CO_2) or chemical ($ZnCl_2$ or H_3PO_4) activation to enhance porosity (Anjum et al., 2017; Gaur & Shankar, 2008; Gawande & Kaware, 2016). Its micro-, meso-, and microporous structure makes it ideal for water purification, air filtration, and the treatment of industrial waste (e.g., Cr removal).

Cr(VI) adsorption on AC occurs via electrostatic attraction, reduction, and complexation, heavily influenced by acidic oxygen-containing groups (carboxyl, phenolic, and lactonic), which both attract anionic Cr species ($HCrO_{4-}$, $CrO_4{}^{2-}$) and reduce Cr^{6+} to Cr^{3+} via electron transfer (Jiang et al., 2014; Li et al., 2010; Valentín-Reyes et al., 2019; Xu et al., 2019). Oxidized AC shows a ~ 50% Cr^{6+} reduction, whereas heat-treated AC with basic sites offers 48 mg g^{-1} adsorption and a ~ 35% reduction (Valentín-Reyes et al., 2019). Desorption using NaOH-NaCl confirms that unreduced Cr^{6+} remains electrostatically bound, while high surface area (1350 $m^2\ g^{-1}$) and microporosity (~18–20 Å) improve kinetics (Gul et al., 2022; Nabais et al., 2009). Removal efficiency depends on pH, adsorbent dose, and pollutant concentration, with removal rates of 75–96% under optimal conditions (Gul et al., 2022).

Aegle marmelos-based AC achieved 82.3 mg L^{-1} uptake at acidic pH (Galán et al., 2005), and bone-derived biochar (92 mg L^{-1}) retains Cr(VI) through electrostatic and ligand exchange (Gode & Pehlivan, 2003). Other efficient adsorbents include cellulose/chitosan (56 mg L^{-1}; Islam et al., 2017), CNC/clay (100 mg L^{-1}; Pattanapipitpaisal et al., 2001), clay–alumina membranes (91.44 mg L^{-1}; Hokkanen et al., 2016), MWCNTs modified with surfactants (98 + mg L^{-1}; Islam et al., 2021), and magnetic chitosan-based hybrids like Chi@Fe_3O_4 and Chi@Fe_3O_4GO (up to 142.3 mg L^{-1}; Rahaman et al., 2021), offering rapid Cr^{6+} removal and magnetic recovery. Doped porous carbons, such as FeNi@HPC, also show potential, albeit with a moderate capacity (30 mg L^{-1}; Choudhury et al., 2018).

Zeolites

Natural zeolites, particularly clinoptilolite and heulandite types, exhibit high efficiency in removing Cr(VI), with reported removal rates of up to 92%, primarily through mechanisms such as ion exchange, surface adsorption, and complexation at active sites (Rangasamy et al., 2014). These zeolites possess a three-dimensional aluminosilicate framework characterized by negatively charged sites, which are balanced by exchangeable cations such as Na^+, K^+, or Ca^{2+}. During Cr removal, these cations are replaced by chromate (CrO_4^{2-}) or dichromate ($Cr_2O_7^{2-}$) ions, effectively trapping Cr within the zeolite structure (Santiago, 1995).

The high cation exchange capacity, especially in zeolites with lower Si/Al ratios (Loizidou et al., 1992; Rangasamy et al., 2014), enhances this process by providing more active sites for Cr(VI) interaction. Additionally, the porous, honeycomb-like structure of zeolite offers a large surface area that facilitates adsorption (Shi et al., 2018; Yang et al., 2023). Cr removal efficiency improves under lower pH conditions due to the protonation of functional groups on the zeolite surface, which strengthens electrostatic attraction with the negatively charged Cr(VI) species. Several factors influence the adsorption capacity of zeolites, including particle size, surface area, porosity, and the Si/Al ratio. When compared to synthetic variants, such as faujasite-type zeolites or organo-modified zeolites, natural zeolites stand out as a more cost-effective and environmentally friendly method for Cr remediation. In addition to zeolites, a variety of low-cost adsorbents, including natural materials and agricultural by-products, have been employed for heavy metal removal. Recently explored options include pulp, clay, clinoptilolite, zeolite, plant leaves, activated C, charcoal, and green walnut shells (Table 7).

7.3 Phytoremediation

Phytoremediation is a cost-effective and environmentally sustainable strategy for remediating heavy metal-contaminated environments (Sinha et al., 2018). Through mechanisms such as phytoextraction, phytostabilization, and rhizofiltration, selected plant species absorb, sequester, transform, or immobilize metal pollutants (Ullah et al., 2023). This process relies on manipulating redox conditions, altering contaminant solubility through environmental changes, or employing complexation to enhance the removal or immobilization of pollutants from the environment (Anju, 2017; Salt et al., 1995). Although hyperaccumulator plants typically tend to grow slowly and require extended periods for effective remediation, they possess well-developed and deep root systems that enable access to contaminants in deeper soil layers, making them particularly effective for heavy metal cleanup (Komives & Gullner, 2006).

Several plant species have been identified as potential Cr hyperaccumulators, including *Cannabis sativa*, *Allium griffithianum*, and *Catharanthus roseus* (Sajid et al., 2020); *Spartina argentinensis* (Redondo-Gómez et al., 2011); *Amaranthus dubius* (Mellem et al., 2012); and *Convolvulus arvensis* (Gardea-Torresdey et al., 2004). Chromium uptake typically occurs during plant transpiration, where soluble Cr(VI) species (e.g., $HCrO_{4-}$) are absorbed by roots and subsequently reduced to Cr(III) within plant tissues through mechanisms such as cell wall immobilization, vacuolar sequestration, and antioxidant production (Han et al., 2023). The efficiency of phytoremediation is influenced by several factors, including plant biomass, chlorophyll content, and the use of soil amendments (Revathi et al., 2011). To enhance

Table 7 Biosorption performance of plant-based materials for Cr(VI) removal: source, modification, and capacity

Material type	Source	Scientific/specific form	Removal (mg g^{-1})	References
Fiber composite	Banana And Areca	Chitosan-coated fiber blend	75 (mg L^{-1})	(Ali et al., 2016)
Modified waste	Banana peel	Acrylonitrile-grafted banana peel	96 (mg L^{-1})	(Daneshvar et al., 2002)
Composite material	SCM-CC	Composite (chemically treated)	99.9 (mg L^{-1})	(Begum et al., 2020)
Industrial by-product	Paper Industry	Lignin	13.5	(Dakiky et al., 2002)
Coir	Coconut	Raw coir	6.3	(Agarwal et al., 2006)
Leaf	*Tradescantia pallida*	Leaf	94 (mg L^{-1})	(Uysal & Ar, 2007)
	Rubber	Leaf	96 (mg L^{-1})	(Rengaraj et al., 2001)
Plant waste	Sugarcane	Bagasse (second entry, untreated)	23	(Pourfadakari et al., 2017)
	Sunflower Stem	Pretreated	5.4	(Chakraborty et al., 2021)
Agricultural waste	Sugarcane	Bagasse	13.4	(Yasmeen et al., 2016)
	Rice	Bran	285.7	(Wan et al., 2018)
	Wheat	Bran	35	(Wang et al., 2019)
	Olive	Cake	35.4	(Singh et al., 2005)
By-product (Agro)	Soybean	Soya cake	0.3	(Meena et al., 2008)
Sawdust	Babool (*Acacia arabica*)	Pretreated sawdust	111.6	(Dupont & Guillon, 2003)
	Pine wood	Modified sawdust	30.5	(Krishnani et al., 2009)
	Generic	–	100 (mg L^{-1})	(Altun et al., 2016)
Shell	Almond	Crushed shell	22.2	(Dupont et al., 2004)
	Almond	Shell	10.6	(Singh et al., 2005)
	Hazelnut	Shell	8.3	(Jain et al., 2009)
	Walnut	Green walnut shell	95 (mg L^{-1})	(Faust & Aly, 2018)
Husk	Rice	Husk	38.4	(Gonzalez et al., 2008)
	Rye	Husk	80 (mg L^{-1})	(Ahalya et al., 2005)

(continued)

Table 7 (continued)

Material type	Source	Scientific/specific form	Removal (mg g^{-1})	References
	Bengal Gram	Husk	91.6	(Celik et al., 2004)
Leaves	London Plane Tree	Leaf	68	(Ahmed et al., 2019)
Seed	*Caryota urens*	Seeds	52.6	(Pehlivan & Altun, 2008)
Straw composite	Rice	A-RS/PVA (Aminated Straw–Grafted PVA)	140.4	(Raji & Anirudhan, 1998)
Straw	Paddy	Straw	21.5	(Sinha et al., 2015)
Wood	Rubber	Polyacrylamide-Grafted Sawdust	22.6	(Aoyama, 2003)

the process, the application of plant-growth-promoting bacteria (PGPB) has shown promise, as these microbes can improve Cr uptake by modifying root architecture and alleviating Cr toxicity (Ahemad, 2015).

Plants such as *Phragmites australis* (reed) have demonstrated notable Cr removal efficiency under experimental conditions. Ranieri et al. (2016) reported 55% removal of Cr from contaminated irrigation water, while Mandi et al. (2009) observed 99% removal efficiency, with Cr accumulation reaching 1690 mg kg^{-1} in roots from tannery wastewater. In addition, aquatic macrophytes such as *Pistia stratiotes* and *Eichhornia crassipes* have demonstrated a high Cr uptake potential, making them suitable for use in constructed wetlands. Their performance can be further improved by applying nitrogen and phosphorus fertilizers (Luca et al., 2014). The Cr remediation capabilities of various plant species, along with their respective bioconcentration and translocation factors, are summarized in Table 8.

7.4 Bioremediation

Bioremediation is a sustainable and cost-effective approach that employs the metabolic capabilities of microorganisms to transform, detoxify, or immobilize Cr. Microbial genera such as *Pseudomonas*, *Bacillus*, and *Shewanella* are well known for their chromate reductase activity, facilitating Cr(VI) reduction under both aerobic and anaerobic conditions (Cheung & Gu, 2007; Viti et al., 2014). These microorganisms can utilize Cr(VI) as a terminal electron acceptor during respiration, resulting in its reduction and subsequent precipitation. The mechanisms involved are multifaceted, encompassing enzymatic, physicochemical, and cellular processes. A primary mechanism is enzymatic reduction, wherein membrane-bound or cytosolic chromate reductases catalyze electron transfer from intracellular donors such as NADH (Mala et al., 2020) to Cr(VI), effectively detoxifying it. In addition to enzymatic activity,

Table 8 Cr(VI) uptake, accumulation, and translocation performance in remediating plants

Plant species	Cr(VI) dose applied	Key observations	References
Tagetes erecta L	2 mg L^{-1}	Achieved exceptional Cr(VI) removal efficiency (94%) within 35 days	Chitraprabha and Sathyavathi (2018)
Convolvulus arvensis L	80 mg L^{-1}	Exhibited a very high BCF of 161.1, showing strong accumulation capability	Gardea-Torresdey et al. (2004)
Spartina argentinensis	10 mmol L^{-1}	Demonstrated efficient metal transport (TF = 1.9), suitable for wetland remediation	Redondo-Gómez et al. (2011)
Amaranthus dubius	25 ppm	Displayed moderate accumulation (BCF = 2.6) and translocation (TF = 1.1)	Mellem et al. (2012)
Zea mays (maize)	16.8 g kg^{-1}	Very low BCF (0.013) but exceptionally high TF (2.67), suggesting shoot-based accumulation	Abou-Shanab et al. (2008)
Cicer arietinum (Chickpea)	10 mg/L and 60 mg/kg	Showed 50% removal efficiency at low concentration; higher dose led to reduced mobility (TF < 0.14)	Velez et al. (2017)
Mahalaxmi rice	47.75 ppm	Achieved 17.9% Cr(VI) removal over a prolonged period (120 days)	Pattnaik et al. (2020)

Note BCF—Bioconcentration Factor: metal concentration in plant tissue versus soil/solution (accumulation capacity); TF—Translocation Factor: metal concentration in shoots versus roots (translocation efficiency)

biosorption plays a significant role; microbial cell walls contain functional groups (–COOH, –OH, $-NH_2$, and phosphate) that interact with Cr(VI) species via electrostatic interactions, ion exchange, or complexation (Tripathi, 2002). This passive biosorption process is particularly effective under acidic conditions, where Cr predominantly exists as anionic species, such as $CrO_4{}^{2-}$ or $HCrO_{4-}$. For instance, *Bacillus subtilis* has shown a Cr removal efficiency of 48 mg L^{-1} (Quiton et al., 2018), while *Staphylococcus xylosus* and *Pseudomonas sp.* achieved 143 mg g^{-1} and 95 mg g^{-1}, respectively (Chug et al., 2016). Environmental factors, such as pH (optimal between 6.0 and 8.0), temperature, redox potential, and availability of electron donors, significantly influence Cr(VI) biotransformation. Reducing agents, like Fe^{2+} or sulfides, enhance microbial reduction through synergistic abiotic interactions.

Fungal strains also demonstrate strong detoxification potentials; *Aspergillus niger* removed 11.8 mg g^{-1} of Cr(VI) (Pena-Castro et al., 2004), *Fusarium verticillioides* reached up to 6400 mg g^{-1} (Chen et al., 2019), and *Penicillium janthinellum* showed 58.5 mg g^{-1} removal under highly acidic conditions (Barkhordar et al., 2004). Recently, microbial consortia have garnered attention for their enhanced efficiency in treating tannery wastewater and industrial effluents by harnessing synergistic interactions among bacteria, fungi, and algae, thereby improving metabolic diversity and environmental adaptability. Algal species also contribute to bioremediation through biosorption and the release of oxygen. Green algae, such as *Cladophora glomerata* and *Spyrogyra sp.*, have been shown to remove 66.6 mg g^{-1} (Losada et al., 2018) and 90 mg L^{-1} (Onyancha et al., 2008) of Cr(VI), respectively. As research advances, next-generation strategies such as genetically engineered strains, nanobiotechnology, and immobilized biocatalysts are emerging further to enhance the scope and effectiveness of Cr bioremediation. Representative examples of organisms and their Cr removal efficiencies are summarized in Table 9.

Table 9 Comparative Cr(VI) biosorption efficiencies of fungal, bacterial, and algal biosorbents

Biosorbent type	Common name	Scientific name	Removal capacity (mg g^{-1})	References
Fungi	Black mold	*Aspergillus niger*	11.79–96.30	(Mondal et al., 2017; Pena-Castro et al., 2004)
	Fusarium fungus	*Fusarium verticillioides*	6400	(Chen et al., 2019)
	Penicillium mold	*Penicillium janthinellum*	58.6	(Barkhordar & Ghias, 2004)
Bacteria	Staph bacteria	*Staphylococcus xylosus*	143	(Chug et al., 2016)
	Pseudomonad bacteria	*Pseudomonas sp.*	95	(Chug et al., 2016)
	Bacillus bacterium	*Bacillus subtilis*	14.54–48.00	(Quiton et al., 2018; Emran et al., 2019)
Cyanobacteria	Blue-green algae	*Spirulina sp.*	90.91	(Gupta et al., 2009)
		Nostoc calcicola HH-12	12.23	(Al-Homaidan et al., 2018)
Microalgae	Green microalga	*Scenedesmus sp.*	98.63	(Suganya et al., 2016)
	Filamentous green algae	*Spyrogyra sp.*	90	(Onyancha et al., 2008)
	Brown seaweed	*Sargassum sp.*	39.61–66.66	(Jayakumar et al., 2015; Krishnani et al., 2008)

7.5 Nanotechnology-Based Remediation

Nanotechnology offers a promising approach for the remediation of Cr(VI) through the use of advanced nanomaterials with high surface areas, superior sorption capacity, and vigorous redox activity. These materials facilitate both the reduction of Cr(VI) to the less toxic Cr(III) and the subsequent immobilization or adsorption of the reduced species, thereby enhancing Cr sequestration from aqueous and terrestrial environments. Among them, nanoscale zero-valent iron (nZVI) has been extensively studied for its strong reductive properties, where Cr(VI) is effectively reduced to Cr(III) via electron transfer from the iron core. The reaction proceeds as follows:

$$Cr_2O_7^{2-} + 3\,Fe^0 + 14\,H^+ \rightarrow 2\,Cr^{3+} + 3\,Fe^{2+} 7\,H_2O$$

The resulting Cr(III) typically precipitates as insoluble $Cr(OH)_3$, leading to effective immobilization. Coating nZVI with organic stabilizers such as polyacrylic acid, carboxymethyl cellulose, or starch improves its stability, dispersion, and reactivity in environmental matrices (Wang & Zhang, 1997).

Magnetic iron oxide nanoparticles (Fe_3O_4) also exhibit a high affinity for both Cr(VI) and Cr(III). Their functionalization with ligands such as amino or thiol groups enhances selective adsorption and enables easy magnetic recovery, thereby reducing secondary pollution. Adsorption mechanisms involve inner-sphere complexation and electrostatic interactions. Carbon nanotubes (CNTs) and graphene oxide (GO), particularly when functionalized with amine and hydroxyl groups, offer excellent Cr(VI) removal performance due to their high surface area and rich surface chemistry. These materials enhance Cr(VI) uptake under acidic conditions via hydrogen bonding, π–π interactions, and electrostatic attraction. CNT-GO hybrids embedded with metal oxides further improve reduction kinetics and capacity (Zhao et al., 2011). Metal–organic frameworks, especially zirconium-based variants like UiO-66 and UiO-66-NH_2, exhibit high Cr(VI) adsorption capacities, ranging from 32 mg g^{-1} to over 200 mg g^{-1}, through electrostatic interactions and hydrogen bonding, with some systems achieving rapid removal within minutes.

8 Conclusion

Chromium, though essential for various industrial applications, poses a significant environmental and public health threat due to the high toxicity and mobility of its hexavalent form, Cr(VI), and its complex redox behavior. The dynamic interconversion between Cr(III) and Cr(VI), regulated by pH, redox potential, organic matter, and microbial activity, plays a critical role in determining its toxicity and persistence across ecosystems. Industrial hotspots such as Hazaribagh (Bangladesh), Kanpur (India), and Qujing (China) underscore the global burden of Cr contamination, particularly in low- and middle-income countries with limited regulatory oversight.

While conventional chemical treatments, such as reduction and precipitation, are widely used, biological approaches, including biosorption, microbial transformation, and phytoremediation, offer more sustainable and cost-effective alternatives. Addressing Cr pollution requires a systems-level strategy that integrates environmental chemistry, toxicology, public health, and policy to ensure long-term protection of ecosystems and human well-being.

References

Abbott, A. P., Alaysuy, O., Antunes, A. P. M., Douglas, A. C., Guthrie-Strachan, J., & Wise, W. R. (2015). Processing of leather using deep eutectic solvents. *ACS Sustainable Chemical Engineering, 3*, 1241–1247.

Abou-Shanab, R. A., Ghanem, K., Ghanem, N., & Al-Kolaibe, A. (2008). The role of bacteria on heavy-metal extraction and uptake by plants growing on multi-metal-contaminated soils. *World Journal of Microbiology and Biotechnology, 24*(2), 253–262.

Agarwal, G. S., Bhuptawat, H. K., & Chaudhari, S. (2006). Biosorption of aqueous chromium (VI) by *Tamarindus indica* seeds. *Bioresource Technology, 97*, 949–956.

Agrawal, A. (2012). Toxicity and fate of heavy metals with particular reference to developing foetus. *Advances in Life Sciences, 2*(2), 29–38.

Aguilar-Ascón, E., Marrufo-Saldaña, L., & Neyra-Ascón, W. (2019). Reduction of total chromium levels from raw tannery wastewater via electrocoagulation using response surface methodology. *Journal of Ecological Engineering, 20*(11), 217–224.

Ahalya, N., Kanamadi, R. D., & Ramachandra, T. V. (2005). Biosorption of chromium (VI) from aqueous solutions by the husk of Bengal gram (*Cicer arientinum*). *Electronic Journal of Biotechnology, 8*(3), 258–264.

Ahemad, M. (2015). Enhancing phytoremediation of chromium-stressed soils through plant-growth-promoting bacteria. *Journal of Genetic Engineering and Biotechnology, 13*(1), 51–58.

Ahmed, N., Islam, M. N., Hossain, M. Y., Rahman, A., & Sultana, A. (2019). Modified coconut coir to remove hexavalent chromium from aqueous solution. *Bangladesh Journal of Scientific and Industrial Research, 54*, 89–98.

Al-Homaidan, A. A., Al-Qahtani, H. S., Al-Ghanayem, A. A., Ameen, F., & Ibraheem, I. B. (2018). Potential use of green algae as a biosorbent for hexavalent chromium removal from aqueous solutions. *Saudi Journal of Biological Sciences, 25*(8), 1733–1738.

Ali, M., & Khan, M. J. (2013). Geophysical hunt for chromite in ophiolite. *International Journal Economic and Environment Geology, 4*, 22–28.

Ali, A., Saeed, K., & Mabood, F. (2016). Removal of chromium (VI) from aqueous medium using chemically modified banana peels as efficient low-cost adsorbent. *Alexandria Engineering Journal, 55*, 2933–2942.

Altun, T., Parlayıcı, Ş, & Pehlivan, E. (2016). Hexavalent chromium removal using agricultural waste "rye husk." *Desalination and Water Treatment, 57*, 17748–17756.

Anju, M. (2017). Biotechnological strategies for remediation of toxic metal(loid)s from environment. In S. Gahlawat, R. Salar, P. Siwach, J. Duhan, S. Kumar, & P. Kaur (Eds.), *Plant biotechnology: Recent advancements and developments* (pp. 315–359). Springer.

Anjum, H., Chemat, F., Gnanasundaram, N., Arunagiri, A., & Thanabalan, M. (2017). Impact of surface modification of activated carbon on BTEX removal from aqueous solutions: A review. In T. K. Sen (Ed.), *Air, gas, and water pollution control using industrial and agricultural solid wastes adsorbents* (pp. 293–312). CRC Press.

Aoyama, M. (2003). Removal of Cr(VI) from aqueous solution by London plane leaves. *Journal of Chemical Technology and Biotechnology, 78*, 601–604.

Argast, S. (1995). Detrital origin of fuchsite-bearing quartzites in the western Dharwar craton, Karnataka,India. *Journal Geological Society of India, 45*(5), 559–575.

Austrheim, H., Corfu, F., & Renggli, C. J. (2021). From peridotite to fuchsite bearing quartzite via carbonation and weathering: With implications for the Pb budget of continental crust. *Contributions to Mineralogy and Petrology, 176*, 94. https://doi.org/10.1007/s00410-021-01851-z

Avudainayagam, S., Megharaj, M., Owens, G., Kookana, R. S., Chittleborough, D., & Naidu, R. (2003). Chemistry of chromium in soils with emphasis on tannery waste sites. In G. W. Ware (Ed.), *Reviews of environmental contamination and toxicology* (pp. 53–91). Springer.

Bagdon, R. E., & Hazen, R. E. (1991). Skin permeation and cutaneous hypersensitivity as a basis for making risk assessments of chromium as a soil contaminant. *Environmental Health Perspectives, 92*, 111–119.

Banu, S. K., Stanley, J. A., Sivakumar, K. K., Arosh, J. A., Taylor, R. J., & Burghardt, R. C. (2017). Chromium VI-induced developmental toxicity of placenta is mediated through spatiotemporal dysregulation of cell survival and apoptotic proteins. *Reproductive Toxicology, 68*, 171–190.

Banu, S. K., Stanley, J. A., Taylor, R. J., Sivakumar, K. K., Arosh, J. A., Zeng, L., Pennathur, S., & Padmanabhan, V. (2018). Sexually dimorphic impact of chromium accumulation on human placental oxidative stress and apoptosis. *Toxicological Sciences, 161*, 375–387.

Bao, Z., Al, T., Couillard, M., Poirier, G., Bain, J., Shrimpton, H. K., Finfrock, Y. Z., Lanzirotti, A., Paktunc, D., & Saurette, E. (2021). A cross scale investigation of galena oxidation and controls on mobilization of lead in mine waste rock. *Journal of Hazardous Materials, 412*, Article 125130. https://doi.org/10.1016/j.jhazmat.2021.125130

Barkhordar, B., & Ghias, A. M. (2004). Comparision of Langmuir and Freundlich equilibriums in Cr, Cu and Ni adsorption by *Sargassum. Iranian Journal of Environmental Health, Science, and Engineering, 1*(2), 58–64.

Barnes, S.-J., Mansur, E. T., Maier, W. D., & Prevec, S. A. (2023). A comparison of trace element concentrations in chromite from komatiites, picrites, and layered intrusions: Implications for the formation of massive chromite layers. *Canadian Journal of Earth Sciences, 60*, 97–132.

Barnes, S.-J., Mansur, E. T., Pagé, P., Meric, J., & Arguin, J.-P. (2020). Major and trace element compositions of chromites from the Stillwater, Bushveld and Great Dyke intrusions compared with chromites from komatiites, boninites and large igneous provinces. *EGU General Assembly 2020*, Online, 4–8 May 2020, EGU2020–9768. https://doi.org/10.5194/egusphere-egu2020-9768

Barnhart, J. (1997). Occurrences, uses, and properties of chromium, regulatory toxicology and pharmacology. *Regulatory Toxicology and Pharmacology, 26*(1), S3–S7.

Bartlett, R. J., & Kimble, J. M. (1976). Behavior of chromium in soils: I. Trivalent forms1. *Journal of Environmental Quality, 5*(4), 379–383.

Beacom, S. E., Hardesty, D. W., & Doty, W. R. (1964). Some physical characteristics of nickel and chromium deposits and their relation to corrosion resistance. *Transactions of the IMF, 42*(1), 77–85.

Begum, H. A., Haque, A. M., Islam, M. D., Hasan, M. M., Ahmed, S., Razzak, M., & Khan, R. A. (2020). Analysis of the adsorption of toxic chromium (VI) by untreated and chitosan treated banana and areca fiber. *Journal of Textile Science and Technology, 6*, 81–106.

Bolan, S., Padhye, L. P., Jasemizad, T., Govarthanan, M., Karmegam, N., Wijesekara, H., et al. (2024). Impacts of climate change on the fate of contaminants through extreme weather events. *Science of the Total Environment, 909*, Article 168388. https://doi.org/10.1016/j.scitotenv.2023.168388

Bouhadi, M., Daoui, O., El Hajjouji, H., Elkhattabi, S., Chtita, S., Talbi, M., & Fougrach, H. (2023). Physiological and molecular modeling investigations of the relationship between sulfate and chromium VI uptake in *Vicia faba* L. *Biocatalysis and Agricultural Biotechnology, 47*, Article 102554. https://doi.org/10.1016/j.bcab.2022.102554

Bourotte, C., Bertolo, R., Almodovar, M., & Hirata, R. (2009). Natural occurrence of hexavalent chromium in a sedimentary aquifer in Urânia, State of São Paulo, Brazil. *Anais Da Academia Brasileria De Ciências, 81*, 227–242.
Brown, E., Taylor, M., & Liu, C.-K. (2016). Soluble collagen approach to a combination tannage mechanism. *Journal of American Leather Chemists Association, 111*, 141–147.
Cao, X., Guo, J., Mao, J., & Lan, Y. (2011). Adsorption and mobility of Cr (III)–organic acid complexes in soils. *Journal of Hazardous Materials, 192*(3), 1533–1538.
Celik, A., Dost, K., & Sezer, H. (2004). An investigation of chromium (VI) ion removal from wastewaters by adsorption on residual lignin. *Fresenius Environmental Bulletin, 13*(2), 124–127.
Chakraborty, R., Verma, R., Asthana, A., Vidya, S. S., & Singh, A. K. (2021). Adsorption of hazardous chromium(VI) ions from aqueous solutions using modified sawdust: Kinetics, isotherm and thermodynamic modelling. *International Journal of Environmental Analytical Chemistry, 101*, 911–928.
Chen, H., Dou, J., & Xu, H. (2018). The effect of low-molecular-weight organic-acids (LMWOAs) on treatment of chromium-contaminated soils by compost-phytoremediation: Kinetics of the chromium release and fractionation. *Journal of Environmental Sciences, 70*, 45–53.
Chen, H. Y., Lu, Y. N., Yin, P. C., Li, X., & Shan, Y. (2019). Exploring the mechanisms of biosorption of Cr (Vi) by marine-derived *Penicillium janthinellum* P1. *International Journal of Agriculture and Biology, 22*, 913–920.
Cheung, K. H., & Gu, J.-D. (2007). Mechanism of hexavalent chromium detoxification by microorganisms and bioremediation application potential: A review. *International Biodeterioration and Biodegradation, 59*, 8–15.
Chirwa, E. M. N., & Molokwane, P. E. (2011). Biological Cr(VI) reduction: Microbial diversity, kinetics and biotechnological solutions to pollution. In A. Sofo (Eds.), *Biodiversity*. InTech Cambridge. https://doi.org/10.5772/24311
Chitraprabha, K., & Sathyavathi, S. (2018). Phytoextraction of chromium from electroplating effluent by *Tagetes erecta* (L.). *Sustainable Environment Research, 28*(3), 128–134.
Choudhury, P. R., Majumdar, S., Sahoo, G. C., Saha, S., & Mondal, P. (2018). High pressure ultrafiltration CuO/hydroxyethyl cellulose composite ceramic membrane for separation of Cr(VI) and Pb(II) from contaminated water. *Chemical Engineering Journal, 336*, 570–578.
Chrysochoou, M., Theologou, E., Bompoti, N., Dermatas, D., & Panagiotakis, I. (2016). Occurrence, origin and transformation processes of geogenic chromium in soils and sediments. *Current Pollution Reports, 2*, 224–235.
Chug, R., Gour, V. S., Mathur, S., & Kothari, S. L. (2016). Optimization of extracellular polymeric substances production using *Azotobacter beijreinckii* and *Bacillus subtilis* and its application in chromium(VI) removal. *Bioresource Technology, 214*, 604–608.
Claudio, L. (2007). Standing on principle: The global push for environmental justice. *Environmental Health Perspectives, 115*, A500–A503.
Coles, C. A., Arisi, J. A., Organ, M., & Veinott, G. I. (2014). Leaching of chromium, copper, and arsenic from CCA-treated utility poles. *Applied and Environmental Soil Science, 2014*, 1–11.
Cook, R. B. (2007). Crocoite: Dundas, Tasmania, Australia. *Rocks and Minerals, 82*, 50–54.
Cooper, M. A., Ball, N. A., Hawthorne, F. C., Paar, W. H., Roberts, A. C., & Moffatt, E. (2011). Georgerobinsonite, $Pb_4(CrO_4)_2(OH)_2FCl$, a new chromate mineral from the Mammoth–St. Anthony mine, Tiger, Pinal County, Arizona: Description and crystal structure. *The Canadian Mineralogist, 49*, 865–876.
Córdova Bravo, H. M., Vargas Parker, R., Cesare Coral, M. F., Flores del Pino, L., & Visitación Figueroa, L. (2014). Tratamiento de las aguas residuales del proceso de curtido tradicional y alternativo que utiliza acomplejantes de cromo. *Revista De La Socieded Química Del Perú, 80*, 183–191.
Costa, M., & Klein, C. B. (2006). Toxicity and carcinogenicity of chromium compounds in humans. *Critical Reviews in Toxicology, 36*, 155–163.

Crane, M., Leverett, P., Shaddick, L., Williams, P., Kloprogge, J. T., & Frost, R. (2001). The $PbCrO_4$-$PbSO_4$ system and its mineralogical significance. *Neues Jahrbuch fur Mineralogie, Monatshefte, 11*(11), 505–519.

Dakiky, M., Khamis, M., Manassra, A., & Mer'Eb, M. (2002). Selective adsorption of chromium(VI) in industrial wastewater using low-cost abundantly available adsorbents. *Advances in Environmental Research, 6*, 533–540.

Daneshvar, N., Salari, D., & Aber, S. (2002). Chromium adsorption and Cr(VI) reduction to trivalent chromium in aqueous solutions by soya cake. *Journal of Hazardous Materials, 94*, 49–61.

Das, A. P., & Mishra, S. (2008). Hexavalent chromium (VI): Environment pollutant and health hazard. *Journal of Environmental Research and Development, 2*, 386–392.

Datta, J. K., Bandhyopadhyay, A., Banerjee, A., & Mondal, N. K. (2011). Phytotoxic effect of chromium on the germination, seedling growth of some wheat (*Triticum aestivum* L.) cultivars under laboratory condition. *Journal of Agricultural Technology, 7*, 395–402.

de Melo, B. A. G., Motta, F. L., & Santana, M. H. A. (2016). Humic acids: Structural properties and multiple functionalities for novel technological developments. *Materials Science and Engineering: C, 62*, 967–974.

Di Luca, G. A., Hadad, H. R., Mufarrege, M. M., Maine, M. A., & Sanchez, G. C. (2014). Improvement of Cr phytoremediation by *Pistia stratiotes* in presence of nutrients. *International Journal of Phytoremediation, 16*(2), 167–178.

Dong, X., Ma, L. Q., & Li, Y. (2011). Characteristics and mechanisms of hexavalent chromium removal by biochar from sugar beet tailing. *Journal of Hazardous Materials, 190*, 909–915.

Dotaniya, M. L., Thakur, J. K., Meena, V. D., Jajoria, D. K., & Rathor, G. (2014). Chromium pollution: A threat to environment-a review. *Agricultural Reviews, 35*, 153–157.

Dupont, L., & Guillon, E. (2003). Removal of hexavalent chromium with a lignocellulosic substrate extracted from wheat bran. *Environmental Science and Technology, 37*, 4235–4241.

Dupont, L., Bouanda, J., Ghanbaja, J., Dumonceau, J., & Aplincourt, M. (2004). Use of analytical microscopy to analyze the speciation of copper and chromium ions onto a low-cost biomaterial. *Journal of Colloid and Interface Science, 279*, 418–424.

Emran, F. K., Walli, H. A., D'asheesh, T. I. A., Shinjar, F. J., & Abdulameer, M. A. (2019). Chromium bioremoval by *Staphylococcus sp.* and *Pseudomonas sp.* isolated from industrial waste water in Baghdad, Iraq. *Journal of Physics: Conference Series, 1294*, 072014. https://doi.org/10.1088/1742-6596/1294/7/072014

Fanfani, L., & Zanazzi, P. F. (1967). Structural similarities of some secondary lead minerals. *Mineralogical Magazine and Journal of the Mineralogical Society, 36*(280), 522–529.

Faust, S. D., Aly, O. M. (2018). *Chemistry of water treatment* (2nd edn). CRC Press. https://doi.org/10.1201/9781315139265

Fendorf, S., Wielinga, B. W., & Hansel, C. M. (2000). Chromium transformations in natural environments: The role of biological and abiological processes in chromium(VI) reduction. *International Geology Review, 42*, 691–701.

Finley, B. L., Kerger, B. D., Katona, M. W., Gargas, M. L., Corbett, G. C., & Paustenbach, D. J. (1997). Human ingestion of chromium (VI) in drinking water: Pharmacokinetics following repeated exposure. *Toxicology and Applied Pharmacology, 142*, 151–159.

Focardi, S., Pepi, M., & Focardi, S. E. (2013). Microbial reduction of hexavalent chromium as a mechanism of detoxification and possible bioremediation applications. In R. Chamy & F. Rosenkranz (Eds.), *Biodegradation-life of science*. IntechOpen. https://doi.org/10.5772/56365

Friedrich, M. J. (2016). Industrial pollution threatens millions in developing world. *Journal of American Medical Association, 316*, 2473. https://doi.org/10.1001/jama.2016.17431

Friess, S. L. (1989). Carcinogenic risk assessment criteria associated with inhalation of airborne particulates containing chromium(VI/III). *Science of the Total Environment, 86*(1–2), 109–112.

Frost, R. L. (2004). Raman microscopy of selected chromate minerals. *Journal of Raman Spectroscopy, 35*, 153–158.

Furukawa, Y., Watkins, J. L., Kim, J., Curry, K. J., & Bennett, R. H. (2009). Aggregation of montmorillonite and organic matter in aqueous media containing artificial seawater. *Geochemical Transactions, 10*, 2. https://doi.org/10.1186/1467-4866-10-2

Galán, B., Castañeda, D., & Ortiz, I. (2005). Removal and recovery of Cr(VI) from polluted ground waters: A comparative study of ion-exchange technologies. *Water Research, 39*, 4317–4324.

Gao, Y., & Xia, J. (2011). Chromium contamination accident in China: Viewing environment policy of China. *Environmental Science and Technology, 45*, 8605–8606.

Gao, Z., Geng, Y., Xiao, S., & Zhuang, M. (2022). Mapping the global anthropogenic chromium cycle: Implications for resource efficiency and potential supply risk. *Environmental Science and Technology, 56*, 10904–10915.

Gardea-Torresdey, J. L., Peralta-Videa, J. R., Montes, M., De la Rosa, G., & Corral-Diaz, B. (2004). Bioaccumulation of cadmium, chromium and copper by *Convolvulus arvensis* L.: Impact on plant growth and uptake of nutritional elements. *Bioresource Technology, 92*(3), 229–235.

Gasik, M., Dashevskii, V., & Bizhanov, A. (2020). Metallurgy of chromium ferroalloys. In C. P. Bergmann (Ed.), *Ferroalloys* (pp. 125–159). Springer.

Gaur, V., & Shankar, P. A. (2008). Surface modification of activated carbon for the removal of water impurities. *Water Conditioning and Purification International Magazine*, 1–5.

Gawande, P. R., & Kaware, J. P. (2016). Preparation and activation of activated carbon from waste materials-a review. *International Journal for Research in Applied Science and Engineering Technology, 4*, 1–4.

Genchi, G., Lauria, G., Catalano, A., Carocci, A., & Sinicropi, M. S. (2021). The double face of metals: The intriguing case of chromium. *Applied Sciences, 11*(2), 638. https://doi.org/10.3390/app11020638

Gheju, M., & Balcu, I. (2017). Assisted green remediation of chromium pollution. *Journal of Environmental Management, 203*, 920–924.

Gilpavas, E., Dobrosz-Gómez, I., & Gómez-García, M. Á. (2011). The removal of the trivalent chromium from the leather tannery wastewater: The optimisation of the electro-coagulation process parameters. *Water Science and Technology, 63*, 385–394.

Gode, F., & Pehlivan, E. (2003). A comparative study of two chelating ion-exchange resins for the removal of chromium(III) from aqueous solution. *Journal of Hazardous Materials, 100*, 231–243.

Goetz, M. E., & Luch, A. (2008). Reactive species: A cell damaging rout assisting to chemical carcinogens. *Cancer Letters, 266*, 73–83.

Gomez, K., Quevedo, N., & Molina, L. (2021). Use of the biochar obtained by slow pyrolysis from *Ulex europaeus* in the removal of total chromium from the Bogotá-colombia River water. *Chemical Engineering Transactions, 86*, 289–294.

Gonzalez, M. H., Araújo, G. C., Pelizaro, C. B., Menezes, E. A., Lemos, S. G., De Sousa, G. B., & Nogueira, A. R. A. (2008). Coconut coir as biosorbent for Cr(VI) removal from laboratory wastewater. *Journal of Hazardous Materials, 159*, 252–256.

Gowd, S. S., Reddy, M. R., & Govil, P. K. (2010). Assessment of heavy metal contamination in soils at Jajmau (Kanpur) and Unnao industrial areas of the Ganga Plain, Uttar Pradesh, India. *Journal of Hazardous Materials, 174*, 113–121.

Griffin, R. A., Au, A. K., & Frost, R. R. (1977). Effect of pH on adsorption of chromium from landfill-leachate by clay minerals. *Journal of Environmental Science and Health. Part a: Environmental Science and Engineering, 12*, 431–449.

Gujar, A. R., Ambre, N. V., Iyer, S. D., Mislankar, P. G., & Loveson, V. J. (2010). Placer chromite along south Maharashtra, central west coast of India. *Current Science, 99*(4), 492–499.

Gul, A., Ma'amor, A., Khaligh, N. G., & Julkapli, N. M. (2022). Recent advancements in the applications of activated carbon for the heavy metals and dyes removal. *Chemical Engineering Research and Design, 186*, 276–299.

Guo, H., Liu, C., Yan, S., Yin, J., & Shan, J. (2024). Source, distribution, and geochemical processes of geogenic high chromium groundwater around the world: A critical review. *Journal of Hydrology, 638*, Article 131480. https://doi.org/10.1016/j.jhydrol.2024.131480

Gupta, V. K., & Rastogi, A. (2009). Biosorption of hexavalent chromium by raw and acid-treated green alga *Oedogonium hatei* from aqueous solutions. *Journal of Hazardous Materials, 163*(1), 396–402.

Gupta, P. K., Nair, V. K., Dalvi, V., Dhali, S., Malik, A., & Pant, K. K. (2024). Field-scale assessment of soil, water, plant, and soil microbiome in and around Rania-Khan Chandpur chromium contaminated site, India. *Journal of Hazardous Materials, 467*, Article 133747. https://doi.org/10.1016/j.jhazmat.2024.133747

Gustafsson, J. P., Persson, I., Oromieh, A. G., Van Schaik, J. W. J., Sjöstedt, C., & Kleja, D. B. (2014). Chromium(III) complexation to natural organic matter: Mechanisms and modeling. *Environmental Science and Technology, 48*, 1753–1761.

Gutterres, M., & Mella, B. (2014). Chromium in tannery wastewater. In S. K. Sharma (Eds.), *Heavy metals in water: Presence, removal and safety*. Royal Society of Chemistry (pp. 315–344). https://doi.org/10.1039/9781782620174

Hai, J., Liu, L., Tan, W., Hao, R., & Qiu, G. (2020). Catalytic oxidation and adsorption of Cr(III) on iron-manganese nodules under oxic conditions. *Journal of Hazardous Materials, 390*, Article 122166. https://doi.org/10.1016/j.jhazmat.2020.122166

Han, L., Gu, H., Lu, W., Li, H., Peng, W. X., Ma, N. L., et al. (2023). Progress in phytoremediation of chromium from the environment. *Chemosphere, 344*, Article 140307. https://doi.org/10.1016/j.chemosphere.2023.140307

Hao, Y., Ma, H., Wang, Q., Zhu, C., & He, A. (2022). Complexation behaviour and removal of organic-Cr (III) complexes from the environment: A review. *Ecotoxicology and Environmental Safety, 240*, Article 113676. https://doi.org/10.1016/j.ecoenv.2022.113676

Hausladen, D. M., Alexander-Ozinskas, A., McClain, C., & Fendorf, S. (2018). Hexavalent chromium sources and distribution in California groundwater. *Environmental Science and Technology, 52*, 8242–8251.

He, Y. T., Chen, & Traina, S. J. (2004). Inhibited Cr(VI) reduction by aqueous Fe(II) under hyperalkaline conditions. *Environmental Science and Technology*, *38*, 5535–5539.

Helene, L. P. I., Moreira, C. A., & Carrazza, L. P. (2016). Applied geophysics on a soil contaminated site by chromium of a tannery in Motuca (SP). *Brazilian Journal of Geophysics, 34*, 309–317.

Hessel, E. V., Staal, Y. C., Piersma, A. H., den Braver-Sewradj, S. P., & Ezendam, J. (2021). Occupational exposure to hexavalent chromium. Part I. Hazard assessment of non-cancer health effects. *Regulatory Toxicology and Pharmacology*, *126*, 105048. https://doi.org/10.1016/j.yrtph.2021.105048

Hokkanen, S., Bhatnagar, A., Repo, E., Lou, S., & Sillanpää, M. (2016). Calcium hydroxyapatite microfibrillated cellulose composite as a potential adsorbent for the removal of Cr(VI) from aqueous solution. *Chemical Engineering Journal, 283*, 445–452.

Hossini, H., Shafie, B., Niri, A. D., Nazari, M., Esfahlan, A. J., Ahmadpour, M., et al. (2022). A comprehensive review on human health effects of chromium: Insights on induced toxicity. *Environmental Science and Pollution Research, 29*, 70686–70705.

Hu, G., Zheng, P., Feng, H., & Jia, G. (2017). Imbalance of oxidative and reductive species involved in chromium (VI)-induced toxic effects. *Reactive Oxygen Species, 3*, 1–11.

In, B. M. (1994). In *Emerging technology for bioremediation of metals* (Vol. 2). CRC Press.

Inegbenebor, A. I., Thomas, J. H., & Williams, P. A. (1989). The chemical stability of mimetite and distribution coefficients for pyromorphite–mimetite solid-solutions. *Mineralogical Magazine, 53*, 363–371.

Islam, L. N., Rahman, A., Mahmud, Z., Nabi, A. N., Hossain, M., & Mohasin, M. (2015). Assessment of physicochemical and biochemical qualities of tannery effluents of Hazaribagh, Dhaka, and comparison with non-tannery water samples. *International Journal of Environment, 4*, 68–81.

Islam, M., Hasan, K. F., Deb, H., Faisal, A. M. M., & Xu, W. (2016). Improving the fastness properties of cotton fabric through the implementation of different mordanting agents dyed with natural dye extracted from Marigold. *American Journal of Polymer Science and Engineering, 4*, 1–17.

Islam, M. M., Khan, M. N., Biswas, S., Choudhury, T. R., Haque, P., Rashid, T. U., & Rahman, M. M. (2017). Preparation and characterization of bijoypur clay-crystalline cellulose composite for application as an adsorbent. *Advances in Materials Science, 2*, 1–7.

Islam, M. M., Islam, M. S., Maniruzzaman, M., Haque, M.M.-U., & Mohana, A. A. (2021). Banana rachis CNC/Clay composite filter for dye and heavy metals adsorption from industrial wastewater. *Engineering Science and Technology, 2*(2), 140–152.

Iyer, M., Anand, U., Thiruvenkataswamy, S., Babu, H. W. S., Narayanasamy, A., Prajapati, V. K., et al. (2023). A review of chromium (Cr) epigenetic toxicity and health hazards. *Science of the Total Environment, 882*, Article 163483. https://doi.org/10.1016/j.scitotenv.2023.163483

Jain, M., Garg, V. K., & Kadirvelu, K. (2009). Chromium(VI) removal from aqueous system using *Helianthus annuus* (sunflower) stem waste. *Journal of Hazardous Materials, 162*, 365–372.

Jang, H., & Topal, E. (2020). Transformation of the Australian mining industry and future prospects. *Mining Technology, 129*, 120–134.

Jayakumar, R., Rajasimman, M., & Karthikeyan, C. (2015). Sorption and desorption of hexavalent chromium using a novel brown marine algae *Sargassum myriocystum*. *Korean Journal of Chemical Engineering, 32*(10), 2031–2046.

Jiang, W., Cai, Q., Xu, W., Yang, M., Cai, Y., Dionysiou, D. D., & O'Shea, K. E. (2014). Cr(VI) adsorption and reduction by humic acid coated on magnetite. *Environmental Science and Technology, 48*, 8078–8085.

Jomova, K., & Valko, M. (2011). Advances in metal-induced oxidative stress and human disease. *Toxicology, 283*, 65–87.

Jordão, C. P., da Silva, A. C., Pereira, J. L., & Brune, W. (1999). Contaminação por crômio de águas de rios proveniente de curtumes em Minas Gerais. *Química Nova, 22*(1), 47–52.

Ju, L., Jiao, Z., Ge, S., Zhan, W., Liu, Y., Ren, Q., Liao, Q., Yang, Z., & Wang, Y. (2022). Formation, stability and mobility of soluble Cr(III) during Cr(VI) reduction by *Pannonibacter phragmitetus* BB. *Environmental Technology and Innovation, 27*, Article 102496. https://doi.org/10.1016/j.eti.2022.102496

Juárez, H. S. (2006). Contaminación del río Rímac por metales pesados y efecto en la agricultura en el cono este de Lima metropolitana. MS Thesis, Universidad Nacional Agraria la Molina, Lima-Perú (pp. 1–88). https://idl-bnc-idrc.dspacedirect.org/server/api/core/bitstreams/7e07c51e-14bb-4970-be4b-af085a03802e/content

Kapoor, R. T., Mfarrej, M. F. B., Alam, P., Rinklebe, J., & Ahmad, P. (2022). Accumulation of chromium in plants and its repercussion in animals and humans. *Environmental Pollution, 301*, Article 119044. https://doi.org/10.1016/j.envpol.2022.119044

Karim, Md. R., Manshoven, S., Islam, Md. R., Gascon, J. A., Ibarra, M., Diels, L., & Rahman, Md. M. (2013). Assessment of an urban contaminated site from tannery industries in Dhaka city, Bangladesh. *Journal of Hazardous Toxic Radioactive Waste, 17*, 52–61.

Katz, S. A., & Salem, H. (1993). The toxicology of chromium with respect to its chemical speciation: A review. *Journal of Applied Toxicology, 13*, 217–224.

Kazakis, N., Kantiranis, N., Voudouris, K. S., Mitrakas, M., Kaprara, E., & Pavlou, A. (2015). Geogenic Cr oxidation on the surface of mafic minerals and the hydrogeological conditions influencing hexavalent chromium concentrations in groundwater. *Science of the Total Environment, 514*, 224–238.

Keim, M. F., & Markl, G. (2015). Weathering of galena: Mineralogical processes, hydrogeochemical fluid path modeling, and estimation of the growth rate of pyromorphite. *American Mineralogist, 100*(7), 1584–1594.

Kerger, B. D., Finley, B. L., Corbett, G. E., Dodge, D. G., & Paustenbach, D. J. (1997). Ingestion of chromium(VI) in drinking water by human volunteers: Absorption, distribution, and excretion of single and repeated doses. *Journal of Toxicology and Environmental Health, 50*, 67–95.

Komives, T., & Gullner, G. (2006). Dendroremediation: The use of trees in cleaning up polluted soils. In M. Mackova, D. N. Dowling & T. Macek (Eds.), *Phytoremediation rhizoremediation* (pp. 23–31). Springer.

Krikorian, R., Eliassen, J. C., Boespflug, E. L., Nash, T. A., & Shidler, M. D. (2010). Improved cognitive-cerebral function in older adults with chromium supplementation. *Nutritional Neuroscience, 13*, 116–122.

Krishnani, K. K., Meng, X., Christodoulatos, C., & Boddu, V. M. (2008). Biosorption mechanism of nine different heavy metals onto biomatrix from rice husk. *Journal of Hazardous Materials, 153*(3), 1222–1234.

Krishnani, K. K., Meng, X., & Dupont, L. (2009). Metal ions binding onto lignocellulosic biosorbent. *Journal of Environmental Science Health, Part A, 44*, 688–699.

Lakshmi, S., & Sundaramoorthy, P. (2010). Effect of chromium on germination and seedling growth of vegetable crops. *Asian Journal of Science and Technology, 1*, 28–31.

Leimbach, M., Tschaar, C., Zapf, D., Kurniawan, M., Schmidt, U., & Bund, A. (2019). Relation between color and surface morphology of electrodeposited chromium for decorative applications. *Journal of Electrochemical Society, 166*, D205.

Levy, B. S., Levenstein, C., & Spangler, E. (1991). Environment and Health in Eastern Europe. *NEW Solutions: A Journal of Environmental and Occupational Health Policy, 1*, 19–27.

Li, J., Lin, Q., & Zhang, X. (2010). Mechanism of electron transfer in the bioadsorption of hexavalent chromium within *Leersia hexandra Swartz* granules by X-ray photoelectron spectroscopy. *Journal of Hazardous Materials, 182*, 598–602.

Li, J., Li, Y., Tian, X., Zou, L., Zhao, X., Wang, S., & Wang, S. (2017). The hardness and corrosion properties of trivalent chromium hard chromium. *Material Science and Applications, 8*, 1014.

Lindsay, D., Jr. (1999). Decorative and hard chromium plating. *Plating and Surface Finishing, 86*, 40–42.

Loizidou, M., Haralambous, K. J., Loukatos, A., & Dimitrakopoulou, D. (1992). Natural zeolites and their ion exchange behavior towards chromium. *Journal of Environmental Science and Health Part a: Environmental Science and Engineering and Toxicology, 27*, 1759–1769.

Lopez-Delacalle, M., Camejo, D. M., García-Martí, M., Nortes, P. A., Nieves-Cordones, M., Martínez, V., Rubio, F., Mittler, R., & Rivero, R. M. (2020). Using tomato recombinant lines to improve plant tolerance to stress combination through a more efficient nitrogen metabolism. *Frontiers in Plant Science, 10*, 1702. https://doi.org/10.3389/fpls.2019.01702

Losada, V. A. R., Bonilla, E. P., Pinilla, L. A. C., & Serrezuela, R. R. (2018). Removal of chromium in wastewater from tanneries applying bioremediation with algae, orange peels and citrus pectin. *Contemporary Engineering Sciences, 11*(9), 433–449.

Losi, M. E., Amrhein, C., & Frankenberger, W. T. (1994). Factors affecting chemical and biological reduction of hexavalent chromium in soil. *Environmental Toxicology and Chemistry, 13*, 1727–1735.

Loyaux-Lawniczak, S., Lecomte, P., & Ehrhardt, J.-J. (2001). Behavior of hexavalent chromium in a polluted groundwater: Redox processes and immobilization in soils. *Environmental Science and Technology, 35*, 1350–1357.

Mala, J. G. S. M., Takeuchi, S., Sujatha, D., & Mani, U. (2020). Microbial chromate reductases: Novel and potent mediators in chromium bioremediation-a review. *Applied Microbiology: Theory and Technology, 1*(1), 32–44.

Mandi, L., Tiglyene, S., & Jaouad, A. (2009). Depuration of tannery effluent by phytoremediation and infiltration percolation under arid climate. *Options Méditerranéennes, 88*, 199–205.

Manian, A. P., Paul, R., & Bechtold, T. (2016). Metal mordanting in dyeing with natural colourants. *Coloration Technology, 132*, 107–113.

Martyn, J. E., & Johnson, G. I. (1986). Geological setting and origin of fuchsite-bearing rocks near Menzies, Western Australia. *Australian Journal of Earth Science, 33*, 1–18.

Masscheleyn, P. H., Pardue, J. H., DeLaune, R. D., & Patrick, W. H., Jr. (1992). Chromium redox chemistry in a lower Mississippi valley bottomland hardwood wetland. *Environmental Science and Technology, 26*, 1217–1226.

Matamet, F. R. M., & Bonotto, D. M. (2013). Evaluation of the chromium contamination at Ribeirão dos Bagres, Franca (SP), Brazil, by the ^{210}Pb method. *Applied Radiation and Isotopes, 82*, 359–369.

Matrosova, E. A., Bobrov, A. V., & Bindi, L. (2020). Basics of geochemistry and mineralogy of chromium. *Geochemistry of chromium in the earth's mantle* (pp. 5–36). Springer.

Matsumoto, S. T., Mantovani, M. S., Malaguttii, M. I. A., Dias, A. L., Fonseca, I. C., & Marin-Morales, M. A. (2006). Genotoxicity and mutagenicity of water contaminated with tannery effluents, as evaluated by the micronucleus test and comet assay using the fish *Oreochromis niloticus* and chromosome aberrations in onion root-tips. *Genetics and Molecular Biology, 29*, 148–158.

Maurice, J. (2001). Tannery pollution threatens health of half-million Bangladesh residents. *Bulletin of the World Health Organization, 79*(1). https://www.scielosp.org/article/bwho/2001.v79n1/78-79/en

Meena, A. K., Kadirvelu, K., Mishra, G. K., Rajagopal, C., & Nagar, P. N. (2008). Adsorptive removal of heavy metals from aqueous solution by treated sawdust (*Acacia arabica*). *Journal of Hazardous Materials, 150*, 604–611.

Mellem, J. J., Baijnath, H., & Odhav, B. (2012). Bioaccumulation of Cr, Hg, As, Pb, Cu and Ni with the ability for hyperaccumulation by *Amaranthus dubius*. *African Journal of Agricultural Research, 7*(4), 591–596.

Minas, F., Chandravanshi, B. S., & Leta, S. (2017). Chemical precipitation method for chromium removal and its recovery from tannery wastewater in Ethiopia. *Chemical International, 3*, 291–305.

Mishra, S., & Bharagava, R. N. (2016). Toxic and genotoxic effects of hexavalent chromium in environment and its bioremediation strategies. *Journal of Environmental Science and Health, 34*, 1–32.

Mondal, N. K., Samanta, A., Dutta, S., & Chattoraj, S. (2017). Optimization of Cr (VI) biosorption onto Aspergillus niger using 3-level Box-Behnken design: Equilibrium, kinetic, thermodynamic and regeneration studies. *Journal of Genetic Engineering and Biotechnology, 15*, 151–160.

Monga, A., Fulke, A. B., & Dasgupta, D. (2022). Recent developments in essentiality of trivalent chromium and toxicity of hexavalent chromium: Implications on human health and remediation strategies. *Journal of Hazardous Materials Advances, 7*, Article 100113. https://doi.org/10.1016/j.hazadv.2022.100113

Morais, S., Fonseca, H. M., Oliveira, S. M., Oliveira, H., Gupta, V. K., Sharma, B., & de Lourdes Pereira, M. (2021). Environmental and health hazards of chromated copper arsenate-treated wood: A review. *International Journal of Environmental Research and Public Health, 18*, 5518. https://doi.org/10.3390/ijerph18115518

Murray, K. J., & Tebo, B. M. (2007). Cr(III) Is indirectly oxidized by the Mn(II)-oxidizing bacterium *Bacillus sp.* Strain SG-1. *Environmental Science and Technology, 41*, 528–533.

Nabais, J. V., Gomes, J. A., Carrott, P. J. M., Laginhas, C., & Roman, S. (2009). Phenol removal onto novel activated carbons made from lignocellulosic precursors: Influence of surface properties. *Journal of Hazardous Materials, 167*, 904–910.

Narayani, M., & Shetty, K. V. (2013). Chromium-resistant bacteria and their environmental condition for hexavalent chromium removal: A Review. *Critical Reviews in Environmental Science and Technology, 43*, 955–1009.

Nickens, K. P., Patierno, S. R., & Ceryak, S. (2010). Chromium genotoxicity: A double-edged sword. *Chemico-Biological Interactions, 188*, 276–288.

Onyancha, D., Mavura, W., Ngila, J. C., Ongoma, P., & Chacha, J. (2008). Studies of chromium removal from tannery wastewaters by algae biosorbents, *Spirogyra condensata* and *Rhizoclonium hieroglyphicum*. *Journal of Hazardous Materials, 158*(2–3), 605–614.

Ozdener, Y., Aydin, B., Fatma Aygün, S., & Yürekli, F. (2011). Effect of hexavalent chromium on the growth and physiological and biochemical parameters on *Brassica oleracea* L. var. acephala DC. *Acta Biologica Hungarica, 62*, 463–476.

Oze, C., Fendorf, S., Bird, D. K., & Coleman, R. G. (2004). Chromium geochemistry in serpentinized ultramafic rocks and serpentine soils from the Franciscan complex of California. *American Journal of Science, 304*, 67–101.

Oze, C., Bird, D. K., & Fendorf, S. (2007). Genesis of hexavalent chromium from natural sources in soil and groundwater. *Proceeding of the National Academy of Science, 104*, 6544–6549.

Papassiopi, N., Vaxevanidou, K., Christou, C., Karagianni, E., & Antipas, G. S. E. (2014). Synthesis, characterization and stability of Cr (III) and Fe (III) hydroxides. *Journal of Hazardous Materials, 264*, 490–497.

Pariser, H. H., Backeberg, N. R., Masson, O. C. M., & Bedder, J. C. M. (2018). Changing nickel and chromium stainless steel markets-a review. *Journal of the Southern African Institute of Mining and Metallurgy, 118*, 563–568.

Pattanapipitpaisal, P., Brown, N. L., & Macaskie, L. E. (2001). Chromate reduction by Microbacterium liquefaciens immobilised in polyvinyl alcohol. *Biotechnology Letters, 23*, 61–65.

Pattnaik, S., Dash, D., Mohapatra, S., Pattnaik, M., Marandi, A. K., Das, S., & Samantaray, D. P. (2020). Improvement of rice plant productivity by native Cr(VI) reducing and plant growth promoting soil bacteria *Enterobacter cloacae*. *Chemosphere, 240*, Article 124895. https://doi.org/10.1016/j.chemosphere.2019.124895

Paustenbach, D. J., Sheehan, P. J., Pauli, J. M., Wisser, L. M., & Finley, B. L. (1992). Review of the allergic contact dermatitis hazard posed by chromium-contaminated soil: Identifying a "safe" concentration. *Journal of Toxicology and Environmental Health, 37*, 177–207.

Paustenbach, D., Finley, B., Mowat, F., & Kerger, B. (2003). Human health risk and exposure assessment of chromium(VI) in tap water. *Journal of Toxicology and Environmental Health, 66*, 1295–1339.

Pehlivan, E., & Altun, T. (2008). Biosorption of chromium(VI) ion from aqueous solutions using walnut, hazelnut and almond shell. *Journal of Hazardous Materials, 155*, 378–384.

Pena-Castro, J. M., Martınez-Jerónimo, F., Esparza-Garcıa, F., & Canizares-Villanueva, R. O. (2004). Heavy metals removal by the microalga *Scenedesmus incrassatulus* in continuous cultures. *Bioresource Technology, 94*(2), 219–222.

Pinto, B. S., Luque, J. C., Valverde, G. R., Salazar, R. J., Fuentes, J. P., & Salas, J. V. (2017). Estudio de la contaminación por cromo (Cr) en el río chili y parque industrial de río seco (PIRS), Arequipa-Perú 2015–2016. *Veritas, 16*(1), 43–46.

Pokhrel, G. R., & Pokhre, G. (2022). The effect of chromium on human-health: A review. *BMC Journal of Scientific Research, 5*, 27–35.

Ponting, J., Kelly, T. J., Verhoef, A., Watts, M. J., & Sizmur, T. (2021). The impact of increased flooding occurrence on the mobility of potentially toxic elements in floodplain soil–A review. *Science of the Total Environment, 754*, Article 142040. https://doi.org/10.1016/j.scitotenv.2020.142040

Pourfadakari, S., Jorfi, S., Ahmadi, M., & Takdastan, A. (2017). Experimental data on adsorption of Cr(VI) from aqueous solution using nanosized cellulose fibers obtained from rice husk. *Data in Brief, 15*, 887–895.

Prasad, S., Yadav, K. K., Kumar, S., Gupta, N., Cabral-Pinto, M. M., Rezania, S., Radwan, N., & Alam, J. (2021). Chromium contamination and effect on environmental health and its remediation: A sustainable approaches. *Journal of Environmental Management, 285*, Article 112174. https://doi.org/10.1016/j.jenvman.2021.112174

Protsenko, V. S., Gordiienko, V. O., Danilov, F. I., & Kwon, S. C. (2011). Preparation and characterization of nanocrystalline hard chromium coatings using eco-friendly trivalent chromium bath. *Journal of Chemistry, 8*, 1925–1929.

Puzon, G. J., Roberts, A. G., Kramer, D. M., & Xun, L. (2005). Formation of soluble organo–chromium(III) complexes after chromate reduction in the presence of cellular organics. *Environmental Science and Technology, 39*, 2811–2817.

Qian, L., Zhang, W., Yan, J., Han, L., Gao, W., Liu, R., & Chen, M. (2016). Effective removal of heavy metal by biochar colloids under different pyrolysis temperatures. *Bioresource Technology, 206*, 217–224.

Quiton, K. G., Doma Jr, B., Futalan, C. M., & Wan, M. W. (2018). Removal of chromium (VI) and zinc (II) from aqueous solution using kaolin-supported bacterial biofilms of Gram-negative *E.*

coli and Gram-positive *Staphylococcus epidermidis*. *Sustainable Environment Research, 28*(5), 206–213.

Raase, P., Raith, M., Ackermand, D., Viswanatha, M. N., & Lal, R. K. (1983). Mineralogy of chromiferous quartzites from South India. *Journal Geological Society of India, 24*(10), 502–521.

Rafiqullah, I. M., Hossain, A. M., Ilias, M., & Hoq, M. M. (2008). Chromium (VI) reducing native microorganisms for remediation of chromium eco-toxicity in environment of Bangladesh. *Bangladesh Journal of Scientific Industrial Research, 43*, 455–466.

Rahaman, M. H., Islam, M. A., Islam, M. M., Rahman, M. A., & Alam, S. N. (2021). Biodegradable composite adsorbent of modified cellulose and chitosan to remove heavy metal ions from aqueous solution. *Current Research in Green and Sustainable Chemistry, 4*, Article 100119. https://doi.org/10.1016/j.crgsc.2021.100119

Rajapaksha, A. U., Alam, M. S., Chen, N., Alessi, D. S., Igalavithana, A. D., Tsang, D. C., & Ok, Y. S. (2018). Removal of hexavalent chromium in aqueous solutions using biochar: Chemical and spectroscopic investigations. *Science of the Total Environment, 625*, 1567–1573.

Raji, C., & Anirudhan, T. S. (1998). Batch Cr (VI) removal by polyacrylamide-grafted sawdust: Kinetics and thermodynamics. *Water Research, 32*, 3772–3780.

Randive, K. R., Korakoppa, M. M., Muley, S. V., Varade, A. M., Khandare, H. W., Lanjewar, S. G., Tiwari, R. R., & Aradhi, K. K. (2015). Paragenesis of Cr-rich muscovite and chlorite in green-mica quartzites of Saigaon-Palasgaon area, Western Bastar Craton, India. *Journal of Earth System and Science, 124*, 213–225.

Rangasamy, S., Alagirisamy, B., & Santiago, M. (2014). Effect of natural zeolite material on hexavalent chromium adsorption. *Journal of Industrial Pollution Control, 30*(1), 79–84.

Ranieri, E., Fratino, U., Petrella, A., Torretta, V., & Rada, E. C. (2016). *Ailanthus altissima* and *Phragmites australis* for chromium removal from a contaminated soil. *Environmental Science and Pollution Research, 23*, 15983–15989.

Rao, J. C., Zhang, X. X., Qin, B., & Fung, K. K. (2003). High-resolution transmission electron microscopy study of epitaxial passive films on nanocubes of chromium. *Philosophical Magazine Letters, 83*(6), 395–401.

Rastogi, S. K., Pandey, A., & Tripathi, S. (2008). Occupational health risks among the workers employed in leather tanneries at Kanpur. *Indian Journal of Occupational and Environmental Medicine, 12*, 132–135.

Ratasuk, N., & Nanny, M. A. (2007). Characterization and quantification of reversible redox sites in humic substances. *Environmental Science and Technology, 41*, 7844–7850.

Ray, A., & Jankar, J. S. (2022). A comparative study of chromium: Therapeutic uses and toxicological effects on human health. *Journal of Pharmacology and Pharmacotherapeutics, 13*, 239–245.

Redondo-Gómez, S., Mateos-Naranjo, E., Vecino-Bueno, I., & Feldman, S. R. (2011). Accumulation and tolerance characteristics of chromium in a cordgrass Cr-hyperaccumulator, *Spartina argentinensis*. *Journal of Hazardous Materials, 185*(2–3), 862–869.

Rengaraj, S., Yeon, K.-H., & Moon, S.-H. (2001). Removal of chromium from water and wastewater by ion exchange resins. *Journal of Hazardous Materials, 87*, 273–287.

Revathi, K., Haribabu, T. E., & Sudha, P. N. (2011). Phytoremediation of chromium contaminated soil using sorghum plant. *International Journal of Environmental Sciences, 2*(2), 417–428.

Rivera, H., Chira, J., Zambrano, K., & Petersen, P. (2007). Dispersión secundaria de los metales pesados en sedimentos de los ríos Chillón, Rímac y Lurín Departamento de Lima. *Revista Del Instituto De Investigaciones FIGMMG, 10*, 19–25.

Rock, M. L., James, B. R., & Helz, G. R. (2001). Hydrogen peroxide effects on chromium oxidation state and solubility in four diverse, chromium-enriched soils. *Environmental Science and Technology, 35*, 4054–4059.

Saha, R., Nandi, R., & Saha, B. (2011). Sources and toxicity of hexavalent chromium. *Journal of Coordination Chemistry, 64*, 1782–1806.

Sajad, M. A., Khan, M. S., Bahadur, S., Naeem, A., Ali, H., Batool, F., et al. (2020). Evaluation of chromium phytoremediation potential of some plant species of Dir Lower, Khyber Pakhtunkhwa, Pakistan. *Acta Ecologica Sinica, 40*(2), 158–165.

Salt, D. E., Blaylock, M., Kumar, N. P., Dushenkov, V., Ensley, B. D., Chet, I., & Raskin, I. (1995). Phytoremediation: A novel strategy for the removal of toxic metals from the environment using plants. *Biotechnology, 13*(5), 468–474.

Santiago, I. (1995). Removal of hexavalent chromium from water using tailored zeolites (pp. 37–43). New Mexico State University. https://p2infohouse.org/ref/28/27809.pdf

Serita, F., Homma, K., Fukuda, K., Sawatari, K., Suzuki, Y., & Toya, T. (1990). Development of an inhalation system of high melting point metal fumes and its use for exposure of rats to chromium and nickel fumes. *Industrial Health, 28*, 185–197.

Shadreck, M., & Mugadza, T. (2013). Chromium, an essential nutrient and pollutant: A review. *African Journal of Pure and Applied Chemistry, 7*, 310–317.

Sharma, G., Sharma, S., Kumar, A., Lai, C. W., Naushad, Mu., Shehnaz, Iqbal, J., & Stadler, F. J. (2022). Activated carbon as superadsorbent and sustainable material for diverse applications. *Adsorption Science and Technology, 2022*, 1–21. https://doi.org/10.1155/2022/4184809

Sheeja, J. P. (2016). Removal of chromium with the complexing agents from industrial effluents. *Oriental Journal of Chemistry, 32*, 2209–2213.

Shelnutt, S. R., Goad, P., & Belsito, D. V. (2007). Dermatological toxicity of hexavalent chromium. *Critical Reviews in Toxicology, 37*, 375–387.

Shi, J., Yang, Z., Dai, H., Lu, X., Peng, L., Tan, X., Shi, L., & Fahim, R. (2018). Preparation and application of modified zeolites as adsorbents in wastewater treatment. *Water Science Technology, 2017*(3), 621–635.

Shin, D. Y., Lee, S. M., Jang, Y., Lee, J., Lee, C. M., Cho, E.-M., & Seo, Y. R. (2023). Adverse human health effects of chromium by exposure route: A comprehensive review based on toxicogenomic approach. *International Journal of Molecular Sciences, 24*, 3410. https://doi.org/10.3390/ijms24043410

Singh, K. K., Rastogi, R., & Hasan, S. H. (2005). Removal of Cr(VI) from wastewater using rice bran. *Journal of Colloid and Interface Science, 290*, 61–68.

Singh, M., Verma, K. K., & Verma, C. L. (2013). An approach to develop a model for describing the influence of fluoride-contaminated irrigation water on physiological responses in poplar (*Populus deltoides* clone S_7C_{15}). *Acta Physiologiae Plantarum, 35*, 3357–3364.

Singh, V., Singh, N., Verma, M., Kamal, R., Tiwari, R., Sanjay Chivate, M., Rai, S. N., Kumar, A., Singh, A., & Singh, M. P. (2022). Hexavalent-chromium-induced oxidative stress and the protective role of antioxidants against cellular toxicity. *Antioxidants, 11*(12), 2375. https://doi.org/10.3390/antiox11122375

Sinha, V., Pakshirajan, K., & Chaturvedi, R. (2015). Evaluation of Cr(VI) exposed and unexposed plant parts of *Tradescantia pallida* (Rose) D. R. Hunt. for Cr removal from wastewater by biosorption. *International Journal of Phytoremediation, 17*, 1204–1211.

Sinha, V., Pakshirajan, K., & Chaturvedi, R. (2018). Chromium tolerance, bioaccumulation and localization in plants: An overview. *Journal of Environmental Management, 206*, 715–730.

Sinha, R., Kumar, R., Sharma, P., Kant, N., Shang, J., & Aminabhavi, T. M. (2022). Removal of hexavalent chromium via biochar-based adsorbents: State-of-the-art, challenges, and future perspectives. *Journal of Environmental Management, 317*, Article 115356. https://doi.org/10.1016/j.jenvman.2022.115356

Southwood, M. J., & Viljoen, E. A. (1986). Lead chromate minerals from the Argent lead-silver mine, Transvaal, South Africa: Crocoite, vauquelinite, and a possible second occurrence of embreyite. *Mineralogical Magazine, 50*, 728–730.

Srivastava, S. R. (2013). Chromium and human health-an overview. *Progressive Research, 8*(1), 1–9.

Standeven, A. M., & Wetterhahn, K. E. (1989). Chromium(VI) toxicity: Uptake, reduction, and DNA damage. *Journal of American College of Toxicology, 8*, 1275–1283.

Steblez, W. G. (1994). The chromium resources of Albania. *International Geology Review, 36*, 785–795.
Stern, A. H., Bagdon, R. E., Hazen, R. E., & Marzulli, F. N. (1993). Risk assessment of the allergic dermatitis potential of environmental exposure to hexavalent chromium. *Journal of Toxicology and Environmental Health, 40*, 613–641.
Suganya, E., Rangabhashiyam, S., Lity, A. V., & Selvaraju, N. (2016). Removal of hexavalent chromium from aqueous solution by a novel biosorbent *Caryota urens* seeds: Equilibrium and kinetic studies. *Desalination and Water Treatment, 57*(50), 23940–23950.
Sun, J., Luo, Y., Ye, J., Li, C., & Shi, J. (2022). Chromium distribution, leachability and speciation in a chrome plating site. *Processes, 10*(1), 142. https://doi.org/10.3390/pr10010142
Teklay, A. (2016). Physiological effect of chromium exposure: A review. *International Journal of Food Science, Nutrition and Dietetics, S7*(001), 1–11.
Tripathi, A. G., A. (2002). Bioremediation of toxic chromium from electroplating effluent by chromate-reducing *Pseudomonas aeruginosa* A2Chr in two bioreactors. *Applied Microbiology and Biotechnology*, *58*, 416–420.
Uddin, S., Afroz, H., Hossain, M., Briffa, J., Blundell, R., & Islam, M. R. (2023). Heavy metals/metalloids in food crops and their implications for human health. In *Heavy metal toxicity and tolerance in plants: A biological, omics, and genetic engineering approach* (pp. 59–86). https://doi.org/10.1002/9781119906506.ch3
Ullah, S., Liu, Q., Wang, S., Jan, A. U., Sharif, H. M. A., Ditta, A., et al. (2023). Sources, impacts, factors affecting Cr uptake in plants, and mechanisms behind phytoremediation of Cr-contaminated soils. *Science of the Total Environment, 899*, Article 165726. https://doi.org/10.1016/j.scitotenv.2023.165726
Uysal, M., & Ar, I. (2007). Removal of Cr(VI) from industrial wastewaters by adsorption: Part I: Determination of optimum conditions. *Journal of Hazardous Materials, 149*, 482–491.
Valavanidis, A., Vlachogianni, T., Fiotakis, K., & Loridas, S. (2013). Pulmonary oxidative stress, inflammation and cancer: Respirable particulate matter, fibrous dusts and ozone as major causes of lung carcinogenesis through reactive oxygen species mechanisms. *International Journal of Environmental Research and Public Health, 10*, 3886–3907.
Valentín-Reyes, J., García-Reyes, R. B., García-González, A., Soto-Regalado, E., & Cerino-Córdova, F. (2019). Adsorption mechanisms of hexavalent chromium from aqueous solutions on modified activated carbons. *Journal of Environmental Management, 236*, 815–822.
Valko, M., Morris, H., & Cronin, M. T. D. (2005). Metals, toxicity and oxidative stress. *Current Medicinal Chemistry, 12*(10), 1161–1208.
Velez, P. A., Talano, M. A., Paisio, C. E., Agostini, E., & González, P. S. (2017). Synergistic effect of chickpea plants and *Mesorhizobium* as a natural system for chromium phytoremediation. *Environmental Technology, 38*(17), 2164–2172.
Viti, C., Marchi, E., Decorosi, F., & Giovannetti, L. (2014). Molecular mechanisms of Cr(VI) resistance in bacteria and fungi. *FEMS Microbiology Reviews, 38*, 633–659.
Vorpahl, K. W., Jordan, P. T., & Mathews, E. J. (1976). Chrome alloy welding fume study. *American Industrial Hygiene Association Journal, 37*, 566–569.
Wan, Z., Li, M., Zhang, Q., Fan, Z., & Verpoort, F. (2018). Concurrent reduction-adsorption of chromium using m-phenylenediamine-modified magnetic chitosan: Kinetics, isotherm, and mechanism. *Environmental Science and Pollution Research, 25*, 17830–17841.
Wang, C. B., & Zhang, W. X. (1997). Synthesizing nanoscale iron particles for rapid and complete dechlorination of TCE and PCBs. *Environmental Science & Technology, 31*(7), 2154–2156.
Wang, Z., Chen, J., Chai, L., Yang, Z., Huang, S., & Zheng, Y. (2011). Environmental impact and site-specific human health risks of chromium in the vicinity of a ferro-alloy manufactory, China. *Journal of Hazardous Materials, 190*(1–3), 980–985.
Wang, Y., Su, H., Gu, Y., Song, X., & Zhao, J. (2017). Carcinogenicity of chromium and chemoprevention: A brief update. *OncoTargets and Therapy, 10*, 4065–4079.

Wang, Z., Lin, F., Huang, L., Chang, Z., Yang, B., Liu, S., Zheng, M., Lu, Y., & Chen, J. (2019). Cyclodextrin functionalized 3D-graphene for the removal of Cr (VI) with the easy and rapid separation strategy. *Environmental Pollution, 254*, Article 112854.

Witt, K. L., Stout, M. D., Herbert, R. A., Travlos, G. S., Kissling, G. E., Collins, B. J., & Hooth, M. J. (2013). Mechanistic insights from the NTP studies of chromium. *Toxicologic Pathology, 41*, 326–342.

Wong, V., Armknecht, S., & Zhitkovich, A. (2012). Metabolism of Cr(VI) by ascorbate but not glutathione is a low oxidant-generating process. *Journal of Trace Elements in Medicine and Biology, 26*, 192–196.

Wu, B. (2017). Relationship between the chromium or chromium compounds on immune functions in animals. *Journal of Veterinary Medicine and Health, 1*, 1–3.

Wu, Y., Deng, B., Xu, H., & Kornishi, H. (2005). Chromium(III) oxidation coupled with microbially mediated Mn(II) oxidation. *Geomicrobiology Journal, 22*, 161–170.

Xu, X., Huang, H., Zhang, Y., Xu, Z., & Cao, X. (2019). Biochar as both electron donor and electron shuttle for the reduction transformation of Cr (VI) during its sorption. *Environmental Pollution, 244*, 423–430.

Xu, T., Nan, F., Jiang, X., Tang, Y., Zeng, Y., Zhang, W., & Shi, B. (2020). Effect of soil pH on the transport, fractionation, and oxidation of chromium(III). *Ecotoxicology and Environmental Safety, 195*, Article 110459. https://doi.org/10.1016/j.ecoenv.2020.110459

Xu, Z.-R., Cai, M.-L., Chen, S.-H., Huang, X.-Y., Zhao, F.-J., & Wang, P. (2021). High-affinity sulfate transporter *Sultr1;2* is a major transporter for Cr(VI) uptake in plants. *Environmental Science and Technology, 55*, 1576–1584.

Yana, R. F. Q., Quispe, G. B., Betancur, H. N. C., Cáceres, S. H., Mendoza, A. P. C., & Miranda, P. S. Y. (2019). Concentración de metales pesados: Cromo, cadmio y plomo en los sedimentos superficiales en el río Coata, Perú. *Revista Boliviana De Química, 36*, 83–90.

Yang, Z., Zhang, X., Jiang, Z., Li, Q., Huang, P., Zheng, C., Liao, Q., & Yang, W. (2021). Reductive materials for remediation of hexavalent chromium contaminated soil–a review. *Science of the Total Environment, 773*, Article 145654. https://doi.org/10.1016/j.scitotenv.2021.145654

Yang, X., Zhong, H., Zhang, W., Liu, Y., Sun, N., Kuang, R., Wang, C., Zhan, A., Zhang, J., Tang, Q., & Li, Z. (2023). Progress in adsorptive removal of volatile organic compounds by zeolites. *Aerosol and Air Quality Research, 23*, Article 220442. https://doi.org/10.4209/aaqr.220442

Yasmeen, S., Kabiraz, M. K., Saha, B., Qadir, M. D., Gafur, M. D., & Masum, S. (2016). Chromium (VI) ions removal from tannery effluent using chitosan-microcrystalline cellulose composite as adsorbent. *International Research Journal of Pure and Applied Chemistry, 10*, 1–14.

Yatera, K., Morimoto, Y., Ueno, S., Noguchi, S., Kawaguchi, T., Tanaka, F., Suzuki, H., & Higashi, T. (2018). Cancer risks of hexavalent chromium in the respiratory tract. *Journal of UOEH, 40*(2), 157–172.

Yoshinaga, M., Ninomiya, H., Al Hossain, M. A., Sudo, M., Akhand, A. A., Ahsan, N., Alim, M. A., Khalequzzaman, M., Iida, M., & Yajima, I. (2018). A comprehensive study including monitoring, assessment of health effects and development of a remediation method for chromium pollution. *Chemosphere, 201*, 667–675.

Yur'Evich, K. S. (2022). Native gold from the crocoite pit of the Berezovsky ore field (Middle Urals). *Известия Уральского Государственного Горного Университета, 4*(68), 16–25.

Zapana-Huarache, S. V., Romero-Sánchez, C. K., Gonza, A. P. D., Torres-Huaco, F. D., & Rivera, A. M. L. (2020). Chromium (VI) bioremediation potential of filamentous fungi isolated from Peruvian tannery industry effluents. *Brazilian Journal of Microbiology, 51*, 271–278.

Zha, S., Yu, A., Wang, Z., Shi, Q., Cheng, X., Liu, C., Deng, C., Zeng, G., Luo, S., & Zhao, Z. (2024). Microbial strategies for effective hexavalent chromium removal: A comprehensive review. *Chemical Engineering Journal, 489*, Article 151457. https://doi.org/10.1016/j.cej.2024.151457

Zhang, Y., Mansel, B. W., Naffa, R., Cheong, S., Yao, Y., Holmes, G., Chen, H.-L., & Prabakar, S. (2018). Revealing molecular level indicators of collagen stability: Minimizing chrome usage in leather processing. *ACS Sustainable Chemical Engineering, 6*, 7096–7104.

Zhao, G., Ren, X., Gao, X., Tan, X., Li, J., Chen, C., Huang, Y., & Wang, X. (2011). Removal of Pb (II) ions from aqueous solutions on few-layered graphene oxide nanosheets. *Dalton Transactions, 40*(41), 10945–10952.

Zhitkovich, A. (2011). Chromium in drinking water: Sources, metabolism, and cancer risks. *Chemical Research in Toxicology, 24*, 1617–1629.

Zhu, L., Hong, C., Zhang, J., & Qiu, Y. (2023). Long-distance mobilization of chromium(III) in soil associated with submicron Cr_2O_3. *Journal of Hazardous Materials, 445*, Article 130519. https://doi.org/10.1016/j.jhazmat.2022.130519

Source and Distribution of Chromium in Environment

Source and Dispersion of Chromium in the Environment

Varun Dhiman

Abstract The element chromium, which is widely distributed in the Earth's crust and is essential for many industrial uses because of its unique physicochemical characteristics, poses serious threats to ecosystems and human health because of its poisonous hexavalent form (Cr(VI)). This chapter describes both natural and man-made sources of chromium in the environment, as well as its origins and modes of diffusion. As a result of geological processes such as weathering, volcanic activity, and leaching, chromium is released from natural sources, primarily ultramafic and mafic rocks, including chromite ($FeCr_2O_4$). Soil concentrations often range from 10 to 150 mg kg^{-1}, and they are higher in soils produced from serpentines. Chromium levels in the environment are greatly increased by human activities such as chromite mining, leather tanning, electroplating, and combustion processes; industrial effluents frequently contain more than 100 mg L^{-1} of Cr(VI]). Dispersion occurs in terrestrial, marine, and atmospheric systems due to intricate physical, chemical, and biological processes. Wet/dry deposition and particulate matter (PM2.5) are the means of atmospheric transport, whereas the solubility and redox dynamics of Cr(VI) influence aquatic dispersion. In soils, microbial reactions, adsorption, and desorption all affect chromium mobility. This chapter summarizes recent research aimed at elucidating these mechanisms, using examples such as the Sukinda Valley in India, where groundwater Cr(VI) concentrations exceed 1 mg/L^{-}. The environmental effects of Cr(VI), a Group 1 carcinogen, include toxicity to aquatic life and soil microbial ecosystems. Regulations and mitigation strategies, including chemical reduction, bioremediation, and phytoremediation, are evaluated together. Advanced speciation techniques and remediation based on nanotechnology are proposed as future research goals to fill in knowledge gaps. This analysis highlights the necessity of multidisciplinary approaches to reduce chromium contamination, safeguarding public health and ecosystems.

Keywords Chromium speciation · Environmental dispersion · Anthropogenic pollution · Toxicity and health risks · Remediation strategies

---

V. Dhiman (✉)
Department of Environment Science, Govt. College Una, Una, Himachal Pradesh, India
e-mail: varundhiman79@gmail.com

N. Kumar (ed.), *Chromium Toxicity*, https://doi.org/10.1007/978-3-032-14323-5_2

1 Introduction

With an average concentration of about 100 μg/g, chromium, a naturally occurring transition metal, is the twenty-first most prevalent element in the Earth's crust (Stefánsson et al., 2015). It is essential in industries including the production of stainless steel, leather tanning, electroplating, paint, pigment, and refractory because to its special qualities, which include corrosion resistance, thermal stability, and glossy finish (Sueker, 1964). As a petrogenetic proxy, the spinel mineral chromite ($FeCr_2O_4$), which is mostly recovered from ultramafic and mafic rocks, is a vital component of industrial economies around the world (Lenaz et al., 2004). Environmental impact of industrial chromium discharge on aquatic ecosystems and subsurface geology is illustrated in Fig. 1.

However, compared to the less hazardous trivalent form (Cr(III)), the high solubility, mobility, and toxicity of Cr(VI) provide serious environmental issues. According to DesMarias and Costa (2019), Cr(VI), a Group 1 carcinogen according to the International Agency for Research on Cancer, is associated with ecological disturbances, reproductive toxicity, and respiratory malignancies. To effectively reduce its effects and protect ecosystems and human health, it is essential to comprehend the sources, dispersion processes, and environmental fate of chromium.

Given their disparate chemical tendencies and toxicological profiles, chromium's Cr(III) and Cr(VI) oxidation states are the most environmentally significant. In minerals, soils, and sediments, Cr(III) predominates and forms low-bioavailability complexes with organic matter or insoluble hydroxides. On the other hand, Cr(VI), which is usually found as chromate (CrO_4^{2-}) or dichromate ($Cr_2O_7^{2-}$) ions, is very soluble and mobile in aquatic environments that range from neutral to alkaline, which

Fig. 1 Environmental impact of industrial chromium discharge on aquatic ecosystems and subsurface geology

makes it easier for it to migrate between environmental compartments (Hao et al., 2022). The intricacy of chromium's environmental behavior is highlighted by the fact that the transition between these states is dependent on redox conditions, pH, and catalysts such as manganese oxides, which can convert Cr(III) to Cr(VI) in aerobic environments.

Complex physical, chemical, and biological mechanisms regulate the dispersion of chromium in the atmosphere, water, and land. Particulate matter, such as PM2.5, is a form of chromium that is carried through the atmosphere. Its concentrations near industrial sources range from 1 to 100 ng m^{-3}. Water chemistry determines aquatic dispersion; in neutral to alkaline environments, Cr(VI) remains highly mobile while Cr(III) precipitates as insoluble compounds. Adsorption to clay minerals, organic matter concentration, microbial activity, and redox changes all affect the mobility and bioavailability of chromium in soils and sediments. Together, these processes determine how chromium behaves in the environment and how it may affect ecosystems and human populations.

The effects of chromium pollution on the environment and human health are severe. Even at concentrations as low as 0.1 mg L^{-1}, Cr(VI) is hazardous to aquatic life and can upset food webs and biodiversity. Elevated chromium levels in terrestrial areas reduce soil fertility by affecting plant development and soil microbial populations. Serious health concerns are associated with human exposure to Cr(VI) through contaminated soil, water, or air, as demonstrated at Hinkley, California, where groundwater Cr(VI) levels reached 0.58 mg/L^{-1}. Effective mitigation techniques and a thorough understanding of the sources and dispersion of chromium are necessary to address these issues.

With an emphasis on its sources, dispersion processes, and effects, this chapter summarizes current studies to give a thorough picture of the environmental dynamics of chromium. It aims to: (1) determine the sources of chromium, both natural and man-made, including geological, industrial, and biological contributions; (2) investigate the physical, chemical, and biological processes that control its distribution throughout environmental matrices; and (3) assess the health and ecological effects of chromium pollution, as well as mitigation and remediation techniques. Natural and man-made sources, dispersion mechanisms, health and environmental impacts, mitigation strategies, and future research needs are all covered in this chapter, which is backed up by recent findings and figures. The Sukinda Valley and Hinkley contamination incidents are two examples of case studies that offer a practical perspective. In an increasingly industrialized world, the chapter seeks to mitigate chromium contamination and safeguard ecosystems and human health by addressing these factors and guiding environmental management measures.

2 Natural Sources of Chromium

As the 21st most abundant metal in the crust of the Earth, chromium is a common element in environmental and geological systems. The main source of chromium is the spinel mineral chromite ($FeCr_2O_4$), which is the main ore used in industrial chromium extraction. Chromium is mostly found in ultramafic and mafic rocks, including peridotites, dunites, and basalts. The distribution of chromium in the environment is greatly influenced by its release into the environment through natural processes such as weathering, volcanic activity, and biological cycles. With the help of quantitative data and a case study to illustrate its inherent environmental dynamics, this part explores the geological sources of chromium, its mobilization through environmental processes, and the biological mechanisms that affect its cycle.

2.1 Geological Origins

The chromite deposits found in ophiolite complexes and layered igneous intrusions are the main source of chromium in ultramafic and mafic lithologies. Major concentrations of chromite with gravity-controlled chromium can be found in places like the Kemi Mine in Finland and the Bushveld Complex in South Africa (Latypov et al., 2017). Serpentine soils, which are made of weathered ultramafic rocks, can have extremely high chromium contents when they are generated from these parent materials.

2.2 Weathering and Mobilization

Chromium is mobilized from geological reservoirs into soils, sediments, and aquatic systems by natural processes as leaching, erosion, and chemical and physical weathering. The comparatively insoluble Cr(III) that is released by chemical weathering of chromite-bearing rocks tends to adsorb onto clay minerals, iron oxides, or organic matter in soils, limiting its mobility. However, in oxidizing conditions, especially in alkaline soils or waterways, manganese oxides or other oxidants can oxidize Cr(III) to Cr(VI), creating highly mobile soluble chromate (CrO_4^{2-}) or dichromate ($Cr_2O_7^{2-}$) ions (Chen et al., 2021).

Another important natural source of chromium is volcanic activity, which releases particles containing chromium into the environment. Long-distance transportation and wet or dry deposition of these particles—which are frequently in the form of fine ash—can lead to the buildup of chromium in soils and water bodies. Research indicates that atmospheric chromium concentrations may be influenced by volcanic eruptions (Shah et al., 2020). Water contamination in aquatic systems is mostly caused by chromium mobilization (e.g., chromite) in natural waters through leaching from chromium-rich rocks and soils (Prasad et al., 2021).

2.3 Biological Cycling

Chromium's bioavailability and cycling in the environment are greatly influenced by biological processes. Some plants may absorb large amounts of chromium from soils, concentrating it in their tissues and changing its distribution in the environment. These plants are referred to as hyperaccumulators. Certain species, including *Salsola kali*, *Alyssum murale*, *Paspalum distichum*, *Pycnandra acuminate*, and *Phyllanthus balgooyi*, can collect heavy metals, including chromium. In addition to lowering soil chromium levels, this phytoextraction method may make it easier for herbivores to reintroduce it into the food chain. The mediating role of microorganisms in chromium redox reactions is crucial. *Pseudomonas* and *P. aeruginosa* are two examples of bacteria that may convert Cr(III) to Cr(VI) under oxidizing circumstances, enhancing the toxicity and mobility of the latter (Ozturk et al., 2012).

By using hydrogenase and cytochrome, sulfate-reducing bacteria such as *Desulfovibrio vulgaris*, on the other hand, can convert Cr(VI) to Cr(III) under reducing conditions, generating insoluble precipitates that restrict its environmental dispersion (Chardin et al., 2002). Environmental elements that act as electron donors, such as pH, oxygen availability, and the presence of organic carbon, are crucial to these microbial processes.

2.4 Case Study

A noteworthy illustration of chromium's natural environmental presence can be found in the serpentine soils of the Jasper Ridge Biological Preserve in the Central Coast Range. The content of chromium in these ultramafic rock-derived soils is 4,760 mg kg^{-1} (Oze et al., 2004). The high concentration of chromium, primarily in the form of Cr(III), supports distinct plant communities that have adapted to environments rich in metals, like the chromium hyperaccumulator *Streptanthus polygaloides*. Cr(III) can, however, be converted to Cr(VI) by localized oxidation processes, especially in regions with alkaline groundwater, which raises the content of Cr(VI) in adjacent water bodies.

3 Anthropogenic Sources of Chromium

Human activity is the main source of increased environmental chromium concentrations, greatly increasing the natural baseline amounts set by geological processes. Industrial operations that emit large amounts of chromium, especially in its hazardous hexavalent form (Cr(VI)), into the atmosphere, water, and land include chromite mining, leather tanning, electroplating, pigment manufacture, and combustion. These actions not only make chromium more hazardous to the environment, but they also make it more mobile and bioavailable because of the molecular characteristics of Cr(VI), which is much more soluble and mobile than the more stable trivalent form (Cr(III)). In order to give a thorough picture of human-driven chromium pollution, this part looks at the main anthropogenic sources of chromium, measures their contributions, and illustrates their effects on the environment using case studies and current research findings.

3.1 *Mining and Metallurgy*

One of the main human-caused sources of environmental chromium is chromite mining and metallurgical processing. South Africa, India, and Kazakhstan mine chromite ($FeCr_2O_4$), the major mineral of chromium, and together they produce more than 90% of the world's chromite (Papp & Lipin, 2001). Chromite ore extraction and processing produce dust, slag, and wastewater, all of which frequently have high concentrations of Cr(VI) because of oxidative conditions during smelting. Chromium ore processing tailings are especially troublesome because they may contain high levels of Cr(VI), which could contaminate groundwater.

For instance, open-pit mining and poor waste management have caused Cr(VI) levels in groundwater leaching to reach 1.2 mg L^{-1}, significantly higher than the WHO's recommended level of 0.05 mg L^{-1} in the Sukinda Valley in Odisha, India, a major chromite mining hub that releases approximately 11.73 tons of chromium (Nayak et al., 2020). These tailings increase the environmental mobility and toxicity of Cr(III) by facilitating its oxidation to Cr(VI) when exposed to air and water.

3.2 *Leather Tanning*

Due to the tanning process's usage of trivalent chromium and chromium(III) sulfate ($Cr_2(SO_4)_3$) to stabilize hides, the leather tanning industry is a significant source of chromium pollution (Codreanu et al., 2024). Every year, millions of tons of chromium salts are consumed worldwide by the leather industry, and residual Cr(VI) from processing oxidation is frequently found in effluents. Untreated or insufficiently treated tannery wastewater may contain amounts of Cr(VI), which can be hazardous to the environment if released into rivers or soils. There have been reports of Cr(VI) levels in surface waters in areas with extensive tanning activities, such as Kanpur and Hazaribagh, Bangladesh, affecting aquatic ecosystems and downstream residents (Sharma & Sharma, 2017). Inappropriate disposal of sludge that contains chromium makes soil pollution more worse.

3.3 Electroplating and Pigments

Chromium is used extensively in the creation of pigments for paints, dyes, and ceramics as well as in electroplating to create coatings that are resistant to corrosion. Cr(VI) compounds, including chromic acid (H_2CrO_4), are used in electroplating operations. These compounds are likely to be released into the environment through wastewater or inappropriate waste disposal, which greatly contributes to local water pollution. Similarly, waste from the manufacturing of chromium-based pigments, including chrome yellow ($PbCrO_4$), adds Cr(VI) to soils and waterways. In areas with inadequate regulatory supervision, such portions of Southeast Asia, incorrect pigment waste disposal has resulted in soil chromium levels that have an impact on groundwater quality and agricultural production (Zhang et al., 2022).

3.4 Combustion Processes

Chromium is released into the environment as fly ash and particulate particles during combustion activities, such as burning coal and burning trash. Chromium is present in trace amounts in coal, especially in some geological formations. During combustion, this element is volatilized and released as tiny particles, such as PM2.5 (Bartoňová & Raclavská, 2022). These particles have a great range and help deposit chromium in soils and waterways through the atmosphere. Fly ash contains chromium, which contributes to atmospheric emissions from trash incineration, especially from industrial and municipal garbage that contains chromium-based products.

3.5 Quantitative Overview and Global Impact

According to estimates, industrial operations emit millions of tons of chromium into the environment each year, making up a significant portion of the world's anthropogenic chromium emissions (Tchounwou et al., 2012). Every year, the leather tanning sector alone is thought to be responsible for 40% of the world's chromium pollution, with untreated effluents frequently having high concentrations of Cr(VI) (Ricky et al., 2022). Every year, more than 10 million tons of chromium-containing tailings are produced by chromite mining activities in nations like South Africa and India, which contaminate nearby land and water.

Anthropogenic sources of Cr(VI) release greatly increase the metal's toxicity and environmental mobility, causing soil, water, and air contamination on a large scale. Ecosystems and human health are at long-term risk due to the persistence of Cr(VI) in aquatic systems and its potential to cause cancer. The dispersion processes that enable the transfer of chromium across environmental compartments and the methods to lessen its effects will be examined in the subsequent parts of this chapter.

4 Dispersion Mechanisms of Chromium

A complex interaction of physical, chemical, and biological processes controls the dispersion of chromium in the environment, affecting its destiny, transformation, and movement through terrestrial, aquatic, and atmospheric compartments. The oxidation state of chromium is closely related to these mechanisms, since trivalent chromium (Cr(III)) and hexavalent chromium (Cr(VI)) display different spectral characteristics and opposing chemical tendencies (Dvoynenko et al., 2021). Thermodynamic stability limits the mobility of Cr(III), which is characterized by low solubility and strong adsorption to soil particles or organic matter, while Cr(VI), which is present as highly soluble chromate (CrO_4^{2-}) or dichromate ($Cr_2O_7^{2-}$)ions, is easily moved in water and air (Zhu et al., 2023). Chemical reactions showing chromium transformations in the environment is presented in Fig. 2.

4.1 Atmospheric Dispersion

According to Assefa et al. (2024), chromium enters the atmosphere through both natural processes like volcanic emissions and lava and man-made ones such debris incineration, coal combustion, and industrial emissions. According to Moonwiriyakit et al. (2024), hexavalent chromium is mostly linked to particulate matter once it is in the air. Fine particles (PM2.5) are very mobile because they may travel great distances and stay suspended for long periods. The amount of chromium in the atmosphere varies greatly between metropolitan locations close to industrial sources, like smelters or power plants, and rural areas, where background levels are lower. Weather conditions that affect the amount of dispersion and deposition, such as wind speed, turbulence, and humidity, have an impact on the movement of these

Fig. 2 Chemical reactions showing chromium transformations in the environment

$$Cr_2O_7^{2-} + 6\,Fe^{2+} + 14\,H^+ \longrightarrow 2\,Cr^{3+} + 6\,Fe^{3+} + 7\,H_2O$$

$$CrO_4^{2-} + MnS + H_2O \longrightarrow Cr(OH)_3 + Mn^{2+} + SO_4^{2-}$$

particles. Through both wet and dry deposition, atmospheric chromium is transferred to terrestrial and aquatic systems.

In precipitation, chromium-containing particles are scavenged and transported to soils, rivers, and lakes, resulting in wet deposition. Research indicates that wet deposition in industrial areas can have a substantial impact on surface environments by contributing chromium. The gravity settling of particles onto surfaces, known as dry deposition, is most noticeable close to emission sources, when soil chromium buildup occurs. Deposition efficiency is influenced by surface properties, wind patterns, and particle size; finer particles allow for longer-range transport and more widespread environmental pollution. Conceptual framework for the atmospheric emission, transformation, and deposition of chromium species (Cr(III) and Cr(VI)) is illustrated in Fig. 3.

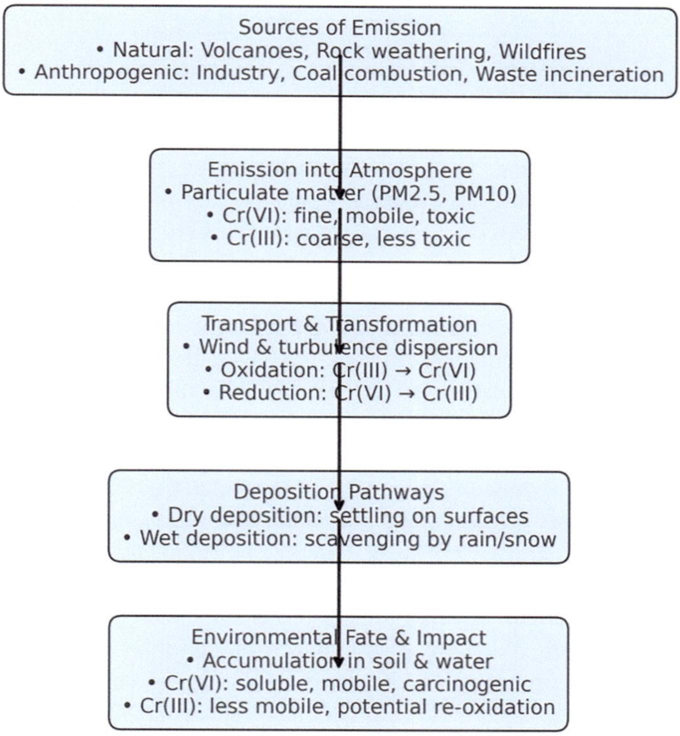

Fig. 3 Conceptual framework for the atmospheric emission, transformation, and deposition of chromium species (Cr(III) and Cr(VI))

4.2 Aquatic Dispersion

Cr(VI) is extremely mobile in neutral to alkaline waters (pH6) and exists as soluble $CrO_4{}^{2-}$ or $Cr_2O_7{}^{2-}$ ions. The degree of aquatic dispersion is noticeable in areas where human activity is substantial. Leaching from mine tailings in the Sukinda Valley, India, a major global center for chromite mining, has caused groundwater Cr(VI) concentrations to significantly exceed the WHO drinking water recommendation of 0.05 mgl^{-1}. In a similar vein, industrial wastewater that is dumped into large rivers, like the Ganges in India, has higher levels of Cr(VI), which affects ecosystems downstream and the people who depend on these water supplies for agriculture and drinking. Because of the high solubility of Cr(VI) and the absence of natural attenuation in oxidizing or alkaline aquatic environments, these increased concentrations can be transported over tens to hundreds of kilometers.

4.3 Soil and Sediment Dispersion

The adsorption, desorption, and redox mechanisms that control chromium dispersion in terrestrial environments are greatly impacted by the environmental factors and the makeup of the soil. Because clay minerals, iron oxides, and organic matter strongly absorbed Cr(III), it had little mobility and accumulated in soil horizons. But under oxidizing circumstances, especially when manganese oxides (MnO_2) are present, Cr(III) can be oxidized to Cr(VI), which greatly increases its solubility and groundwater leaching potential (Guha et al., 2001). Acidic soils ($pH < 6$) encourage Cr(III) retention by precipitation and adsorption, while alkaline soils ($pH > 7$) boost Cr(VI) mobility, making this change more noticeable.

Chromium is accumulated by sedimentation and precipitation processes in aquatic environments, where sediments act as sinks for the metal. Cr(III) usually precipitates as hydroxides in rivers and lakes or forms stable complexes in sediment layers by binding to organic materials. Cr(VI) can, however, be mobile in sediment pore fluids, especially in aerobic settings, which makes it easier for it to return to the water column (Hao et al., 2022). In order to control the dispersion of chromium in soils and sediments, microbial activity is essential. Although this process is frequently sluggish in natural environments because there aren't many electron donors available, anaerobic bacteria like *Desulfovibrio vulgaris* can convert Cr(VI) to Cr(III), creating insoluble precipitates that restrict mobility. On the other hand, Cr(III) oxidation can be catalyzed by aerobic microorganisms, improving dispersion in oxidative conditions.

4.4 *Case Study: Chromium Dispersion in the Ganges River*

India's Ganges River serves as an example of the intricate chromium dispersion in aquatic systems caused by human activity. Kanpur's leather tanning and electroplating industries discharge Cr(VI) into the river as industrial effluents. These high concentrations threaten aquatic ecosystems, especially fish populations and benthic creatures, and have an impact on water quality as they are carried downstream over hundreds of kilometers. Because of reductive processes in anaerobic sediment layers, sediments along the river serve as both sinks and secondary sources of chromium, with the majority of chromium found in sediment samples as Cr(III) (Sharma et al., 2012). The environmental fate of chromium in major river systems is dynamic, as evidenced by the high mobility of Cr(VI) in neutral to alkaline river waters, which favors its dispersion, and the ability of sediment-bound Cr(III) to remobilize under changing redox conditions.

5 Environmental and Health Impacts

The high toxicity, solubility, and mobility of chromium in environmental systems, especially in its hexavalent form (Cr(VI)), make it a serious threat to ecosystems and human health. However, under certain circumstances, trivalent chromium (Cr(III)), despite being less hazardous and comparatively immobile, can exacerbate environmental stress. Chromium pollution has ecological effects that disturb both terrestrial and aquatic ecosystems, impacting food webs, biodiversity, and soil function (Sharma et al., 2022). According to the International Agency for Research on Cancer (IARC), Cr(VI) is a Group 1 carcinogen that can cause serious health effects, including as carcinogenicity and mutagenicity, when humans are exposed to it through eating, inhalation, or skin contact. This part explores the negative effects of chromium contamination on the environment and human health, highlighting the urgent need for efficient mitigation techniques to safeguard ecosystems and public health by combining quantitative data, current research, and a case study.

5.1 *Ecological Impacts: Aquatic Ecosystems*

Cr(VI) is extremely poisonous to a wide variety of aquatic creatures, such as fish, algae, and invertebrates, which causes food chain disruptions and a decline in biodiversity. According to research, exposure to Cr(VI) causes zebrafish (*Danio rerio*) and rainbow trout (*Oncorhynchus mykiss*) to have aberrant development, reproductive failure, and higher mortality rates. For example, exposure to 0.2 mg L^{-1}, Cr(VI) can decrease fish spawning success by as much as 50%, which can have a cascade effect on ecosystem stability and predator–prey dynamics (Dang et al., 2023).

Particularly vulnerable are algae, which are important primary producers. Growth suppression has been seen at Cr(VI) concentrations as low as 0.05 mg L^{-1}, which lowers primary productivity and modifies aquatic food webs. The decreased survival and reproduction of invertebrates, such as mollusks and crustaceans, exacerbates ecological imbalances. Because of its high solubility and mobility in neutral to alkaline fluids, Cr(VI) can persist and travel long distances in rivers, lakes, and groundwater, which increases its ecological impact. Concentrations of Cr(VI) cause fish populations and benthic ecosystems to drop in contaminated systems such as the Ganges River in India; sediment-bound Cr(III) may serve as a secondary source in response to shifting redox conditions (Paul, 2017). These consequences highlight the necessity of strict regulations on chromium emissions in order to protect aquatic environments.

5.2 Terrestrial Ecosystems

Elevated chromium levels, especially Cr(VI), hurt microbial populations, plant growth, and soil fertility in terrestrial ecosystems. Chromium inhibits photosynthesis, root elongation, seed germination, and ultra-structural changes in plant cells; it is frequently found close to industrial areas like tanneries or mining operations (Sharma et al., 2020). For instance, jeopardizes food security and agricultural productivity. Additionally, soil microbial communities—which are essential for nutrient cycling—are disrupted by chromium poisoning. Concentrations of Cr(VI) more than 100 mg kg^{-1} can reduce microbial biomass by as much as 50%, which hinders processes like organic matter decomposition and nitrogen fixation.

Native plants like *Streptanthus polygaloides* have developed tolerance mechanisms in areas with naturally high chromium levels, such as serpentine soils in California's Coast Range, but anthropogenic inputs worsen ecological stress and cause long-term degradation of soil functionality. These effects demonstrate the wider ramifications of chromium pollution on the sustainability of agriculture and terrestrial ecosystems.

5.3 Human Health Impacts

Ingestion of tainted water or food, skin contact with contaminated soils or materials, and inhalation of airborne particles are the main ways that humans are exposed to Cr(VI). Cr(VI) is categorized as a category 1B carcinogen by the European Union, and there is strong evidence that inhalation exposure causes lung cancer, nasal tumors, and other respiratory illnesses (den Braver-Sewradj et al., 2021). Concentrations of Cr(VI) in the air are a major concern for occupational exposure in industries such as chromite mining, leather tanning, and electroplating. One important exposure route,

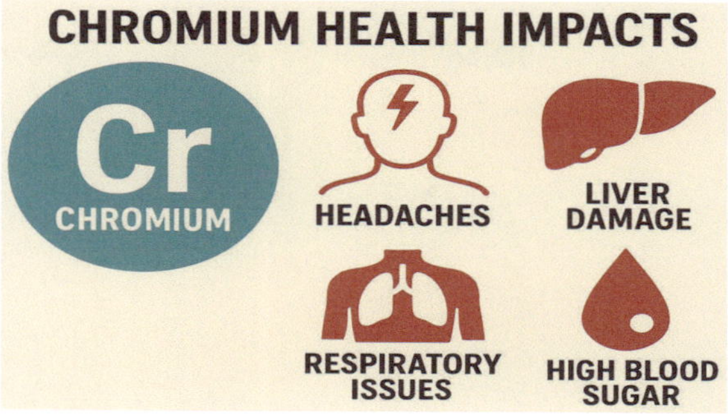

Fig. 4 Systemic toxicity of chromium exposure

especially in industrial areas, is drinking water tainted with Cr(VI). Systemic toxicity of chromium exposure is illustrated in Fig. 4.

A higher risk of cancer, liver and kidney damage, and gastrointestinal issues is linked to long-term exposure to Cr(VI) levels exceeding 0.05 mg L^{-1}. Especially for personnel who handle chromium salts, dermal contact with materials containing Cr(VI), such as leather goods or industrial waste, can result in skin ulcers and allergic contact dermatitis. Because Cr(VI) may penetrate cell membranes and cause abnormal transcription and DNA replication, it has the capacity to cause genetic abnormalities and cancer (Nickens et al., 2010). Communities close to industrial sites are disproportionately exposed to these health concerns, which are especially noticeable in developing nations with lax enforcement of regulations. Comparative toxicological profile of chromium (III) and chromium (VI): exposure routes, health effects, and regulatory standards is tabulated in Table 1.

5.4 *Case Study: Hinkley, California*

A notable example of the detrimental effects of Cr(VI) on human health and the environment is the contamination incident in Hinkley, California (Roberts, 2021). Pacific Gas and Electric Company contaminated groundwater with Cr(VI) concentrations surpassing the U.S. Environmental Protection Agency's maximum contamination level of 0.58 mg/L by discharging wastewater containing Cr(VI) into unlined ponds during the 1950s and 1980s (Izbicki et al., 2023). Due to increased rates of cancer, respiratory problems, and dermatological illnesses reported by locals exposed through drinking water, a high-profile legal battle and significant cleanup measures resulted. The long-term health hazards of Cr(VI) contamination are highlighted by this example, as is the significance of regulatory supervision in averting such occurrences.

Table 1 Comparative toxicological profile of chromium (III) and chromium (VI): exposure routes, health effects, and regulatory standards

Chromium type	Exposure routes	Health consequences	Symptoms observed	Impacted systems	Regulatory limits	Vulnerable groups
Chromium (III)	Inhalation of dust, ingestion via food or supplements	Low toxicity; supports metabolic functions in trace amounts. High doses may lead to minor irritation or digestive upset	Mild skin rash, stomach discomfort, occasional nausea	Skin, digestive system	OSHA limit: 0.5 mg/m^3 (inhalable); Dietary guideline: 20–35 µg/day (adults)	Workers in tanning or metallurgy, individuals taking high-dose supplements
Chromium(VI)	Inhalation of fine particles, skin contact with solutions, and ingestion of tainted water	Highly hazardous; causes cancer (lungs), severe respiratory issues, skin allergies, and organ damage	Nasal damage, breathing difficulties, skin ulcers, kidney issues, and liver abnormalities	Lungs, skin, kidneys, liver, and genetic material	OSHA limit: 0.005 mg/m^3 (action level: 0.0025 mg/m^3); EPA water standard: 0.1 mg/L (total chromium)	Factory workers (e.g., plating, welding), and residents near industrial zones
Both types	Long-term exposure through air, water, or skin	Cumulative effects include oxidative damage and potential systemic toxicity	Chronic fatigue, skin lesions, and possible reproductive issues (primarily Cr(VI))	Multiple systems (lungs, kidneys, liver, reproductive)	NIOSH recommendation for Cr(VI): 0.0002 mg/m^3 (inhalation)	Communities near emission sources, workers with mixed exposures

6 Mitigation and Remediation Strategies

An integrated strategy combining source management, cutting-edge treatment technology, and strong regulatory frameworks is needed to mitigate chromium pollution, especially that of the extremely hazardous and transportable hexavalent chromium (Cr(VI)). Compared to the more stable trivalent chromium (Cr(III)), the environmental persistence and health hazards of Cr(VI) call for specific measures to cut emissions, clean up polluted areas, and stop additional effects on the environment and human health. With the use of contemporary studies and quantitative data, this part assesses important mitigation and remediation techniques, such as source control, chemical remediation, bioremediation, and phytoremediation. Additionally included are regulatory actions and their function in reducing chromium contamination, highlighting the necessity of international collaboration in tackling this widespread environmental issue.

6.1 Source Control

One of the most important methods for lowering environmental inputs is to avert chromium pollution at its source. Emissions of Cr(VI) can be considerably reduced by using cleaner production methods, such as replacing chromium salts in leather tanning with vegetable tannins, glutaraldehyde, or compounds based on titanium (Zhang et al., 2016). For example, some facilities have seen a 90% reduction in effluent Cr(VI) contents since implementing chromium-free tanning techniques. Cr(VI) leaking into groundwater can be avoided in chromite mining by using appropriate waste management techniques, such as lining tailings ponds and neutralizing alkaline wastes.

The use of UG2 mining techniques and closed-loop water systems has decreased the amount of chromium discharge from mining activities in areas such as the Bushveld Complex in South Africa (Mooney, 1997). These actions are economical in reducing long-term environmental harm and remedial expenses, even though they demand an upfront investment.

6.2 Chemical Remediation

Wastewater and soils contaminated with Cr(VI) are commonly treated using chemical remediation procedures, especially reduction-precipitation. This procedure entails precipitating Cr(III) as insoluble hydroxides ($Cr(OH)_3$) after reducing Cr(VI) to Cr(III) with reducing agents like ferrous sulfate ($FeSO_4$), sodium metabisulfite, or zero-valent iron (Azis et al., 2021). For instance, ferrous sulfate treatment can meet strict regulatory standards by lowering Cr(VI) contents in wastewater from 100 mg L^{-1} to less than 0.05 mg L^{-1} (Qin et al., 2005).

This technique works well for industrial tanning and electroplating effluents, but it needs precise pH control to work as best it can. In contaminated groundwater, emerging technologies like nanotechnology-based adsorbents (such as iron oxide nanoparticles) improve removal efficiency and reduce Cr(VI) by more than 95% (Aragaw et al., 2021).

6.3 Bioremediation

Microorganisms that can convert Cr(VI) to Cr(III) are used in bioremediation to provide an economical and sustainable remediation method. Under specific circumstances (such as anaerobic settings with organic carbon sources), bacterial species like *Bacillus subtilis* and *Pseudomonas aeruginosa* have demonstrated potential in lowering Cr(VI) concentrations in soils and water by up to 80% to 95% (Ruparelia & Patel, 2024). In chromium-contaminated soils next to tanneries, for example, bioremediation experiments have decreased Cr(VI) levels. However, site-specific tailoring is required since several parameters, including soil pH, oxygen availability, and the presence of electron donors, limit the effectiveness of bioremediation.

6.4 Phytoremediation

In order to remove chromium from contaminated soils and lessen its environmental impact, phytoremediation uses hyperaccumulator plants like *Leersia hexandra* and *Alyssum murale*. By accumulating Cr(VI) concentrations in their tissues, these plants can efficiently extract chromium from soil matrices (Oubohssaine & Dahmani, 2024). Despite its cost-effectiveness and sustainability, slow growth rates and the requirement for appropriate disposal of chromium-laden biomass to stop re-entry into the environment are drawbacks of phytoremediation.

6.5 Regulatory Frameworks

Controlling chromium pollution and guaranteeing adherence to environmental regulations depend on regulatory structures. While the European Union enforces a stricter restriction of 0.05 mg L^{-1} for Cr(VI), the U.S. Environmental Protection Agency (EPA) sets a maximum contamination threshold for total chromium in drinking water at 0.1 mg L^{-1}. The Central Pollution Control Board's standards in nations like India require industrial emissions to have Cr(VI) effluent limits of 0.1 mg/L, although enforcement is still uneven. Reducing chromium contamination requires strict monitoring, public awareness campaigns, and global standardization, especially in developing countries with substantial industrial activity.

7 Future Directions and Research Needs

Critical research gaps still exist despite tremendous progress in our understanding of chromium's environmental behavior, especially the differential dynamics of trivalent (Cr(III)) and hexavalent (Cr(VI)) forms. The creation of all-encompassing plans to effectively anticipate, control, and mitigate chromium pollution is hampered by these inadequacies. Improving speciation methods, developing remediation solutions, and addressing the long-term destiny of chromium under shifting environmental conditions are critical. Furthermore, addressing the transboundary nature of chromium dispersion and its effects requires interdisciplinary and international cooperation. With a focus on creative methods to improve our comprehension and management of chromium in the environment, this section highlights important research requirements and future directions. It is backed by recent literature and quantitative findings.

7.1 Long-Term Fate of Chromium

Chromium's long-term fate in intricate environmental systems is still poorly understood, especially in light of climate change, which modifies soil chemistry, precipitation patterns, and redox conditions. For example, studies have shown that elevated precipitation and flooding can boost the leaching of Cr(VI) from contaminated soils, leading to an increase in groundwater concentrations of Cr(VI). Long-term droughts, on the other hand, might concentrate chromium in soils, boosting its bioavailability (Wei et al., 2021). Research is required to use dynamic models that include hydrological, climatic, and geochemical aspects to model the long-term behavior of chromium under changing climatic conditions. Cr(VI) mobility in susceptible areas could be predicted by such models.

7.2 Speciation Dynamics

Chromium speciation (Cr(III) vs. Cr(VI)) in environmental matrices must be accurately and continuously monitored in order to forecast its mobility and toxicity. Atomic absorption spectroscopy is one of the current techniques that frequently lacks the sensitivity to identify speciation in complicated matrices or detect low concentrations. Cutting-edge analytical methods like inductively coupled plasma mass spectrometry (ICP-MS) and X-ray absorption spectroscopy (XAS) show promise for in situ speciation for Cr(VI) in various samples (Cuello et al., 2016). Future studies should concentrate on creating affordable, portable instruments for real-time speciation in soils, water, and air so that pollution hazards can be quickly assessed. In industrial areas, where effluents contain higher quantities of Cr(VI), these instruments might be especially helpful.

7.3 Emerging Remediation Technologies

Although phytoremediation, bioremediation, and chemical remediation have demonstrated effectiveness, their scalability and efficiency are still constrained. In lab conditions, nanotechnology-based adsorbents, including iron oxide nanoparticles, have shown an efficiency of over 95% in removing Cr(VI); nevertheless, stability and cost concerns limit field usage (Kuruthukulangara et al., 2024). To maximize nanomaterials for widespread usage, research is required, with an emphasis on environmental safety and recyclability. Furthermore, *Alyssum murale* and other hyperaccumulator plants may be genetically modified to improve chromium uptake and hence increase removal rates. Likewise, using synthetic biology techniques to increase the reduction of microbial Cr(VI) by species such as *Bacillus subtilis* may increase the effectiveness of soil bioremediation (Wróbel et al., 2023).

7.4 Interdisciplinary and International Collaboration

Multidisciplinary strategies combining environmental engineering, microbiology, geochemistry, and policy analysis are needed to address chromium pollution. For instance, integrating microbial genomes and geochemical modeling may clarify the processes of Cr(VI) reduction in anaerobic soils, guiding bioremediation tactics. Since chromium pollution frequently spreads across national borders through the air and water, international cooperation is equally important. International frameworks for technology transfer and regulatory harmonization would be advantageous for areas with high levels of industrial activity, such as South Asia and Sub-Saharan Africa. Transboundary Cr(VI) dispersion, which can reach concentrations of 0.5 mg L^{-1} in shared river systems like the Ganges, could be addressed by collaborative research and data sharing made possible by initiatives such as the United Nations Environment Program.

8 Conclusion

The presence of chromium in the environment, which is caused by a complex interaction between natural and man-made sources, poses serious problems because of its toxicity and chemical flexibility, especially in its hexavalent form (Cr(VI)). Natural soil concentrations of chromium, the 21st most abundant element in the Earth's crust, come from geological causes such as volcanic eruptions and ultramafic rocks. However, man-made operations that greatly increase ambient chromium levels include chromite mining, leather tanning, electroplating, and combustion. Physical, chemical, and biological processes regulate the dispersion of chromium across atmospheric, aquatic, and terrestrial compartments. Given its high solubility and mobility, Cr(VI) can be transported over great distances, as demonstrated in places like the Sukinda Valley, India, where groundwater levels of Cr(VI) can reach 1.2 mg L^{-1}. Significant ecological and health effects result from Cr(VI), which damages soil fertility and microbial populations and disturbs aquatic food webs at concentrations of more than 0.1 mg L^{-1}. The contamination event in Hinkley, California, when groundwater Cr(VI) reached 0.58 mg L^{-1}, shows that human exposure through ingestion, inhalation, or skin contact is associated with serious health concerns, such as lung cancer and gastrointestinal issues. Combining source management (such as chromium-free tanning), chemical remediation (such as reduction-precipitation reaching 0.05 mg L^{-1} Cr(VI) in wastewater), bioremediation, and phytoremediation is necessary for effective mitigation. Research gaps remain despite advancements, especially with regard to real-time speciation dynamics, the long-term fate of chromium under climate change, and the scalability of new technologies such as adsorbents based on nanotechnology. To create comprehensive solutions, interdisciplinary methods that integrate environmental engineering, microbiology, and geochemistry are crucial, as is international cooperation. To manage chromium pollution, it will be essential to keep developing analytical methods like X-ray absorption spectroscopy and ecologically sound restoration approaches. Society can guarantee the sustainable management of this vital component by tackling these issues, shielding ecosystems and public health from the widespread dangers of Cr(VI) contamination.

References

Aragaw, T. A., Bogale, F. M., & Aragaw, B. A. (2021). Iron-based nanoparticles in wastewater treatment: A review on synthesis methods, applications, and removal mechanisms. *Journal of Saudi Chemical Society, 25*, Article 101280. https://doi.org/10.1016/j.jscs.2021.101280

Assefa, H., Singh, S., Olu, F. E., Dhanjal, D. S., Mani, D., Khan, N. A., Singh, J., & Ramamurthy, P. C. (2024). Advances in adsorption technologies for hexavalent chromium removal: Mechanisms, materials, and optimization strategies. *Desalination and Water Treatment, 319*, Article 100576. https://doi.org/10.1016/j.dwt.2024.100576

Azis, M. Y., Amedyan, N., Hanefiatni, & Suprabawati, A. (2021). Study of reducing Chromium (VI) to Chromium (III) ion using reduction and coagulation methods for electroplating industrial

waste. *Journal of Physics: Conference Series, 1763*, 012042. https://doi.org/10.1088/1742-6596/1763/1/012042
Bartoňová, L., & Raclavská, H. (2022). Behavior of Cr during coal combustion: An overview. *Fuel, 322*, Article 124210. https://doi.org/10.1016/j.fuel.2022.124210
Chardin, B., Dolla, A., Chaspoul, F., Fardeau, M., Gallice, P., & Bruschi, M. (2002). Bioremediation of chromate: Thermodynamic analysis of the effects of Cr(VI) on sulfate-reducing bacteria. *Applied Microbiology and Biotechnology, 60*, 352–360. https://doi.org/10.1007/s00253-002-1091-8
Chen, K., Bocknek, L., & Manning, B. (2021). Oxidation of Cr(III) to Cr(VI) and Production of Mn(II) by Synthetic Manganese(IV) Oxide. *Crystals, 11*. https://doi.org/10.3390/cryst11040443
Codreanu (Manea), A. M. N., Stefan, D. S., Kim, L., & Stefan, M. (2024). Depollution of polymeric leather waste by applying the most current methods of chromium extraction. *Polymers, 16*. https://doi.org/10.3390/polym16111546
Cuello, S., Entwisle, J., Benning, J., Liu, C., Coburn, S., McAdam, K. G., Braybrook, J., & Goenaga-Infante, H. (2016). Complementary HPLC-ICP-MS and synchrotron X-ray absorption spectroscopy for speciation analysis of chromium in tobacco samples. *Journal of Analytical Atomic Spectrometry, 31*, 1818–1829. The Royal Society of Chemistry. https://doi.org/10.1039/C5JA00442J
Dang, K. D., Ho, C. N., Van, H. D., Dinh, S. T., Nguyen, Q. T., Nguyen, T. T., Kien, X. T., et al. (2023). Hexavalent chromium inhibited zebrafish embryo development by altering apoptosis- and antioxidant-related genes. *Current Issues in Molecular Biology, 45*, 6916–6926. https://doi.org/10.3390/cimb45080436
den Braver-Sewradj, S. P., van Benthem, J., Staal, Y. C. M., Ezendam, J., Piersma, A. H., & Hessel, E. V. S. (2021). Occupational exposure to hexavalent chromium. Part II. Hazard assessment of carcinogenic effects. *Regulatory Toxicology and Pharmacology, 126*, 105045. https://doi.org/10.1016/j.yrtph.2021.105045
DesMarias, T. L., & Costa, M. (2019). Mechanisms of chromium-induced toxicity. *Mechanistic Toxicology, 14*, 1–7. https://doi.org/10.1016/j.cotox.2019.05.003
Dvoynenko, O., Lo, S. H., Chen, Y. J., Chen, G. W., Tsai, H. M., Wang, Y. L., & Wang, J. K. (2021). Speciation analysis of Cr(VI) and Cr(III) in water with surface-enhanced Raman spectroscopy. *ACS Omega*, 6, 2052–2059. American Chemical Society. https://doi.org/10.1021/acsomega.0c05020
Guha, H., Saiers, J. E., Brooks, S., Jardine, P., & Jayachandran, K. (2001). Chromium transport, oxidation, and adsorption in manganese-coated sand. *Journal of Contaminant Hydrology, 49*, 311–334. https://doi.org/10.1016/S0169-7722(00)00199-6
Hao, Y., Ma, H., Wang, Q., Zhu, C., & He, A. (2022). Complexation behaviour and removal of organic-Cr(III) complexes from the environment: A review. *Ecotoxicology and Environmental Safety, 240*, Article 113676. https://doi.org/10.1016/j.ecoenv.2022.113676
Izbicki, J. A., Blaine McCleskey, R., Burton, C. A., Clark, D. A., & Smith, G. A. (2023). *Groundwater chemistry and hexavalent chromium* (pp. 2330–7102). US Geological Survey.
Kuruthukulangara, N., Anjali, V. A., Nair, A. G., & Asharani, I. V. (2024). Environmentally friendly iron oxide nanoparticles: Synthesis, characterization, and adsorption performance for Cr (VI) ion removal. *Inorganic Chemistry Communications, 168*, Article 112760. https://doi.org/10.1016/j.inoche.2024.112760
Latypov, R., Chistyakova, S., & Mukherjee, R. (2017). A novel hypothesis for origin of massive chromitites in the bushveld igneous complex. *Journal of Petrology, 58*, 1899–1940. https://doi.org/10.1093/petrology/egx077
Lenaz, D., Skogby, H., Princivalle, F., & Hålenius, U. (2004). Structural changes and valence states in the $MgCr_2O_4$–$FeCr_2O_4$ solid solution series. *Physics and Chemistry of Minerals, 31*, 633–642. https://doi.org/10.1007/s00269-004-0420-0
Mooney, D. G. (1997). *An evaluation of small scale open cast mining of UG2 in the bushveld complex*. Rhodes University.

Moonwiriyakit, A., Dinsuwannakol, S., Sontikun, J., Timpratueang, K., Muanprasat, C., & Khemawoot, P. (2024). Fine particulate matter PM2.5 and its constituent, hexavalent chromium induce acute cytotoxicity in human airway epithelial cells via inflammasome-mediated pyroptosis. *Environmental Toxicology and Pharmacology*, 107, 104416. https://doi.org/10.1016/j.etap.2024.104416

Nayak, S., Rangabhashiyam, S., Balasubramanian, P., & Kale, P. (2020). A review of chromite mining in Sukinda Valley of India: impact and potential remediation measures. *International Journal of Phytoremediation, 22*, 804–818. Taylor & Francis. https://doi.org/10.1080/15226514.2020.1717432

Nickens, K. P., Patierno, S. R., & Ceryak, S. (2010). Chromium genotoxicity: A double-edged sword. *Cell Death or Survival: THe Double-Edged Sword of Environmental Toxicity, 188*, 276–288. https://doi.org/10.1016/j.cbi.2010.04.018

Oubohssaine, M., & Dahmani, I. (2024). Phytoremediation: Harnessing plant power and innovative technologies for effective soil remediation. *Plant Stress, 14*, Article 100578. https://doi.org/10.1016/j.stress.2024.100578

Oze, C., Fendorf, S., Bird, D. K., & Coleman, R. G. (2004). Chromium geochemistry in serpentinized ultramafic rocks and serpentine soils from the Franciscan complex of California. *American Journal of Science, 304*, 67–101. American Journal of Science.

Ozturk, S., Kaya, T., Aslim, B., & Tan, S. (2012). Removal and reduction of chromium by *Pseudomonas* spp. and their correlation to rhamnolipid production. *Journal of Hazardous Materials, 231–232*, 64–69. https://doi.org/10.1016/j.jhazmat.2012.06.038

Papp, J. F., & Lipin, B. R. (2001). Chromium and chromium alloys. In *Kirk-othmer encyclopedia of chemical technology*. https://doi.org/10.1002/0471238961.0308181523051920.a01.pub2

Paul, D. (2017). Research on heavy metal pollution of river Ganga: A review. *Annals of Agrarian Science, 15*, 278–286. https://doi.org/10.1016/j.aasci.2017.04.001

Prasad, S., Yadav, K. K., Kumar, S., Gupta, N., Cabral-Pinto, M. M. S., Rezania, S., Radwan, N., & Alam, J. (2021). Chromium contamination and effect on environmental health and its remediation: A sustainable approaches. *Journal of Environmental Management, 285*, Article 112174. https://doi.org/10.1016/j.jenvman.2021.112174

Qin, G., McGuire, M. J., Blute, N. K., Seidel, C., & Fong, L. (2005). Hexavalent chromium removal by reduction with ferrous sulfate, coagulation, and filtration: A pilot-scale study. *Environmental Science & Technology, 39*, 6321–6327. American Chemical Society. https://doi.org/10.1021/es050486p

Ricky, R., Shanthakumar, S., Ganapathy, G. P., & Chiampo, F. (2022). Zero liquid discharge system for the tannery industry—An overview of sustainable approaches. *Recycling, 7*. https://doi.org/10.3390/recycling7030031

Roberts, W. C. (2021). Facts and ideas from anywhere. *Baylor University Medical Center Proceedings, 34*, 339–344. Taylor & Francis. https://doi.org/10.1080/08998280.2021.1874825

Ruparelia, J. R., & Patel, H. K. (2024). Harnessing the chromium reduction potential of Pseudomonas aeruginosa JRHM33: A comprehensive study on bioinformatics, phenotype microarray, and CCD-RSM optimization. *Heliyon, 10*, Article e35650. https://doi.org/10.1016/j.heliyon.2024.e35650

Shah, D., Karibayev, M., Adotey, E. K., & Torkmahalleh, M. A. (2020). Impact of volatile organic compounds on chromium containing atmospheric particulate: Insights from molecular dynamics simulations. *Scientific Reports, 10*, 17387. https://doi.org/10.1038/s41598-020-74522-x

Sharma, P., Bihari, V., Agarwal, S. K., Verma, V., Kesavachandran, C. N., Pangtey, B. S., Mathur, N., Singh, K. P., Srivastava, M., & Goel, S. K. (2012). Groundwater contaminated with hexavalent chromium [Cr (VI)]: A health survey and clinical examination of community inhabitants (Kanpur, India). *PLOS ONE, 7*, e47877. Public Library of Science. https://doi.org/10.1371/journal.pone.0047877

Sharma, P. K., & Sharma, R. K. (2017). The Ganga Jal pollution status: A brief review. https://doi.org/10.17577/IJERTV6IS070017

Sharma, A., Kapoor, D., Wang, J., Shahzad, B., Kumar, V., Bali, A. S., Jasrotia, S., Zheng, B., Yuan, H., & Yan, D. (2020). Chromium bioaccumulation and its impacts on plants: An overview. *Plants, 9*. https://doi.org/10.3390/plants9010100

Sharma, P., Singh, S. P., Parakh, S. K., & Tong, Y. W., & (2022). Health hazards of hexavalent chromium (Cr (VI)) and its microbial reduction. *Bioengineered, 13*, 4923–4938. Taylor & Francis. https://doi.org/10.1080/21655979.2022.2037273

Stefánsson, A., Gunnarsson, I., Kaasalainen, H., & Arnórsson, S. (2015). Chromium geochemistry and speciation in natural waters, Iceland. *A Special Issue to Honor the Geochemical Contributions of D. Kirk Nordstrom, 62*, 200–206. https://doi.org/10.1016/j.apgeochem.2014.07.007

Sueker, J. K. (1964). 5—Chromium. In R. D. Morrison, & B. L. Murphy (Eds.),*Environmental forensics* (pp. 81–95). Academic Press, Burlington. https://doi.org/10.1016/B978-012507751-4/50027-6

Tchounwou, P. B., Yedjou, C. G., Patlolla, A. K., & Sutton, D. J. (2012). Heavy metal toxicity and the environment. In A. Luch (Ed.), *Molecular, clinical and environmental toxicology: Environmental toxicology* (Vol. 3, pp. 133–164). Springer Basel. https://doi.org/10.1007/978-3-7643-8340-4_6

Wei, Y., Xu, X., Zhao, L., Chen, X., Qiu, H., Gao, B., & Cao, X. (2021). Migration and transformation of chromium in unsaturated soil during groundwater table fluctuations induced by rainfall. *Journal of Hazardous Materials, 416*, 126229. https://doi.org/10.1016/j.jhazmat.2021.126229

Wróbel, M., Śliwakowski, W., Kowalczyk, P., Kramkowski, K., & Dobrzyński, J. (2023) Bioremediation of heavy metals by the Genus *Bacillus. International Journal of Environmental Research and Public Health, 20*. https://doi.org/10.3390/ijerph20064964

Zhang, C., Lin, J., Jia, X., & Peng, B. (2016). A salt-free and chromium discharge minimizing tanning technology: The novel cleaner integrated chrome tanning process. *Journal of Cleaner Production, 112*, 1055–1063. https://doi.org/10.1016/j.jclepro.2015.07.155

Zhang, D., Zhang, X., Zhang, Z., Zhang, X., Qiu, F., Liu, Z., & Li, W. (2022). Treatment of methylene blue wastewater with nano-$PbCrO_4$ photocatalyst prepared from chromite ore processing residue. *Journal of Cleaner Production, 379*, Article 134352. https://doi.org/10.1016/j.jclepro.2022.134352

Zhu, L., Hong, C., Zhang, J., & Qiu, Y. (2023). Long-distance mobilization of chromium(III) in soil associated with submicron Cr_2O_3. *Journal of Hazardous Materials, 445*, Article 130519. https://doi.org/10.1016/j.jhazmat.2022.130519

Regulations and Policies for Chromium Emissions to the Environment

Rysul Hasan Hredoy, Farjana Jannat Shimo, Md. Kamruzzaman, Md. Shihab Uddine Khan, Md. Sadiqul Amin, Sadia Afroz Ritu, and Shihab Uddin

Abstract Chromium (Cr), especially in its hexavalent form [Cr(VI)], is a hazardous pollutant of global concern due to its carcinogenicity, persistence, and widespread industrial use. This chapter critically examines the evolution of Cr emission regulations, from early workplace safety measures to major interventions such as the U.S. Clean Air and Water Acts, the EU's REACH Regulation, and the IARC's classification of Cr(VI) as a Group 1 carcinogen. It explores policy challenges, including species-specific thresholds, enforcement disparities, and economic resistance across regional, national, and international frameworks. Sector-specific controls in leather tanning, electroplating, and textiles are reviewed, along with landmark legal cases like the Erin Brockovich litigation and India's tannery reforms, illustrating the role of judicial activism in strengthening accountability. The analysis highlights emerging strategies such as green chemistry, AI-based environmental monitoring, and circular economy integration as pathways for closing regulatory gaps. Overall, this chapter underscores the need for harmonized standards, strict enforcement, and multi-stakeholder collaboration to mitigate the transboundary risks of Cr pollution.

Keywords Chromium emissions · Cr(VI) regulation · Industrial pollution · Polluter pays principle · Environmental policy · Bioremediation

---

R. H. Hredoy
Department of Environmental Science, Bangladesh University of Professionals, Dhaka, Bangladesh

F. J. Shimo · Md. Kamruzzaman · Md. S. Amin
Soil, Water and Environment Discipline, Khulna University, Khulna, Bangladesh

F. J. Shimo · Md. Kamruzzaman · S. Uddin (✉)
Department of Soil Science, Faculty of Agriculture, Bangladesh Agricultural University, Mymensingh, Bangladesh
e-mail: shihab.ss@bau.edu.bd

Md. S. U. Khan
Agricultural Research Station, Bangladesh Agricultural Research Institute, Satkhira, Bangladesh

S. A. Ritu
Department of Crop Botany, Faculty of Agriculture, Bangladesh Agricultural University, Mymensingh, Bangladesh

N. Kumar (ed.), *Chromium Toxicity*, https://doi.org/10.1007/978-3-032-14323-5_3

1 Introduction

Chromium (Cr) is a trace element extensively used in industrial processes such as leather tanning, electroplating, and stainless-steel manufacturing (Losi et al., 1994). However, its widespread industrial application has resulted in substantial environmental contamination, raising serious ecological health concerns. Chromium exists in multiple oxidation states, with hexavalent Cr [Cr(VI)] being highly toxic and mobile, whereas trivalent Cr [Cr(III)] is comparatively less harmful and biologically essential in trace amounts (Shadreck & Mugadza, 2013; Uddin et al., 2023). Industrial wastewater discharge is a major source of Cr pollution, leading to its accumulation in aquatic and terrestrial organisms, where it impairs physiological functions, reduces plant productivity, and disrupts ecosystem balance (Prasad et al., 2021). Cr(VI) is particularly hazardous to aquatic life even at concentrations as low as 5–10 $\mu g\ L^{-1}$ (Eisler, 1985), while both Cr(III) and Cr(VI) can exert toxic, mutagenic, and carcinogenic effects at elevated levels (Kimbrough et al., 1999). In humans, Cr exposure has been linked to cancer, kidney disorders, and liver damage (Cohen et al., 1993; Pandey et al., 2024). To address the problem, numerous remediation strategies have been developed, with biosorption emerging as a cost-effective and sustainable method using agricultural waste, plants, and microorganisms. A thorough understanding of Cr's biochemistry and toxicity is crucial for designing effective treatment solutions and reducing its environmental impact.

Human exposure to Cr, particularly Cr(VI), poses serious health risks. Cr(VI) is classified as a Group 1 carcinogen by the International Agency for Research on Cancer (IARC) and is strongly linked to lung cancer, respiratory ailments, liver and kidney damage, and skin disorders (Hessel et al., 2021; Saha et al., 2011). Workers in industries such as tanning and electroplating are at elevated risk of chronic exposure, often exhibiting respiratory diseases and skin sensitization (Rastogi et al., 2008). Even Cr(III), long considered less hazardous, has been shown to induce oxidative stress and genotoxic effects under certain conditions (Dubey et al., 2022). Communities near industrial zones may also face increased health burdens due to contaminated water sources.

In response, national and international regulatory frameworks have evolved to mitigate Cr emissions. Countries like the USA have established discharge limits under environmental laws such as the Clean Air Act and Clean Water Act (Baral et al., 2006). European nations and South Asian countries have implemented variable national standards to restrict Cr levels in air and water (Vaiopoulou & Gikas, 2020). Despite these efforts, enforcement remains inconsistent, especially in developing regions. Alongside conventional treatments, bioremediation using plants, microbes, and agricultural waste is emerging as a sustainable and cost-effective strategy to manage Cr pollution (Paul et al., 2024). Effective regulation, combined with technological innovation, is essential for reducing ecological damage and safeguarding public health. This chapter critically evaluates the current Cr emission restrictions and policies. Additionally, case studies demonstrating successful regulatory interventions, policy enforcement problems, and the role of stakeholders in mitigating Cr contamination are discussed in detail.

2 Historical Context of Chromium Regulation

2.1 Early Recognition of Chromium Toxicity

The application of Cr dates back more than two millennia, with ancient Chinese artisans utilizing a Cr-based coating to prevent corrosion or rusting (Luo, 1993). Archaeological findings from the tomb of Emperor Qin Shi Huang (259–210 BCE) revealed bronze crossbow bolts and swords coated with a thin layer of Cr oxide, which remarkably resisted corrosion for over 2000 years (McVeigh, 1984; Li et al., 2015; Zhou et al., 2007). Despite this early use, Cr remained largely unrecognized in the scientific community until the work of French chemist Louis Nicolas Vauquelin (1763–1829) in the late eighteenth century. In 1797, Vauquelin successfully isolated Cr from Siberian red lead ore (crocoite), establishing it as a distinct element (Guertin et al., 2004; Vauquelin, 1798).

However, recognition of Cr's toxic effects did not emerge until the Industrial Revolution. By the mid-nineteenth century, occupational illnesses—including chronic skin ulcers, nasal septum perforations, and respiratory ailments—were reported among workers in tanneries, dye works, and electroplating industries. These conditions were later linked to Cr(VI) exposure, particularly in Germany and the UK, where Cr salts were widely used (Langård et al., 1990; Lin et al., 1994).

In the early twentieth century, scientific evidence began associating Cr(VI) with cancer. Notably, German physician Ludwig Teleky (1872–1957) documented lung cancer cases among Cr pigment workers in the 1920s, marking one of the earliest correlations between Cr exposure and malignancy (Teleky, 1936). This link was further substantiated by studies in the 1940s and 1950s, which showed increased lung cancer risks from inhaling chromate dust in electroplating and pigment production settings (Gad, 1989; Mancuso & Hueper, 1951).

2.2 Development of Initial Regulations

The regulation of Cr evolved in two phases: initial occupational health protection and later environmental regulatory control. Table 1 represents the historical timeline of Cr regulatory framework establishment.

Phase I: The first phase (1930s–1960s) focused on occupational exposure limits in response to rising incidences of disease among workers in Cr-intensive industries,

Table 1 Early regulatory actions on chromium (1935–1980)

Country/region	Regulatory authority	Year	Cr(VI) limit	Regulatory focus	References
Germany	TRgA 901	1940	0.1 mg m^{-3} (air)	Occupational exposure	Alderson et al. (1981)
USA	ACGIH	1948	0.1 mg m^{-3} (air)	Industrial hygiene guideline	Mackenzie et al. (1958)
UK	Factory inspectorate	1953	0.5 mg m^{-3} (air)	Occupational safety	Alderson et al. (1981)
USA	OSHA	1971	0.5 mg m^{-3}	Workplace	Bartlett and Vesilind (1998)
USA	EPA (SDWA)	1974	0.1 mg L^{-1} (water)	Drinking water safety	Álvarez-Antonio et al. (2021)
European EEC	Directive 76/464/EEC	1976	Industry-specific	Surface water protection	Prasad et al. (2021)

such as chrome plating, pigment production, and tanning. These efforts marked the first formal recognition of Cr—particularly Cr(VI)—as an occupational hazard.

- Germany was an early leader, establishing workplace exposure guidelines for chromates under the Technische Regeln für Gefahrstoffe (TRgA 901) as early as 1940 (Alderson et al., 1981).
- In the USA, the American Conference of Governmental Industrial Hygienists (ACGIH) introduced a threshold limit value (TLV) of 0.1 mg m^{-3} for Cr(VI) in 1948, followed by the UK in 1953, which adopted exposure standards based on mounting epidemiological evidence linking Cr(VI) to respiratory disease and cancer (Alderson et al., 1981; Mackenzie et al., 1958).
- In 1971, the establishment of the U.S. Occupational Safety and Health Administration (OSHA) led to the codified permissible exposure limit (PEL) of 0.5 mg m^{-3}, making the first enforceable Cr(VI) workplace regulation in the USA (Bartlett & Vesilind, 1998).

Phase II: The second phase (1970s–1980s) emerged amid growing environmental awareness. This phase recognized Cr as a pollutant of concern in water, air, and soil systems.

- Chromium was designated as a priority pollutant under the U.S. Clean Water Act in 1972, promoting control of industrial discharges to aquatic ecosystems (Kimbrough et al., 1999).
- In 1974 Safe Drinking Water Act (SDWA) established a Maximum Contaminant Level (MCL) of 0.1 mg L^{-1} for total Cr in drinking water, although it did not distinguish between Cr(VI) and Cr(III) (Álvarez-Antonio et al., 2021).
- At the European level, the 1976 Directive 76/464/EEC by the European Economic Community (EEC) introduced harmonized effluent limits for hazardous substances, including Cr (Prasad et al., 2021).

Despite these advancements, several critical limitations undermined the effectiveness of early Cr regulations. Foremost among them was the failure to differentiate between Cr(III)—a relatively non-toxic and biologically essential form—and Cr(VI), a highly carcinogenic compound. Most early standards targeted total Cr, overlooking the significantly greater health risks posed by Cr(VI) exposure (Álvarez-Antonio et al., 2021). Enforcement also varied widely across countries. While industrialized nations gradually adopted monitoring and compliance systems, many developing regions lacked the institutional capacity, technical expertise, or financial resources to establish or enforce Cr standards effectively (Prasad et al., 2021). Moreover, economic opposition from industries dependent on Cr-based processes—such as electroplating and leather tanning—slowed the regulatory process due to concerns over compliance costs and required technological upgrades (Kimbrough et al., 1999). These scientific, institutional, and economic challenges collectively delayed the implementation of globally consistent and protective Cr regulation frameworks.

2.3 *Key Milestones in Chromium Regulation History*

The regulation of Cr has evolved significantly over the past several decades, shaped by legislative action, scientific advancements, and growing awareness of the health risks associated with Cr(VI). These key milestones have played a critical role in establishing national and international regulatory frameworks that address the multifaceted challenges posed by Cr pollution—encompassing occupational safety, environmental protection, and public health.

USA: Foundational Policies and Amendments
In the USA, the regulatory journey began with the Clean Water Act of 1972, under which the U.S. Environmental Protection Agency (EPA) identified Cr as a priority pollutant (Baral et al., 2006). This designation initiated federal oversight of industrial discharges into aquatic systems, aiming to protect both water quality and aquatic life (Kimbrough et al., 1999). The Safe Drinking Water Act (1974) subsequently established a MCL of 0.1 mg L^{-1} for total Cr in drinking water. However, although it did not initially differentiate between Cr(VI) and Cr(III) (Bartlett & Vesilind, 1998).

A pivotal development occurred with the enactment of the Resource Conservation and Recovery Act (RCRA) in 1976, which classified Cr-laden industrial wastes as hazardous. This legislation introduced a cradle-to-grave approach to managing hazardous waste, targeting high-risk industries like electroplating, pigment manufacturing, and leather tanning (U.S. EPA, 2016). Concurrently, OSHA set a PEL of 0.5 mg m^{-3} for Cr(VI) in 1971, later revised to 0.005 mg m^{-3} in 2006 in response to updated cancer risk assessments (OSHA, 2006), highlighting the increasing emphasis on worker safety.

European Union: Harmonization through REACH and IED
The European Union's regulatory landscape has emphasized integrated environmental governance since the 1970s (McCormick, 1999). One of the earliest directives, 76/464/EEC, set discharge limits for hazardous substances, including Cr, into water bodies. In the 1990s, these standards were further strengthened. However, specific national discharge limits for Cr(VI) and total Cr still vary, with upper limits typically at 1 mg L^{-1} and 5 mg L^{-1}, respectively (Vaiopoulou & Gikas, 2020).

The Registration, Evaluation, Authorization, and Restriction of Chemicals (REACH) regulation, enacted in 2006, marked a major advance in European chemical policy. Cr(VI) compounds were designated Substances of Very High Concern (SVHC) due to their carcinogenic and mutagenic properties, triggering strict reporting requirements, substitution mandates, and usage restrictions (Prasad et al., 2021). The Industrial Emissions Directive (IED) complemented REACH by enforcing Best Available Techniques (BAT) for industries using Cr, especially in electroplating and metal finishing sectors.

Global Scientific Consensus: IARC Carcinogen Classification
The International Agency for Research on Cancer (IARC) classified Cr(VI) as a Group 1 human carcinogen in 2006, based on substantial evidence linking inhalation exposure to lung cancer, particularly in occupations such as chrome plating and welding (Bencko, 1985; IARC, 1990). This classification promoted regulatory tightening worldwide and reinforced the need for stricter Cr(VI) emission monitoring and occupational exposure limits.

Regional Focus: South Asia and Industrial Pollution
In South Asia, Cr regulation has been significantly influenced by the leather tanning industry. In Bangladesh, tanneries in Dhaka's Hazaribagh area historically discharged untreated Cr-laden effluent into urban waterways, causing severe health and environmental damage. In response, the government relocated tanneries to a centralized site in Savar in 2017, aiming to improve compliance through a centralized effluent treatment plant (CETP). Despite operational challenges, the move represents a major regulatory intervention (Prasad et al., 2021).

In India and Pakistan, Cr-specific discharge standards have been established under national environmental protection laws, particularly targeting the tanning and electroplating sectors. However, enforcement remains uneven due to institutional and infrastructural limitations, highlighting the ongoing need for regulatory strengthening (Khan et al., 2020).

Case Study 1: Cr(VI) Ban in California (2004).

In 2004, California introduced one of the World's strictest Cr(VI) drinking water regulations, posing an interim public health goal of 0.02 μg L^{-1} and a formal MCL of 10 μg L^{-1}—distinct from total Cr. This regulatory initiative was spurred by emerging evidence linking oral Cr(VI) exposure to gastrointestinal cancers, particularly in the context of Hinkley, California, popularized by the film *Erin Brockovich* (Scharfenaker, 2001). Although the MCL was withdrawn in 2017 due to procedural issues in cost–benefit analysis, the move established a global precedent for stringent Cr(VI) regulation.

Case Study 2: Leather Industry Reform in Bangladesh (2017).

The relocation of over 150 tanneries from Hazaribagh to the Savar Tannery Industrial Estate in 2017 was a landmark regulatory action by the Bangladeshi government to address rampant Cr pollution in the Buriganga River. The CEPT, operational since 2020, was designed to improve environmental compliance; however, initial shortcomings in solid waste management, effluent separation, and sludge treatment have been documented. Despite these challenges, this initiative reflects growing regulatory commitment among emerging economies to confront industrial pollution and safeguard public health (Prasad et al., 2021).

3 Global Regulatory Frameworks for Chromium Emissions

Global regulatory frameworks for Cr emissions encompass international, regional, and national laws, guidelines, and standards aimed at controlling Cr release into the environment. These frameworks seek to mitigate health risks and ecological harm by setting permissible exposure limits, restricting hazardous uses, and enforcing monitoring and compliance protocols.

3.1 International Standards and Guidelines

International organizations have developed influential, though non-binding, guidelines, and classifications to support Cr regulation across nations. These standards lay the groundwork for national laws and. Reflect global scientific consensus on Cr toxicity.

World Health Organization (WHO) Guidelines

- Drinking Water: The WHO sets a guideline value of 0.05 mg L^{-1} (50 μg L^{-1}) for total Cr in drinking water, primarily to mitigate the potential carcinogenic effects of Cr (VI) (WHO, 2003).
- Air Quality: While the WHO recognizes Cr(VI) as a hazardous air pollutant, it does not prescribe a specific air quality guideline. Instead, it emphasizes minimizing exposure due to Cr(VI)'s carcinogenic nature.

International Agency for Research on Cancer (IARC) Classification

- Cr(VI) compounds are classified as Group 1 carcinogens ("carcinogenic to humans") based on sufficient evidence in both humans and experimental animals (IARC, 2012).

United Nations Environment Program (UNEP) and Basel Convention

- Under the **Basel Convention (1989)**, Cr-containing industrial waste—especially from leather tanning, electroplating, and pigment production—is categorized as hazardous. The Convention governs the global transboundary movement and safe disposal of such wastes (Peiry, 2011).

3.2 *Regional Regulations*

Regional regulations provide harmonized Cr control mechanisms across multiple countries, often exceeding international standards in stringency. These frameworks are typically enforced by regional governing bodies or economic unions.

USA

- **Environmental Protection Agency (EPA)**:

1. **Clean Air Act (CAA)**: Regulates Cr emissions from industrial sources via National Emission Standards for Hazardous Air Pollutants (NESHAP).
2. **Clean Water Act (CWA)**: Establishes Cr discharge limits into surface waters.
3. **Resource Conservation and Recovery Act (RCRA)**: Designates Cr (VI) waste as hazardous, requiring special handling and disposal.
4. **Comprehensive Environmental Response, Compensation, and Liability Act (CERCLA)**: Includes Cr in the National Priorities List for contaminated site cleanup.

- **Occupational Safety and Health Administration (OSHA)**:
  - Sets a PEL for Cr(VI) at 5 $\mu g\ m^{-3}$ as an 8-h time-weighted average (OSHA, 2006).

European Union (EU)

- Registration, Evaluation, Authorization, and Restriction of Chemicals (REACH):
  - Cr(VI) compounds are SVHC, requiring authorization and often substitution due to their toxic properties.
- Water Framework Directive (WFD):
  - Establishes environmental quality standards (EQS) for total Cr in surface waters.

- **Industrial Emissions Directive (IED):**
  - Mandates the use of BAT to minimize Cr emissions from industrial facilities.

Australia

- **Governed Under the National Environment Protection (Assessment of Site Contamination) Measure** (Abraham et al., 2018):
  - Provides health-based investigation levels for Cr(VI) in soil and groundwater.

Latin America

- Regulations vary by country, for example:
  - Brazil enforces Cr discharge limits through **CONAMA Resolution 357/2005**, which governs water quality and effluent discharge into water bodies.

3.3 National Regulations

National regulations reflect context-specific approaches to Cr management, shaped by industrial practices, public health priorities, and regulatory capacity.

Bangladesh

- The **Department of Environment (DoE)** sets industrial effluent standards:
  - Maximum permissible limit for total Cr in effluent: 0.5 mg L^{-1}.
- The **Environmental Conservation Act (1995)** and **Environmental Conservation Rules (1997)** regulate heavy metals in water and soil.

China

- GB Standards:
  - **GB 5749–2006**: Limits total Cr in drinking water to 0.05 mg L^{-1}.
  - **GB 21900–2008 (Electroplating Emission Standards)**: Enforces strict discharge controls for Cr-bearing wastewater.

India

- **Central Pollution Control Board (CPCB)** guidelines:
  - Maximum allowable Cr (VI) in industrial effluents: 0.1 mg L^{-1}.
  - Ambient Water Quality Standard for total Cr in drinking water: 0.05 mg L^{-1}.

African and Asian Countries

- Many adopt WHO guideline values for Cr in drinking water guidelines.
- Enforcement is often inconsistent, particularly in Cr pollution hotspots such as tanneries in Ethiopia, Kenya, and Pakistan, where weak institutional capacity hampers effective regulation.

Summary of chromium standards is tabulated in Table 2.

Table 2 Summary of chromium standards

Guideline	Medium	Chromium type	Limit	References
WHO guideline	Drinking water	Total Cr	0.05 mg L^{-1}	WHO (2003)
OSHA PEL	Air	Cr(VI)	5 μg m^{-3}	OSHA (2006)
EU water framework directive EQS	Inland water	Total Cr	4.7 μg L^{-1}	Directive (2000)
Bangladesh effluent standard	Water	Total Cr	0.5 mg L^{-1}	DoE (2015)
China GB 5749	Water	Total Cr	0.05 mg L^{-1}	China GB Standards (2006)
EPA (USA)	Air	Cr(VI)	0.006 mg m^{-3}	EPA (2024)
CPCB (India)	Water effluent	Total Cr	2 mg L^{-1}	CPCB, India (2012)
National effluent standards	Effluent	Cr(VI)	0.5 mg L^{-1}	MoE, Japan

4 Policy Measures for Chromium Mitigation

Chromium pollution, particularly from industrial sources, poses significant threats to human health and ecosystems. Effective policy interventions are critical for the prevention, control, and remediation of Cr emissions. Global experiences indicate that an integrated approach—combining preventive strategies with stringent regulatory frameworks—offers the greatest success in mitigating Cr-related risks.

4.1 Preventive Policies

Preventive policies are proactive measures aimed at minimizing Cr release at the source before it enters the environment. These strategies emphasize process redesign, adoption of clean technologies, substitution of hazardous materials, and promotion of industrial innovation to reduce Cr use and emissions.

4.1.1 Best Available Techniques (BAT) in Industries

The Best Available Techniques (BAT) principle is widely applied in industrial policy to reduce environmental emissions, including those from Cr-based processes.

- Definition: BAT refers to the most effective and advanced stage in the development of activities and their operational methods, indicating the practical suitability of specific techniques as a basis for emission limit values (Pinasseau et al., 2018).
- Implementation: BAT is prioritized in industries that are major users of Cr, such as metal processing, leather tanning, electroplating, and dye manufacturing.

Table 3 Best available techniques for chromium mitigation in key industries

Industry sector	Best available techniques (BAT)	Expected outcome
Leather tanning	Cr recycling systems	50–70% reduction in Cr discharge
Electroplating	Membrane filtration; chromium-free coating alternatives	60–90% reduction in Cr emissions
Metal finishing	Enclosed systems; wet scrubbing	Improved worker safety and 80% emission control
Pigments and dyes	Cr substitution (e.g., organic pigments)	100% reduction in Cr exposure

Source Pinasseau et al. (2018), (Zhang et al., 2017)

- Examples of BAT Applications:
  - Adoption of closed-loop water systems to reduce wastewater discharge.
  - Substitution of Cr(VI) compounds with less hazardous alternatives.
  - Use of advanced wastewater treatment methods (e.g., membrane filtration, electrochemical reduction).
  - Installation of dust and fume control systems (e.g., scrubbers, bag filters).

Best available techniques for chromium mitigation in key industries are presented in Table 3.

4.1.2 Emission Limits and Effluent Standards

Emission limits and effluent standards set maximum allowable concentrations of Cr in air, water, and workplace environments. These regulations are sector-specific, targeting industries such as metal plating and tanning, and serve as direct mechanisms to control Cr discharges into the environment. Some examples are

- Air Emission Standards: The U.S. Environmental Protection Agency (EPA) limits Cr(VI) emissions from electroplating operations to 0.006 mg m^{-3} (EPA, 2024).
- Water Effluent Standards: In India, the Central Pollution Control Board (CPCB) sets the permissible limits for total Cr in industrial effluents at 2.0 mg L^{-1} and for Cr(VI) at 0.1 mg L^{-1} (CPCB, India, 2012).
- Occupational Standards: OSHA enforces a permissible exposure limit (PEL) of 5 μg m^{-3} for Cr(VI), based on an 8-h time-weighted average (OSHA, 2006).

4.2 Mitigation Policy

The following flowchart illustrates (Fig. 1) the typical framework adopted for Cr mitigation in regulatory policies.

Fig. 1 Policy framework for chromium mitigation

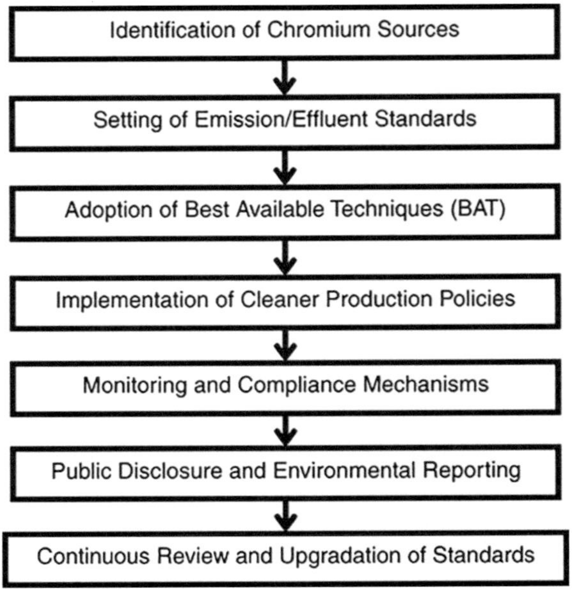

4.2.1 Polluter Pays Principle (PPP)

The Polluter Pays Principle (PPP) is a foundational concept in environmental law, asserting that those responsible for pollution should bear the costs of managing it to human health and the environment. Formally introduced by the OECD in 1972, PPP has since become integral to national and international environmental policies (Munir, 2013). Its roots lie in the early-twentieth-century economic theories, later evolving into a legal doctrine influencing frameworks on taxation, civil liability, and state aid (de Sadeleer et al., 2014; Joseph, 2014). The principle plays redistributive, preventive, and curative roles by assigning financial responsibility for pollution to its source. Although some exceptions exist—such as government subsidies for pollution control—these may dilute the strict application of PPP and raise concerns over fairness and regulatory consistency (Joseph, 2014). In practice, the principle has been expanded to cover costs related to accidental pollution, including prevention and remediation, with critics warning that liability caps may undermine legal precedent and fundamental rights. PPP also holds increasing relevance in transboundary pollution disputes, including cases involving Cr-laden effluents, underscoring the need for enforceable regional agreements to ensure accountability across borders.

4.3 Remediation Policies

Chromium contamination from industrial activities has prompted the development of remediation policies worldwide, aimed at environmental restoration and enforcing polluter accountability. These policies range from direct cleanup mandates, like the U.S. Superfund program, to economic and legal instruments like the Polluter Pays Principle (Baral et al., 2006). The particularly toxic and carcinogenic nature of Cr(VI) has led to stringent regulatory controls, especially in industries such as metal finishing (Baral et al., 2006; Prasad et al., 2021). Although command-and-control approaches have achieved reductions in Cr discharge, full compliance remains difficult, promoting the adoption of stricter standards and complementary voluntary measures (Baral et al., 2006). Remediation methods include a range of physical and chemical techniques. However, biological strategies—such as bioremediation using microorganisms and phytoremediation—are increasingly favored for their cost-effectiveness and environmental compatibility (Das & Mathew, 2011). These nature-based solutions rely on mechanisms like phytoextraction, rhizofiltration, and phyto-detoxification to remove or neutralize Cr in contaminated environments (Das & Mathew, 2011).

4.3.1 Soil and Groundwater Remediation Mandates (E.G., Superfund Sites in the USA)

Chromium contamination in soil and groundwater poses serious environmental and public health risks, necessitating effective remediation strategies. In the USA, the Superfund program has addressed such risks through site-specific interventions, including the cleanup of the United Chrome Products plant in Corvallis, Oregon, using pump-and-treat systems (McKinley & Pratt, 1992). Effective remediation requires understanding both physical processes—such as advection and dispersion—and chemical interactions, including redox transformations and adsorption/desorption dynamics (Palmer & Wittbrodt, 1991). Cr(VI) is of particular concern due to its toxicity and mobility (Prasad et al., 2021). A variety of remediation techniques have been developed, including nanoscale zero-valent iron adsorption, electrocoagulation, and bioremediation, which typically aim to reduce Cr(VI) to the less toxic and less soluble Cr(III) form (Acharyya et al., 2023). The efficiency of these methods depends on factors such as pH, temperature, Cr(VI) concentration, contact time, and the presence of competing ions.

Cr-contaminated soils present unique challenges for risk assessment and remediation. The complex speciation and reactivity of Cr compounds influence their environmental behavior, toxicity, and treatment efficacy (Palmer & Wittbrodt, 1991; Uddin et al., 2023). Common remediation approaches include source removal, groundwater treatment, and in situ chemical reduction of Cr(VI) to less toxic Cr(III) (James, 1996). Risk assessment is complicated by uncertainties in chronic toxicity data—particularly for oral exposure—and by difficulties distinguishing between effects of Cr(VI) and Cr(III) (LaVelle, 1991). Developing soil cleanup standards thus requires considering multiple toxicity endpoints, including liver, kidney, and blood effects from ingestion; lung cancer from inhalation; and allergic contact dermatitis from dermal exposure (Felter & Dourson, 1997). Despite these uncertainties, soil Cr concentrations between 130 and 450 ppm have been proposed as protective thresholds, balancing health risks with technical feasibility and socio-economic considerations (Felter & Dourson, 1997).

Globally, similar frameworks have emerged. In the European Union, the Environmental Liability Directive (2004/35/EC) obliges polluters to prevent and remediate damage to soil and water. However, implementation has faced legal and technical hurdles (Pouikli, 2016). In rapidly industrializing countries such as India and China, remediation projects are increasingly focused on industrial hotspots, though regulatory enforcement and industrial capacity vary (McAllister et al., 2010; Chiang & Gu, 2015). Internationally, sustainable remediation frameworks have been developed to integrate environmental, social, and economic dimensions. While these frameworks perform well in addressing triple-bottom-line concerns, they often fall short of meeting all proposed sustainability principles, particularly in managing trade-offs (Ridsdale & Noble, 2016).

4.3.2 Extended Producer Responsibility (EPR)

Extended Producer Responsibility (EPR) expands PPP by holding manufacturers accountable for the entire lifecycle of their products, including post-consumer waste management and pollution mitigation (Agamuthu & Victor, 2011). It aims to enhance environmental and social outcomes by extending producer obligations to stages such as product take-back, recycling, and final disposal. While traditionally applied to solid waste streams—such as electronics and plastics—EPR is increasingly relevant in sectors that involve hazardous chemical additives, including Cr use in leather, textiles, and electroplating. ERP policies can take various forms, including product take-back requirements, advance disposal fees, and deposit-refund systems. However, the framework itself does not prescribe specific instruments for different product categories or contexts. Policymakers must therefore consider factors such as waste volume, toxicity, and prevailing disposal practices when designing EPR mechanisms (Walls, 2003). Importantly, although EPR primarily addresses end-of-life impacts, it can also influence upstream decisions, promoting safer design and material substitution across the product lifecycle (Curtis et al., 2014).

In the EU, EPR operates in tandem with REACH regulations, incentivizing the replacement of Cr(VI) with less hazardous alternatives and promoting responsible management of residual waste. EPR has become integral to circular economy strategies, fostering "design-for-environment" approaches and penalizing non-compliance. Emerging economies like India and Bangladesh are also beginning to integrate EPR principles into their environmental governance frameworks. For example, India's Draft National Framework for Chemical Safety proposes that industries track and reduce hazardous inputs, including Cr, across supply chains. Nevertheless, enforcement remains a challenge, particularly for small and medium enterprises (SMEs).

5 Industry-Specific Emission Guidelines

Chromium emissions—particularly Cr(VI), a highly toxic and carcinogenic form—primarily originate from industries such as leather tanning, electroplating, textile dyeing, and chromate manufacturing. These emissions pose significant ecological and public health risks, entering the environment through solid waste, airborne particles, and industrial wastewater (WHO, 2003; EPA, 2024). In response, national and international regulatory bodies have established sector-specific occupational exposure limits and emission control guidelines to mitigate these risks.

5.1 *Leather Tanning Industry*

Cr(III) sulfate is extensively used in the leather tanning industry due to its effectiveness in stabilizing collagen fibers in hides. However, under certain processing conditions—such as burning or improper waste disposal—Cr(III) can oxidize to the more toxic Cr(VI) form (CPCB, India, 2012). As a result, many countries enforce strict regulations on Cr discharges from tanneries. For example, the Central Pollution Control Board of India prescribes maximum permissible limits of 0.1 mg L^{-1} for Cr(VI) and 2 mg L^{-1} for total Cr in treated tannery effluents. The Department of Environment (DoE) in Bangladesh sets comparable thresholds, emphasizing control at both in-plant treatment and final discharge stages (DoE, 2024). To minimize Cr pollution, modern tanneries are encouraged to adopt cleaner production (CP) techniques that reduce the quantity and toxicity of Cr-bearing waste. These methods include chrome recovery units, process optimization, and enzyme-assisted tanning processes (Oruko et al., 2021).

5.2 *Electroplating Industry*

Chromium electroplating is a major source of Cr(VI) emissions. Chromic acid (H_2CrO_4) is commonly used in plating baths, and the resulting fumes and effluents often contain high concentrations of Cr(VI). The U.S. EPA's Electroplating Effluent Guidelines (40 CFR Part 413) set specific discharge limits for total Cr, particularly for indirect dischargers sending waste to Publicly Owned Treatment Works. These include daily maximum values of 2.77 mg L^{-1} for total Cr and 0.1 mg L^{-1} for Cr(VI)

(EPA, 2024). To mitigate emissions and environmental contamination, electroplating facilities are required to implement scrubbers, fume suppressants, and closed-loop rinse systems. Integrated waste management systems should also include regular leak monitoring, chrome mist eliminators, and rinse water pH control (OSHA, 2006).

5.3 *Textile and Dyeing Industry*

Cr-based dyes and mordants are employed in the textile and dyeing industries, particularly for blended and synthetic fabrics. While these compounds enhance dye fixation and fastness, their use results in significant effluent generation. Consequently, regulations require the removal of Cr prior to discharge through Effluent Treatment Plants (ETPs) utilizing filtration, chemical precipitation, and, in some cases, advanced oxidation or membrane technologies (Alderson et al., 1981). In the European Union, industries must comply with the Water Framework Directive (2000/60/EC) and the REACH Regulation, both of which restrict hazardous substances such as Cr(VI) in consumer textiles. Voluntary environmental standards, including ZDHC and OEKO-TEX, further promote the adoption of Cr-free dye formulations.

5.4 *Wastewater Discharge Standards*

Strict regulations on Cr discharge in industrial effluents have been established in several countries to protect environmental health (Table 4). Although these limits vary depending on the receiving water bodies and industrial sectors, they generally reflect international best practices.

5.5 *Occupational Exposure Limits (OELs)*

Workers exposed to Cr—particularly Cr(VI)—face serious long-term health risks, including lung cancer, nasal ulcers, asthma, and dermatitis (ILO, 2021; OSHA, 2006). To mitigate these risks, various regulatory bodies have established occupational exposure limits (OELs) for airborne Cr(VI), typically defined as time-weighted averages (TWA) over an 8-h workday (Table 5).

Table 4 Chromium standards for wastewater discharge

Country/region	Total Cr limit (mg L^{-1})	Cr(VI) limit (mg L^{-1})	Applicable regulation	References
USA	2.77 (daily max)	0.1	EPA categorical standards	EPA (2024)
EU	1.5–2.0	0.05–0.1	Water framework directive, REACH	Vaiopoulou and Gikas (2020)
India	2.0	0.1	CPCB general standards	The Environment (Protection) Rules, 1986, CPCB, India
Bangladesh	1.0–2.0	0.1	DoE environmental conservation rules	Environmental Conservation Rules 2023, Bangladesh
China	1.5	0.1	GB 8978–1996 Wastewater Discharge Standards	Integrated wastewater discharge standard, 1998, China

Table 5 Chromium standards for workplace

Organization	Cr(VI) exposure limit	Basis	Reference
OSHA (USA)	5 μg m^{-3}	Permissible exposure limit (PEL)	Bartlett and Vesilind (1998)
ACGIH	0.05 mg m^{-3}	Threshold limit value (TLV)	Mackenzie et al. (1958)
EU (Directive 2004/37/EC)	10 μg m^{-3}	Limit for carcinogenic workplace substances	Prasad et al. (2021)

6 Legal Battles and Landmark Cases in Chromium Contamination

Legal battles over Cr contamination have significantly influenced environmental policy, regulatory enforcement, and public awareness of industrial pollution risks. Through class-action lawsuits, public interest litigation, and government enforcement actions, these cases have set important precedents in environmental justice, corporate accountability, and regulatory reform. The following landmark cases illustrate the scope and impact of legal interventions across various jurisdictions and highlight the evolving legal landscape surrounding Cr pollution.

6.1 The Erin Brockovich Case in Hinkley (California, USA)

Case: Anderson et al. versus Pacific Gas and Electric Company (PG&E).

Between 1952 and 1966, PG&E used Cr(VI) as a rust inhibitor at its Hinkley compressor station, disposing of contaminated water in unlined ponds. This diposition led to significant groundwater contamination, which was linked to increased rates of cancer, respiratory illness, and other health problems among residents. The case gained international prominence due to the efforts of legal clerk Erin Brockovich (Sharma & Mitra, 2024).

Outcome: In 1996, PG&E agreed to a $333 million settlement—one of the largest environmental settlements in U.S. history at the time.

Significance: The case became a symbol of grassroots environmental activism and played a major role in shaping California's regulatory approach to Cr(VI) in drinking water. It also exposed systemic shortcomings in corporate accountability and public health safeguards.

6.2 Topock Compressor Station Case in California (USA)

Case: PG&E versus California Department of Toxic Substances Control.

PG&E was implicated in Cr contamination at its Topock facility near the Colorado River, raising concerns about the migration of polluted groundwater toward the river. The site was designated for long-term remediation under both federal and state environmental programs.

Outcome: PG&E agreed to implement a comprehensive cleanup plan under the joint oversight of the U.S. Environmental Protection Agency (EPA) and the California Department of Toxic Substances Control (DTSC).

Significance: This case underscored the importance of the Comprehensive Environmental Response, Compensation, and Liability Act (CERCLA/Superfund) and highlighted effective coordination between federal and state agencies in managing legacy industrial pollution.

6.3 Leather Industry Litigation in Kanpur (India)

Case: M.C. Mehta versus Union of India.

This landmark public interest litigation addressed Cr pollution from leather tanneries discharging effluents into the Ganges River in Kanpur. The Indian Supreme Court established a direct connection between tannery waste and the declining water quality of the river (Dubey & Srivastava, 2023).

Outcome: The court ordered the closure of non-compliant tanneries and mandated the installation of effluent treatment plants (ETPs) as a condition for continued operation.

Significance: The judgment institutionalized the Polluter Pays Principle (PPP) within Indian environmental law and reinforced the judiciary's proactive role in ensuring environmental accountability.

6.4 *Chromium Waste Contamination in Glasgow (Scotland)*

Case: Residents versus Scottish Development Agency (SDA).

In Glasgow, Cr waste from historical industrial operations was discovered in residential areas, prompting legal action against the Scottish Development Agency for its role in managing these industrial legacies (Whalley et al., 1999).

Outcome: The agency was found liable and required to undertake cleanup measures and provide compensation to affected residents.

Significance: This case established a key precedent in the UK by affirming that government agencies can be held accountable for environmental contamination arising from legacy industrial activities, reinforcing the need for institutional responsibility in pollution management.

6.5 *Chromium Contamination in Hudson County (New Jersey, USA)*

Case: Multiple lawsuits against Honeywell International and PPG Industries, Hudson County, New Jersey, contained over 200 Cr-contaminated waste sites, largely stemming from historical chromite ore processing activities (Burke et al., 1991).

Outcome: The companies were ordered to undertake extensive cleanup under state oversight, with PPG Industries committing approximately $600 million to remediation.

Significance: This case represents one of the largest Cr cleanup efforts in the USA, emphasizing the vital role of community advocacy and robust state-level regulatory enforcement in addressing legacy pollution.

6.6 *Sukinda Valley Chromium Contamination in Odisha (India)*

The Sukinda Valley, which hosts one of the world's largest chromite ore reserves, has experienced extensive soil and water contamination due to decades of poorly regulated mining activities (Nayak et al., 2020).

Legal Action: While no single lawsuit defines this case, it has led to multiple public interest litigations (PILs) and interventions by the National Green Tribunal (NGT).

Significance: The Sukinda case illustrates persistent regulatory failures in mineral-rich regions and emphasizes the urgent need for robust judicial and legislative oversight in environmental governance.

6.7 PPG Industries Chromium Cleanup in Jersey City (USA)

PPG Industries was identified as a major source of Cr contamination in Hudson County, New Jersey, with over 200 affected sites.

Legal Action: The New Jersey Department of Environmental Protection (NJDEP) filed lawsuits requiring PPG Industries to fund remediation efforts (PPG Industries Inc. v. USA, 2020).

Outcome: PPG committed to a remediation program estimated at $600 million.

Significance: This case underscored the effectiveness of state-level environmental litigation in holding corporations accountable for widespread environmental damage.

6.8 Leather Industry Pollution in Tamil Nadu (India)

Case: Vellore Citizens' Welfare Forum versus Union of India (1996).

Cr-laden effluents from leather tanneries in Tamil Nadu severely contaminated the Palar River, threatening the health of local communities and damaging agricultural productivity.

Outcome: The Supreme Court of India mandated the installation of Common Effluent Treatment Plants (CETPs) and ordered the closure of tanneries that failed to comply with environmental regulations (John & Rajan, 2022).

Significance: This landmark judgment institutionalized the Precautionary Principle and Polluter Pays Principle within Indian environmental law, reinforcing judicial activism as a key mechanism for combating environmental degradation.

6.9 Chromium Contamination in Willits, California (USA)

The Remco Hydraulics facility in Willits, California, used Cr(VI) in its manufacturing processes, resulting in significant soil and groundwater contamination (Remy et al., 2017).

Legal Action: Residents filed a class-action lawsuit citing health issues, including cancer and reproductive harm.

Outcome: Settlements exceeded $20 million, and the site was designated for cleanup under the Superfund program.

Significance: This case highlights the long-term health, legal, and financial consequences of industrial negligence, underscoring the need for early and robust regulatory frameworks to prevent such outcomes.

7 Policy Innovations and Future Directions

As the industrial transition toward more sustainable practices, both regulatory advancements and technical innovations are becoming increasingly vital for managing Cr contamination. In response to the well-documented environmental and human health risks—particularly those posed by Cr(VI)—governments, industry, and civil society are strengthening regulatory frameworks, adopting voluntary environmental standards, and promoting cleaner production methods. This section critically examines current developments, emerging trends, and strategic directions aimed at enhancing the effectiveness of Cr pollution control.

7.1 Voluntary Environmental Standards and Certifications

Voluntary standards have emerged as effective complements to statutory regulations, offering structured frameworks for industries to enhance environmental and occupational performance. ISO 14001 (Environmental Management Systems) emphasizes environmental compliance, resource efficiency, and pollution prevention, making it particularly relevant for Cr-intensive sectors (Potoski & Prakash, 2013). ISO 9001 (Quality Management Systems) supports the systematic application of process controls to minimize unintended Cr discharges during production. Additionally, ISO 45001 (Occupational Health and Safety Management Systems) addresses workplace hazards such as Cr exposure, thereby strengthening occupational health protections (Dougherty, 2000). These voluntary standards encourage organizations to adopt proactive environmental practices that exceed regulatory requirements, fostering a culture of continuous improvement in environmental management.

7.2 Cleaner Production Technologies

Technological innovation plays a fundamental role in Cr pollution mitigation. Advancements in cleaner production technologies have substantially reduced Cr discharges across multiple sectors. Prominent methods include:

- **Electrocoagulation**: Utilizes electric current to precipitate Cr from wastewater.
- **Membrane Filtration**: Employs semi-permeable membranes for Cr separation in effluents.
- **Bioremediation**: Leverages microbial metabolism to transform Cr(VI) into its less toxic Cr(III) form (Das & Mathew, 2011).

Simultaneously, industries are increasingly adopting Cr-free production techniques. For example, vegetable tanning has gained acceptance in the leather industry as an environmentally preferable alternative to Cr tanning. In the metal finishing sector, non-toxic coatings and trivalent chromium plating are replacing Cr(VI)-based technologies (Acharyya et al., 2023).

7.3 Integration with Circular Economy Principles

The adoption of circular economy principles has led to more resource-efficient strategies for managing Cr throughout its life cycle. By promoting reuse, recycling, and closed-loop technologies, circular economy models aim to decouple economic growth from environmental deterioration. In industrial settings, Cr can be recovered from effluents through methods such as chemical precipitation, ion exchange, and advanced electrochemical techniques, enabling its reintegration into production processes. Closed-loop systems, particularly in the electroplating and leather industries, facilitate the recycling of process water and Cr recovery, thereby reducing environmental emissions and operational costs. These approaches align with broader sustainability objectives focused on resource conservation and waste minimization (Prasad et al., 2021).

7.4 Need for Stricter Enforcement and Harmonized Global Standards

Legal frameworks for Cr management have progressed over time, yet enforcement remains uneven, particularly in developing countries where weak institutional capacity and inadequate treatment policies hinder effective pollution control (Garg et al., 2012). While command-and-control regulations have helped reduce Cr emissions, full industry compliance is still lacking. In response, policy developments have introduced stricter standards and voluntary mechanisms, though with limited impact (Baral et al., 2006). The World Bank's model emphasizes flexible norms and non-state pressures, local enforcement challenges and limited governmental commitment constrain its success. Strengthening regulatory outcomes may require a blend of democratic institutions, legal remedies, environmental activism, and localized, bottom-up approaches tailored to specific contexts (Tang et al., 1998). Efforts to harmonize global standards can draw on existing international frameworks such as the Basel Convention, which provides mechanisms for managing hazardous waste, including Cr residues. Enhanced international cooperation can further align technical standards, monitoring protocols, and enforcement strategies to support more effective and equitable Cr regulation.

7.5 *Role of Green Chemistry and Sustainable Industrial Practices*

Green chemistry and sustainable industrial practices prioritize environmentally friendly production by promoting the use of renewable resources, recyclable catalysts, and energy-efficient methods (Ratti, 2020). Recent advancements—particularly in the development of sustainable catalysts—have demonstrated promising applications in sectors such as ammonia and hydrogen peroxide manufacturing, offering both environmental and economic benefits (Iqbal, 2024; Anastas & Lankey, 2002). By encouraging the adoption of eco-friendly alternatives and the redesign of product life cycles, green chemistry enhances industrial competitiveness while supporting broader sustainability goals. In the context of Cr management, regulatory pressure on hazardous chemical use, combined with incentives for green product innovation, plays a crucial role in driving the transition toward safe and more sustainable practices.

7.6 *Public Awareness and Stakeholder Engagement*

Effective Cr pollution management depends on broad societal participation, including public awareness, transparent corporate practices, and active stakeholder engagement. In areas with legacy contamination—such as industrial zones in Bangladesh, India, and parts of the USA—community involvement is essential for driving change. Governments, environmental NGOs, and civil society organizations play key roles in educating the public about Cr risks and advocating for stronger regulation. Meanwhile, corporate social responsibility (CSR) initiatives can help align industrial activities with community interests, fostering shared accountability and supporting more sustainable and equitable environmental outcomes.

7.7 *Emerging Technologies for Chromium Monitoring and Abatement*

Recent advancements in Cr monitoring and abatement technologies offer promising tools for environmental remediation. Electrochemical sensors and biosensors—particularly those incorporating nanomaterials and biorecognition elements—have demonstrated high sensitivity and specificity in Cr detection (Hilali et al., 2020). A wide range of adsorbent materials, including natural minerals, synthetic polymers, and carbon-based compounds, have proven effective in removing Cr from aqueous environments (Xing et al., 2022). Additionally, membrane filtration, electrochemical precipitation, and phytoremediation are being actively explored as innovative treatment approaches (Hussain & Memon, 2020). The integration of artificial intelligence (AI) in environmental monitoring enables real-time data aggregation across multiple sites. These platforms facilitate predictive analytics, enhancing early warning systems, regulatory oversight, and informed decision-making for pollution control. Emerging solutions such as nanomaterial-enhanced adsorbents and automated remediation systems represent the frontier of the next-generation Cr abatement technologies, offering scalable and efficient pathways for long-term environmental management.

8 Conclusion

The regulation of Cr emissions reflects the global challenge of balancing industrial growth, environmental protection, and public health. Once widely used, Cr has become a regulated pollutant due to mounting scientific evidence and public pressure, leading to evolving legal and technological responses. However, enforcement gaps, especially in developing nations, continue to hinder progress and exacerbate environmental inequality. Emerging strategies—such as ISO 14001, Extended Producer Responsibility, and circular economy models—are integrating sustainability into industry. Technological innovations like electrocoagulation, bioremediation, real-time monitoring, and nanofiltration offer promising solutions. However, without strong enforcement and global standards, these advances may fall short. Effective Cr mitigation demands collaborative efforts among governments, industries, researchers, and communities to ensure both scientific accountability and social equity.

References

Abraham, J., Dowling, K., & Florentine, S. (2018). Assessment of potentially toxic metal contamination in the soils of a legacy mine site in Central Victoria, Australia. *Chemosphere, 192,* 122–132.

Acharyya, S., Das, A., & Thaker, T. P. (2023). Remediation processes of hexavalent chromium from groundwater: A short review. *AQUA—Water Infrastructure, Ecosystems and Society, 72*(5), 648–662.

Agamuthu, P., & Victor, D. (2011). Policy trends of extended producer responsibility in Malaysia. *Waste Management & Research, 29*(9), 945–953.

Alderson, M. R., Rattan, N. S., & Bidstrup, L. (1981). Health of workmen in the chromate-producing industry in Britain. *British Journal of Industrial Medicine, 38*, 117–124.

Álvarez-Antonio, C., Meza-Sánchez, G., Calampa, C., Casanova, W., Carey, C., Alava, F., Rodríguez-Ferrucci, H., & Quispe, A. M. (2021). Seroprevalence of anti-SARS-CoV-2 antibodies in Iquitos, Peru in July and august, 2020: A population-based study. *The Lancet Global Health, 9*(7), e925–e931. https://doi.org/10.1016/S2214-109X(21)00173-X

Anastas, P. T., & Lankey, R. L. (2002). Sustainability through green chemistry and engineering. In *Advancing sustainability through green chemistry and engineering* (823, pp. 1–11). American Chemical Society. https://doi.org/10.1021/bk-2002-0823.ch001.

Baral, A., Engelken, R., & Stewart, P. (2006). Chromium-based regulations applicable to metal finishing Industries in the United States: A policy assessment. *Review of Policy Research, 23*(1), 1–21.

Bartlett, L., & Vesilind, P. (1998). Expediency and human health: The regulation of environmental chromium. *Science and Engineering Ethics, 4*, 191–201.

Bencko, V. (1985). Chromium: A review of environmental and occupational toxicology. *Journal of Hygiene, Epidemiology, Microbiology, and Immunology, 29*(1), 37–46.

Burke, T., Fagliano, J., Goldoft, M., Hazen, R. E., Iglewicz, R., & McKee, T. (1991). Chromite ore processing residue in Hudson County, New Jersey. *Environmental Health Perspectives, 92*, 131–137.

Chiang, S. Y. D., & Gu, Q. (2015). Brownfield sites remediation technology overview, trends, and opportunities in China. *Remediation Journal, 25*(3), 85–99.

Cohen, M. D., Kargacin, B., Klein, C. B., & Costa, M. (1993). Mechanisms of chromium carcinogenicity and toxicity. *Critical Reviews in Toxicology, 23*(3), 255–281.

CPCB, India. (2012). *Indian standard drinking water—specification*. Retrieved on December 9, 2025 from https://cpcb.nic.in/wqm/BIS_Drinking_Water_Specification.pdf

Curtis, C., Collins, S., Cunningham, S., Stigler, P., & Novotny, T. E. (2014). Extended producer responsibility and product stewardship for tobacco product waste. *International Journal of Waste Resources, 4*(3), 157.

Das, N., & Mathew, L. (2011). Chromium pollution and bioremediation: An overview. In M. S. Khan, A. Zaidi, R. Goel, & J. Musarrat (Eds.), *Biomanagement of metal-contaminated soils. Environmental pollution* (pp. 297–321). Springer. https://doi.org/10.1007/978-94-007-1914-9_13

De Sadeleer, N., Morin, J. F., & Orsini, A. (2014). Polluter pays principle. In J.-F. Morin & A. Orsini (Eds.), *Essential concepts of global environmental governance* (pp. 55–156). Routledge.

Directive, W. F. (2000). EU Water framework directive. *EC Directive, 60*.

Dobbie, B., & Green, D. (2015). Australians are not equally protected from industrial air pollution. *Environmental Research Letters, 10*(5), 055001.

DoE. (2015). *Bangladesh standards and guidelines for sludge management*. Department of Environment (DoE). Retrieved from https://doe.portal.gov.bd/sites/default/files/files/doe.portal.gov.bd/publications/2398e6c5_c300_472d_9a0c_0385522748f3/Bangladesh%20Standards%20and%20Guideline%20for%20sludge%20management-%20September%202016.pdf.

DoE. (2024). *Enforcement guidelines*. Retrieved on July 10, 2025 from http://moef.portal.gov.bd/sites/default/files/files/moef.portal.gov.bd/page/ac0ce881_4b1d_4844_a426_1b6ee36d2453/DOE%20Enforcement%20Guideline.pdf

Dougherty, R. (2000). ISO 14001: Real world implementation and status. In C. F. Bonser (Ed.), *Security, trade, and environmental policy: A US/European Union transatlantic agenda* (pp. 225–232). Springer. https://doi.org/10.1007/978-1-4615-4399-2_26

Dubey, I., & Srivastava, M. (2023). A case study on chromium effluent treatment plant of leather processing units. *Chemistry for Technological Advances* (pp. 121–128). Available from https://www.nexgenpublication.com/2022-2023/chemistry-for-technological-advances.php

Dubey, R., Verma, P., & Kumar, S. (2022). Cr (III) genotoxicity and oxidative stress: An occupational health risk for leather tannery workers of south Asian developing countries. *Toxicology and Industrial Health, 38*(2), 112–126.

Eisler, R. (1985). *Cadmium hazards to fish, wildlife, and invertebrates: A synoptic review.* Biological Report 85(1.2). https://pubs.usgs.gov/publication/5200065

EPA. (2024). *Pretreatment standards and requirements-categorical pretreatment standards*. EPA.gov. Retrieved 2025, July 20 from https://www.epa.gov/npdes/pretreatment-standards-and-requirements-categorical-pretreatment-standards

Felter, S. P., & Dourson, M. L. (1997). Hexavalent chromium-contaminated soils: Options for risk assessment and risk management. *Regulatory Toxicology and Pharmacology, 25*(1), 43–59.

Gad, S. (1989). Acute and chronic systemic chromium toxicity. *The Science of the Total Environment, 86*(1–2), 149–157. https://doi.org/10.1016/0048-9697(89)90201-5

Garg, S. K., Tripathi, M., & Srinath, T. (2012). Strategies for chromium bioremediation of tannery effluent. *Reviews of Environmental Contamination and Toxicology, 217*, 75–140.

Guertin, J., Jacobs, J. A., & Avakian, C. P. (2004). *Chromium (VI) handbook* (pp. 1–22). CRC Press.

Hessel, E. V., Staal, Y. C., Piersma, A. H., den Braver-Sewradj, S. P., & Ezendam, J. (2021). Occupational exposure to hexavalent chromium. Part I. Hazard assessment of non-cancer health effects. *Regulatory Toxicology and Pharmacology, 126*, 105048.

Hilali, N., Mohammadi, H., Amine, A., Zine, N., & Errachid, A. (2020). Recent advances in electrochemical monitoring of chromium. *Sensors, 20*(18), 5153. https://doi.org/10.3390/s20185153

Hussain, F. S., & Memon, N. (2020). Materials and technologies for the removal of chromium from aqueous systems. *Sustainable Agriculture Reviews, 40*, 113–177.

IARC Working Group on the Evaluation of Carcinogenic Risks to Humans. (2012). *Arsenic, metals, fibres and dusts*. International Agency for Research on Cancer, Lyon (FR). (IARC Monographs on the Evaluation of Carcinogenic Risks to Humans, No. 100C.). Available from https://www.ncbi.nlm.nih.gov/books/NBK304375/

IARC Working Group on the Evaluation of Carcinogenic Risks to Humans (with International Agency for Research on Cancer). (1990). *Chromium, nickel and welding* (Vol. 49). IARC. Available from: https://www.ncbi.nlm.nih.gov/books/NBK519250/

ILO. (2021). *Exposure to hazardous chemicals at work and resulting health impacts: A global review.* Retrieved on December 06 from https://www.ilo.org/sites/default/files/2024-07/wcms_811455%20%284%29.pdf

Iqbal, M. (2024). Advancements in green chemistry: Developing sustainable catalysts for eco-friendly industrial processes. *Transpublika International Research in Exact Sciences, 3*(2), 70–76. https://doi.org/10.55047/tires.v3i2.1395

James, B. R. (1996). The challenge of remediating chromium-contaminated soil. *Environmental Science & Technology, 30*(6), 248A–251A. https://doi.org/10.1021/es962269h

John, R., & Rajan, A. P. (2022). A real-time investigation on public health consequences of chromium laden effluent from the leather industries. *Indian Journal of Public Health, 66*(1), 27–32.

Joseph, S.-A. (2014). The polluter pays principle and land remediation: A comparison of the United Kingdom and Australian approaches. *Australian Journal of Environmental Law, 1*, 24. Retrieved from: https://classic.austlii.edu.au/au/journals/AUJlEnvLaw/2014/2.html

Khan, Z. I., Ahmad, K., Siddique, S., Ahmad, T., Bashir, H., Munir, M., Mahpara, S., Malik, I. S., Wajid, K., Ugulu, I., Nadeem, M., Noorka, I. R., & Chen, F. (2020). A study on the transfer of chromium from meadows to grazing livestock: An assessment of health risk. *Environmental Science and Pollution Research, 27*(21), 26694–26701. https://doi.org/10.1007/s11356-020-09062-y

Kimbrough, D., Cohen, Y., Winer, A., Creelman, L., & Mabuni, C. (1999). A critical assessment of chromium in the environment. *Critical Reviews in Environmental Science and Technology, 29*, 1–46. https://doi.org/10.1080/10643389991259164

Langård, S., Andersen, A., & Ravnestad, J. (1990). Incidence of cancer among ferrochromium and ferrosilicon workers: An extended observation period. *Occupational and Environmental Medicine, 47*(1), 14–19.

LaVelle, J. M. (1991). Mechanisms of toxicity/carcinogenicity and superfund decisions. *Environmental Health Perspectives, 92*, 127–130.

Li, L., Li, Y., Cao, L., & Yang, C. (2015). Enhanced chromium (VI) adsorption using nanosized chitosan fibers tailored by electrospinning. *Carbohydrate Polymers, 125*, 206–213.

Lin, S. C., Tai, C. C., Chan, C. C., & Wang, J. D. (1994). Nasal septum lesions caused by chromium exposure among chromium electroplating workers. *American Journal of Industrial Medicine, 26*(2), 221–228.

Losi, M. E., Amrhein, C., & Frankenberger, W. T., Jr. (1994). Environmental biochemistry of chromium. *Reviews of Environmental Contamination and Toxicology*, 91–121.

Luo, Z. (1993). *China's Imperial tombs and mausoleums the university of Michigan* (p. 44). Foreign Languages Press.
Mackenzie, R., Byerrum, R., Decker, C., Hoppert, C., & Langham, R. (1958). Chronic toxicity studies. II. Hexavalent and trivalent chromium administered in drinking water to rats. *A.M.A. Archives of Industrial Health, 18*(3), 232–234.
Mancuso, T. F., & Hueper, W. C. (1951). Occupational cancer and other health hazards in a chromate plant: a medical appraisal. 1. Lung cancers in chromate workers. *Industrial Medicine and Surgery, 20*(8), 358–363. CABI Databases.
McAllister, L. K., Van Rooij, B., & Kagan, R. A. (2010). Reorienting regulation: Pollution enforcement in industrializing countries. *Law & Policy, 32*(1), 1–13.
McCormick, J. (1999). Environmental policy. In *Developments in the European Union* (pp. 193–210). Macmillan Education UK.
McKinley, W., & Pratt, R. C. (1992). Cleaning up chromium. *Civil Engineering, 62*(3), 69–71.
McVeigh, J. C. (1984). *Energy around the world: An introduction to energy studies, global resources, needs, utilization* (pp. 23–25). Pergamon Press.
Munir, M. (2013). *History and evolution of the polluter pays principle: How an economic idea became a legal principle?* (SSRN Scholarly Paper No. 2322485). Social Science Research Network. https://doi.org/10.2139/ssrn.2322485.
Nayak, S., Rangabhashiyam, S., Balasubramanian, P., & Kale, P. (2020). A review of chromite mining in Sukinda Valley of India: Impact and potential remediation measures. *International Journal of Phytoremediation, 22*(8), 804–818. https://doi.org/10.1080/15226514.2020.1717432
Oruko, R. O., Ogola, H. J. O., Edokpayi, J. N., Volenzo, T. E., & Odiyo, J. O. (2021). Integration of sustainable development goals into leather tanning industries in Sub-Saharan Africa. In Sustainable development in Africa: Fostering sustainability in one of the world's most promising continents (pp. 575–589). Springer.
OSHA. (2006). Occupational exposure to hexavalent chromium. *Federal Register, 71*(100), 10099–10115.
Palmer, C. D., & Wittbrodt, P. R. (1991). Processes affecting the remediation of chromium-contaminated sites. *Environmental Health Perspectives, 92*, 25–40.
Pandey, K., Saharan, B. S., Kumar, R., Jabborova, D., & Duhan, J. S. (2024). Modern-day green strategies for the removal of chromium from wastewater. *Journal of Xenobiotics, 14*(4), 1670–1696.
Paul, A., Dey, S., Ram, D. K., & Das, A. P. (2024). Hexavalent chromium pollution and its sustainable management through bioremediation. *Geomicrobiology Journal, 41*(4), 324–334.
Pinasseau, A., Zerger, B., Roth, J., Canova, M., & Roudier, S. (2018). *Best available techniques (BAT) reference document for waste treatment industrial emissions directive 2010/75/EU (integrated pollution prevention and control).* JRC Science for Policy Reports. https://doi.org/10.2760/407967.
Potoski, M., & Prakash, A. (2013). Do voluntary programs reduce pollution? Examining ISO 14001's effectiveness across countries. *Policy Studies Journal, 41*(2), 273–294.
Pouikli, K. C. (2016). Overview of the implementation of the directive 2004/35/EC on environmental liability with regard to the prevention and remedying of environmental damage at European level. *Desalination and Water Treatment, 57*(25), 11520–11527.
PPG Industries Inc v. USA (2020). *2020 Decisions. 469. Opinions of the United States Court of appeals for the third circuit.* Available from https://digitalcommons.law.villanova.edu/thirdcircuit_2020/469
Prasad, S., Yadav, K., Kumar, S., Gupta, N., Cabral-Pinto, M., Rezania, S., Radwan, N., & Alam, J. (2021). Chromium contamination and effect on environmental health and its remediation: A sustainable approaches. *Journal of Environmental Management, 285*, 112174. https://doi.org/10.1016/j.jenvman.2021.112174
Rastogi, S. K., Pandey, A., & Tripathi, S. (2008). Occupational health risks among the workers employed in leather tanneries at Kanpur. *Indian Journal Of Occupational And Environmental Medicine, 12*(3), 132–135.

Ratti, R. (2020). Industrial applications of green chemistry: Status, challenges and prospects. *SN Applied Sciences, 2*(2), 263. https://doi.org/10.1007/s42452-020-2019-6
Remy, L. L., Byers, V., & Clay, T. (2017). Reproductive outcomes after non-occupational exposure to hexavalent chromium, Willits California, 1983–2014. *Environmental Health, 16*, 1–15.
Ridsdale, D. R., & Noble, B. F. (2016). Assessing sustainable remediation frameworks using sustainability principles. *Journal of Environmental Management, 184*, 36–44.
Saha, R., Nandi, R., & Saha, B. (2011). Sources and toxicity of hexavalent chromium. *Journal of Coordination Chemistry, 64*(10), 1782–1806.
Scharfenaker, M. A. (2001). Chromium VI: A review of recent developments. *Journal AWWA, 93*(11), 20–26. https://doi.org/10.1002/j.1551-8833.2001.tb09322.x
Shadreck, M., & Mugadza, T. (2013). Chromium, an essential nutrient and pollutant: A review. *African Journal of Pure and Applied Chemistry, 7*(9), 310–317.
Sharma, P., & Mitra, K. (2024). Hexavalent chromium contamination in groundwater: Erin Brockovich promulgating awareness since 2000. *Ground Water, 62*(6), 832–833.
Tang, S. Y., Prakash, V., & Tang, C. P. (1998). Local enforcement of pollution control in developing countries: A comparison of Guangzhou, Delhi, and Taipei. *Journal of Public Policy, 18*(3), 265–282.
Teleky, L. (1936). Krebs bei chromarbeitern. *DMW-Deutsche Medizinische Wochenschrift, 62*(33), 1353–1353.
Peiry, K. K. (2011). *Basel convention on the control of transboundary movements of hazardous wastes and their disposal.* United Nations Environment Programme (UNEP). Retrieved 2025, July 22 from https://legal.un.org/avl/ha/bcctmhwd/bcctmhwd.html
Uddin, S., Afroz, H., Hossain, M., Briffa, J., Blundell, R., & Islam, M. R. (2023). Heavy metals/metalloids in food crops and their implications for human health. In *Heavy metal toxicity and tolerance in plants: A biological, omics, and genetic engineering approach* (59–86). https://doi.org/10.1002/9781119906506.ch3.
US EPA. (2016). *Hazardous waste [collections and lists].* Retrieved on 2025, June 20 from https://www.epa.gov/hw
U.S. Environmental Protection Agency. (2024, April 22). *Chromium compounds: National emission standards for hazardous air pollutants (NESHAP).* Retrieved December 9, 2025, from https://www.epa.gov/stationary-sources-air-pollution/chromium-compounds-national-emission-standards-hazardous-air
Vaiopoulou, E., & Gikas, P. (2020). Regulations for chromium emissions to the aquatic environment in Europe and elsewhere. *Chemosphere, 254*, 126876. https://doi.org/10.1016/j.chemosphere.2020.126876
Vauquelin. (1798). XV. *On the use of the new metal called chrome, the oxyd of chrome, and the chromie acid. The Philosophical Magazine, 2*(5), 74–77. https://doi.org/10.1080/14786449808676883
Walls, M. (2003). *The role of economics in extended producer responsibility: Making policy choices and setting policy goals* (Discussion Paper). https://doi.org/10.22004/ag.econ.10855.
Whalley, C., Hursthouse, A., Rowlatt, S., Iqbal-Zahid, P., Vaughan, H., & Durant, R. (1999). Chromium speciation in natural waters draining contaminated land, Glasgow, UK. *Water, Air, and Soil Pollution, 112*, 389–405.
World Health Organization. (2003). *Chromium in drinking-water: Background document for development of WHO Guidelines for Drinking-water Quality.* Page number? or web address? https://cdn.who.int/media/docs/default-source/wash-documents/wash-chemicals/chromium.pdf?sfvrsn=37abd598_6 Accessed on 2025, July 28.
Xing, X., Alharbi, N. S., Ren, X., & Chen, C. (2022). A comprehensive review on emerging natural and tailored materials for chromium-contaminated water treatment and environmental remediation. *Journal of Environmental Chemical Engineering, 10*(2), 107325. https://doi.org/10.1016/j.jece.2022.107325

Zhang, C., Xia, F., Long, J., & Peng, B. (2017). An integrated technology to minimize the pollution of chromium in wet-end process of leather manufacture. *Journal of Cleaner Production, 154*, 276–283.

Zhou, M., Liu, Y., Zeng, G., Li, X., Xu, W., & Fan, T. (2007). Kinetic and equilibrium studies of Cr (VI) biosorption by dead bacillus licheniformis biomass. *World Journal of Microbiology and Biotechnology, 23*, 43–48.

Exposure to Environmental Chromium—An Ongoing Challenge

Md. Shihab Uddine Khan, Md. Kamruzzaman, Sadia Afroz Ritu, Samia Binta Zaman, Nahid Afridi, Noshin Tabassum Hasan, Shimul Mondal, Mehedi Hasan, and Shihab Uddin

Abstract Chromium is a toxic heavy metal that exists primarily as an essential trace element (Cr(III)) and a highly toxic, mobile, and carcinogenic form (Cr(VI)). Industrial processes, including the leather tanning process, the electroplating process, and the production of dyes and pigments, have substantially increased Cr pollution in soil, water, and air. Cr(VI) persists in the environment, bioaccumulates in living organisms, and enters the food chain, posing severe risks to human and ecosystem health. Although legislative initiatives and advancements in remediation technology

Md. S. U. Khan
Horticulture Division, Agricultural Research Station, Bangladesh Agricultural Research Institute, Satkhira, Bangladesh

Md. Kamruzzaman
Soil, Water and Environment Discipline, Khulna University, Khulna, Bangladesh

Md. Kamruzzaman · S. Uddin (✉)
Department of Soil Science, Faculty of Agriculture, Bangladesh Agricultural University, Mymensingh, Bangladesh
e-mail: shihab.ss@bau.edu.bd

S. A. Ritu
Department of Crop Botany, Bangladesh Agricultural University, Mymensingh, Bangladesh

S. B. Zaman
Faculty of Agriculture, Khulna Agricultural University, Khulna, Bangladesh

N. Afridi
Department of Genetics and Plant Breeding, Bangladesh Agricultural University, Mymensingh, Bangladesh

N. T. Hasan
Floriculture Division, Horticulture Research Centre, Bangladesh Agricultural Research Institute, Gazipur, Bangladesh

S. Mondal
Soil Science Division, Agricultural Research Station, Bangladesh Agricultural Research Institute, Satkhira, Bangladesh

M. Hasan
Department of Agriculture, Gopalganj Science and Technology University, Gopalganj, Bangladesh

N. Kumar (ed.), *Chromium Toxicity*, https://doi.org/10.1007/978-3-032-14323-5_4

have been made, the management of Cr pollution remains a worldwide concern due to insufficient waste treatment techniques, inadequate law enforcement in developing areas, and the complexity of extracting Cr(VI) from natural environments. This chapter clarifies the origins, mechanisms, toxicological effects, and contemporary mitigation efforts associated with Cr exposure. This chapter also explains the challenges of Cr remediation and recommends future work to fill the current gaps. It emphasizes the immediate need for sustainable methods, stronger regulations, and heightened public awareness to combat the ongoing danger of environmental Cr pollution.

Keywords Cr exposure · Industrial use · Vulnerable group · Health effect · Regulatory gap

1 Introduction

Chromium (Cr) is a naturally occurring transition metal found in the Earth's crust, primarily as chromite ore ($FeCr_2O_4$), and in trace amounts in rocks, soil, volcanic ash, and dust. Natural geologic and atmospheric processes, including volcanic eruptions, rock weathering, and erosion, contribute to background levels of Cr in terrestrial and aquatic environments (Shanker, 2019). However, anthropogenic activities, including leather tanning, stainless steel manufacturing, electroplating, pigment and dye production, chromate manufacturing, and wood preservation, have dramatically increased Cr concentrations, often far above natural baselines (Xing et al., 2025; Zhang et al., 2021). In regions with limited environmental regulations, untreated or poorly treated Cr-containing effluents have created severe contamination hotspots, particularly in developing nations where industrial growth outpaces the development of environmental protection infrastructure (Garg et al., 2014). The release of Cr into the environment has prompted the exploration of low-cost removal technologies. Clays, zeolites, industrial by-products, and biosorbents, such as agricultural waste and biomass, have demonstrated high adsorption capacities for Cr and other heavy metals (Bhattacharya et al., 2019). Biosorption, in particular, offers a low-energy, scalable, and eco-friendly option for wastewater treatment (Saleh et al., 2022). Meanwhile, phytoremediation, utilizing mechanisms such as phytoextraction, phytostabilization, and rhizofiltration, is also being investigated (Uddin et al., 2024).

A distinctive feature of Cr is its wide range of oxidation states (– II to + VI), with trivalent [Cr(III)] and hexavalent [Cr(VI)] forms being the most stable and environmentally significant (Uddin et al., 2023). Cr(III) is an essential micronutrient for humans and animals, contributing to glucose metabolism, insulin function, and lipid regulation. In the environment, Cr(III) forms insoluble hydroxide complexes, making it relatively immobile and less bioavailable. Conversely, Cr(VI)—predominantly of anthropogenic origin—is highly toxic and occurs mainly as the soluble oxyanions chromate (CrO_4^{2-}) and dichromate ($Cr_2O_7^{2-}$), whose stability depends on pH and redox conditions (Costa & Klein, 2006). Industrial processes release Cr(VI) into air, water, and soil, where it can contaminate surface and groundwater, infiltrate soils, and bioaccumulate in plants and animals. Its strong oxidative potential enables it to enter cells via sulfate transport channels, where it is reduced to Cr(III), generating reactive oxygen species (ROS). This process induces oxidative stress, DNA damage, and lipid peroxidation, and can result in mutagenesis and carcinogenesis. Chronic exposure to Cr(VI) is associated with dermatitis, respiratory illnesses, liver and kidney damage, reproductive impairment, and various cancers, particularly of the lung and gastrointestinal tract (Costa & Klein, 2006; WHO, 2020). Microbial remediation approaches have shown promise in reducing Cr(VI) in contaminated soils and wastes (Dhal et al., 2013).

Once in the environment, Cr's behavior depends strongly on its oxidation state. Cr(III) remains relatively immobile due to its low solubility and tendency to form stable hydroxides, whereas Cr(VI) is highly mobile under aerobic and neutral to alkaline conditions, increasing the risk of groundwater and surface water contamination (Ali et al., 2023; James, 1994). Plants can absorb Cr from soils and water, with uptake influenced by species, soil properties, and pH. This uptake can lead to bioaccumulation and biomagnification through food chains, exposing humans via crop consumption and animal products (Sharma et al., 2020). Adding complexity, climate change is altering key environmental parameters, such as pH, redox potential, rainfall patterns, and temperature, that influence Cr speciation and mobility, potentially increasing the persistence and toxicity of Cr(VI) (Saha et al., 2011). Furthermore, the coexistence and interconversion of Cr(III) and Cr(VI) in soils and waters are influenced by organic matter and microbial activity, complicating the assessment and remediation of contamination.

Despite growing awareness, the global response to Cr pollution remains inconsistent. While the World Health Organization (WHO) and national regulatory agencies have set permissible limits for Cr in water and food, enforcement is often weak, and legacy contamination from historical industrial activity persists. The urgent need for action has driven research into Cr toxicity mechanisms, environmental fate, and remediation technologies. Conventional removal methods such as chemical precipitation and adsorption are now complemented by emerging techniques, including phytoremediation, microbial bioremediation, and nanotechnology-based approaches (Xing et al., 2025). However, scaling these methods for real-world applications remains a significant challenge. Given Cr's dual nature, as both an essential micronutrient in its trivalent form and a potent environmental toxin in its hexavalent form, addressing its contamination requires interdisciplinary research, strong policy frameworks, and sustainable, locally adaptable remediation strategies. Overall, the persistence of Cr in the environment, its severe toxicological impacts, and its continued release from diverse industrial sources make Cr exposure an ongoing global environmental challenge that demands coordinated and sustained action.

2 Pathways of Exposure

Chromium, particularly Cr(VI), is a significant environmental toxicant due to its high solubility, mobility, and bioavailability (Mishra & Bharagava, 2016). Human exposure to Cr occurs through multiple routes, primarily inhalation, ingestion, and dermal absorption, each associated with distinct environmental and occupational sources (Fig. 1) (Shukla et al., 2024). Characterizing these exposure pathways is essential for understanding risk dynamics and developing targeted mitigation strategies.

2.1 *Inhalation*

Inhalation of airborne Cr compounds is a significant route of exposure, especially in industrialized and urban areas (Rowbotham et al., 2000). Cr(VI) is frequently emitted as fine particulate matter during high-temperature and chemically intensive processes, such as stainless steel welding, electroplating, pigment manufacturing, and leather tanning (Laxmi & Kaushik, 2020; Ma & Garbers-Craig, 2006; Shadreck & Mugadza, 2013). These particulates can remain suspended in the air for extended periods and penetrate deeply into the pulmonary system (Beaver et al., 2009). Once inhaled, Cr(VI) readily crosses epithelial barriers, triggering oxidative stress, DNA damage, and inflammatory responses that contribute to both acute and chronic respiratory disorders (Beaver et al., 2009; Singh et al., 2022a, 2022b).

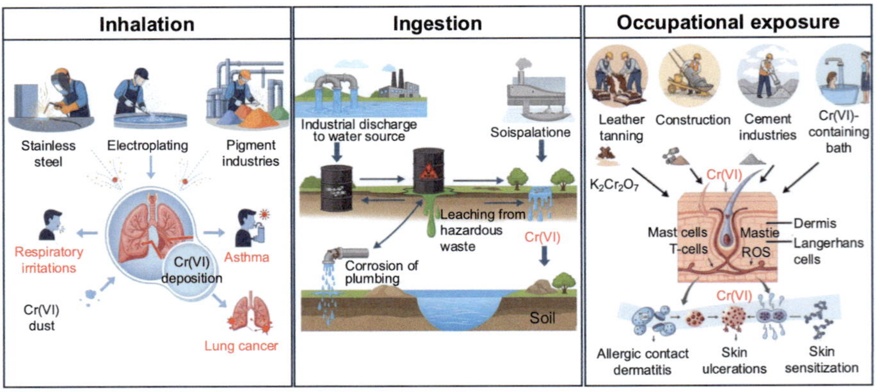

Fig. 1 Human chromium exposure pathways and possible effects

2.2 Ingestion

Oral exposure to Cr primarily occurs through the consumption of contaminated drinking water and food. The high aqueous solubility of Cr(VI) facilitates its leaching into groundwater near industrial discharge points or naturally occurring Cr-rich geogenic formations (Saha et al., 2011; Tumolo et al., 2020). In agricultural systems, the use of contaminated water for irrigation and the application of Cr-laden fertilizers or sludge can lead to bioaccumulation in crops (Dintsi et al., 2023). Chronic ingestion of Cr(VI) has been linked to gastrointestinal irritation, systemic toxicity, and elevated carcinogenic risk, particularly in regions without adequate water treatment infrastructure (Fernandez-Luqueno et al., 2013; Georgaki & Charalambous, 2023).

2.3 Dermal Contact

Dermal exposure results from direct contact with Cr-contaminated media, including soil, water, and consumer or industrial products (Sheehan et al., 1991). This pathway is especially relevant in occupational environments such as chromate manufacturing, construction, and leather processing. Cr(VI) is a potent skin sensitizer that can penetrate damaged skin and, in some cases, enter systemic circulation (Teklay, 2016). Prolonged contact may lead to allergic contact dermatitis, ulcerations, and, under severe exposure, systemic toxicity (Shelnutt et al., 2007). Communities living near Cr-contaminated sites may also experience dermal exposure through routine contact with polluted environments.

2.4 Factors Modulating Exposure

The magnitude of Cr exposure is influenced by physicochemical factors such as pH, redox conditions, and organic matter content, which determine Cr speciation and mobility (Zhu et al., 2025). Anthropogenic inputs, including industrial emissions, improper waste disposal, and inadequate regulatory oversight, exacerbate the environmental persistence of Cr(VI). Socioeconomic determinants, such as limited access to clean water, a lack of personal protective equipment, and insufficient awareness, further increase exposure risks, particularly in low-resource regions. A comprehensive understanding of these variables is essential for accurate exposure modeling, practical risk assessment, and the development of targeted intervention strategies.

3 Vulnerable Populations

Certain population groups face heightened risks from Cr exposure due to occupational hazards, environmental proximity, physiological vulnerabilities, and socioeconomic disadvantage (Iyer et al., 2023; Khan et al., 2024). Identifying these groups is crucial for developing targeted public health policies and effective remediation strategies. Industrial workers are among the most directly affected, especially those in Cr-intensive sectors such as leather tanning, electroplating, chromate manufacturing, welding, and pigment production. They are regularly exposed to Cr(VI) through inhalation of airborne particulates, skin contact, and accidental ingestion, particularly during high-temperature processes and the handling of chemicals (Fig. 2) (Diene et al., 2012; Islam et al., 2022). Occupational risks are exacerbated in settings with poor ventilation, inadequate personal protective equipment (PPE), and weak enforcement of safety regulations, particularly in poorly regulated workplaces, especially in low- and middle-income countries (Georgaki & Charalambous, 2023; Zhitkovich, 2011). Documented health effects include nasal septum perforation, chronic bronchitis, skin lesions, dermatitis, respiratory irritation, and increased lung cancer risk. Even low-dose chronic exposure can cause systemic toxicity and permanent organ damage. Many affected workers come from economically disadvantaged backgrounds, limiting their access to healthcare and compensation (Murthy et al., 2023; Yasmeen & Hafeez, 2023).

Communities near industrial belts, landfill sites, tannery zones, and wastewater discharge points experience chronic low-level exposure through contaminated groundwater, airborne dust, and consumption of crops grown in polluted soils (Junaid et al., 2016). In Bangladesh, for instance, tanneries in Hazaribagh and Savar have historically discharged large amounts of Cr into the Buriganga River and surrounding lands, contaminating water and food chains (Islam et al., 2022). Such populations often lack access to environmental monitoring, clean water, and medical screening, which further amplifies their exposure and health consequences, and face systemic barriers to legal and political advocacy, thereby reinforcing environmental injustice (WHO, 2020).

Children and pregnant women are especially vulnerable due to physiological and developmental factors. Children's higher respiratory rates, lower body mass, and underdeveloped detoxification systems, combined with exploratory behaviors, increase their risk of ingesting or inhaling Cr-contaminated dust and soil (Dutta et al., 2022; Ferguson et al., 2017). Even ingestion of a few milligrams of Cr(VI) can trigger oxidative stress, damage intestinal linings, impair neurological development, and lead to long-term effects such as growth retardation, respiratory diseases, cognitive deficits, and immune suppression (McDermott et al., 2015; Singh et al., 2022a). Pregnant women face both the direct toxic effects and the risk of transplacental Cr transfer, which can accumulate in fetal tissues and disrupt embryonic development, leading to low birth weight, spontaneous abortions, congenital malformations, preterm birth, and endocrine disruption (ATSDR, 2012). Nutritional deficiencies, such as iron deficiency, may further increase Cr absorption and toxicity (Tumolo

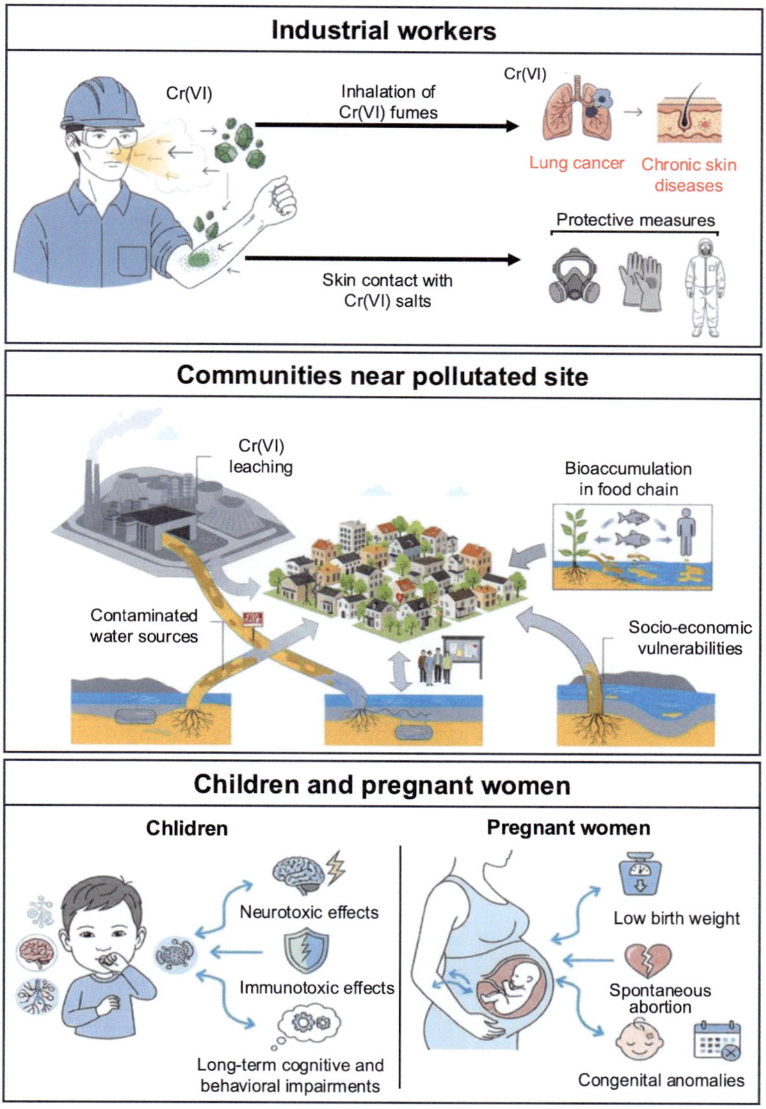

Fig. 2 Vulnerable populations of chromium exposure and health effects

et al., 2020). Marginalized or displaced communities, including refugees and residents of informal settlements, often endure compounded risks due to poorly regulated water and sanitation systems.

4 Health Effects of Chromium Exposure

The health effects of Cr exposure, particularly Cr(VI), are well-documented and multifaceted, ranging from acute toxicity to long-term carcinogenic consequences (Fig. 3). Intracellularly, Cr(VI) is reduced to Cr(III) through metabolic processes, generating reactive intermediates and significant amounts of ROS. These intermediates cause oxidative injury to DNA, including strand breaks and base modifications, leading to mutagenesis and genomic instability (Singh et al., 2022b; Zhitkovich, 2011). Oxidation of proteins can impair enzymatic function and cellular signaling, while damage to cell membranes disrupts cellular integrity. The reduction process relies heavily on cellular reductants like glutathione (GSH), whose depletion exacerbates oxidative stress and weakens antioxidant defenses (Shanker, 2019).

Prolonged Cr(VI) exposure is linked to a wide range of serious health outcomes. Epidemiological studies strongly associate inhalation with increased lung cancer risk, while ingestion of contaminated water is linked to gastrointestinal cancers (Sazakli et al., 2024; Tumolo et al., 2020). Systemic toxicity can manifest as kidney and liver damage, respiratory problems, gastroenteritis, and ulcers (Pokhrel & Pokhre, 2022; Shanker, 2019; Singh et al., 2022a). Dermal contact can cause inflammation, allergic responses, eczema-type rashes, and ulcerative lesions (Singh et al., 2022a). Cr(VI) is widely regarded as the most toxic form of Cr and classified as a class 1 carcinogenic metal (National Toxicology Program, 2021).

Children are especially vulnerable, with exposure linked to impaired neurodevelopment, reduced cognitive function, growth retardation, and higher susceptibility

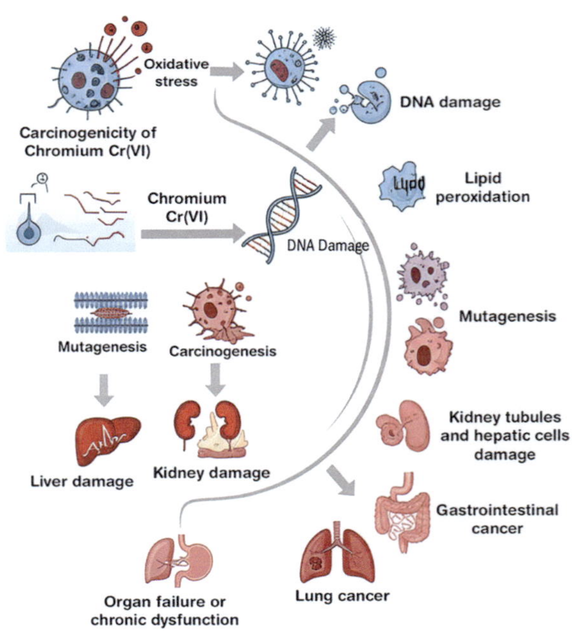

Fig. 3 Human health effects of chromium exposure

to respiratory diseases (McDermott et al., 2015). Prenatal exposure raises concerns over teratogenic effects and impacts on fetal development, necessitating stringent protective measures. Although Cr(III) plays a role in glucose metabolism and generally exhibits low toxicity, environmental and occupational concerns overwhelmingly focus on the hazards of Cr(VI). Chromium can accumulate in vital organs such as the liver and kidneys, causing hepatotoxicity and nephrotoxicity (Chakraborty et al., 2022). Beyond physiological harm, communities in contaminated areas often experience psychological burdens, including anxiety over health outcomes, stigma, property devaluation, and social isolation. These mental health impacts are intensified in regions where industrial pollution intersects with poverty (Di Lisi et al., 2025; Gee & Payne-Sturges, 2004). Thus, the health effects of Cr extend beyond physical illness, affecting overall well-being and community resilience.

5 Environmental Impacts

Chromium contamination imposes significant environmental harm across terrestrial, aquatic, and atmospheric systems, often with persistent and multifaceted ecological consequences (Fig. 4) (Suljević et al., 2024). Although airborne deposition of Cr-containing dust contributes to pollution (Kośla et al., 2019), environmental disruption typically begins in soil, where Cr—especially Cr(VI)—alters structural, chemical, and biological properties (Ao et al., 2022).

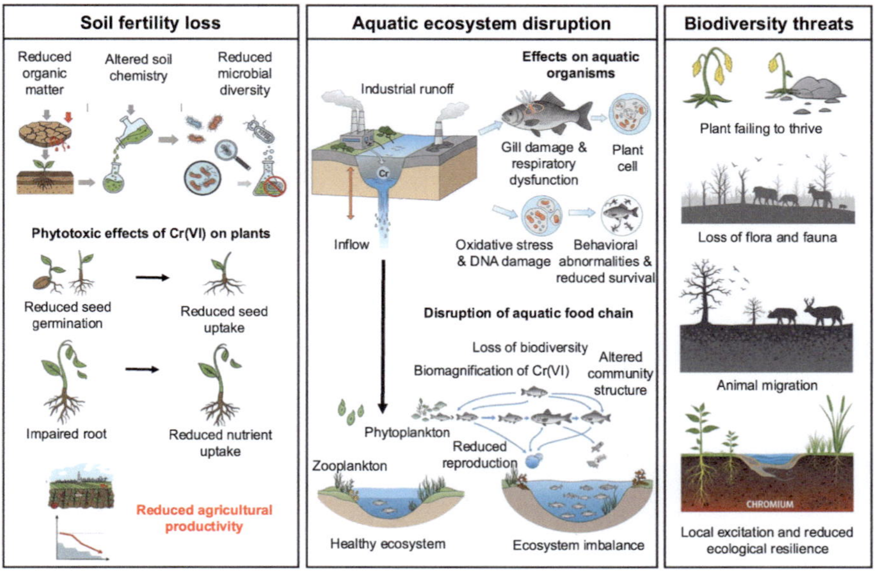

Fig. 4 Environmental impacts of chromium contamination

5.1 Soil Systems

Elevated Cr contamination impairs soil health by degrading soil structure, porosity, and cation exchange capacity (Del Real et al., 2020). It lowers soil pH by reducing alkali-hydrolysable nitrogen (AN) and suppressing activities of essential enzymes like urease and sucrase, thereby disrupting soil nitrogen and carbon cycling (Zhang et al., 2021). These disruptions diminish soil fertility and productivity by limiting nutrient availability, carbon sequestration, and other soil chemical characteristics (Campillo-Cora et al., 2025). Additionally, Cr(VI) exerts toxic effects on microbial communities, reducing diversity, enzymatic activity, and biomass production (Rashid et al., 2023a). The extent of microbial response depends on contamination levels and inherent soil properties (Campillo-Cora et al., 2025). These soil disruptions can cascade into aquatic systems through leaching and runoff, affecting food webs and ecosystem stability.

5.2 Plant Systems

Once mobilized, Cr is absorbed by plants, even at low concentrations. As a non-essential element, Cr interferes with seed germination, shoot/root growth, photosynthetic function, and enzyme activity (Rani et al., 2020; Suljević et al., 2024). Chromium toxicity induces oxidative stress via excess reactive oxygen species (ROS), disrupting redox balance and triggering enzymatic antioxidant responses that may themselves hinder growth (Qin et al., 2024; Srivastava et al., 2021). Physiological effects include chlorosis, inhibited nutrient uptake, reduced water transport, and symptoms such as dwarfing, early senescence, and decreased leaf area (Saud et al., 2022). At higher concentrations, Cr causes irreversible cellular damage, affecting membranes and organelles (Gill et al., 2016). These effects cumulatively reduce biomass and crop yields (Zulfiqar et al., 2023).

5.3 Animal System

Chromium exposure also poses risks to terrestrial animals. In mammals, it is associated with reproductive toxicity (e.g., deteriorated semen quality with several reproductive disorders in males and reduced fertility in females), carcinogenicity, and organ damage (Kośla et al., 2019). For instance, mongooses near Cr-polluted tanneries showed severe hematological changes (e.g., 59.6% reduction in RBCs and 64% in WBCs), liver/kidney dysfunction, histopathological damage exceeding 75% in affected tissues, including hepatocyte shrinkage, vacuolization, sinusoidal expansion, nuclear pyknosis, renal tubule epithelial hypertrophy, lumen reduction, inter-renal cell atrophy, glomerular compression, and hematopoietic tissue necrosis

(Andleeb et al., 2020). In rodents, dietary and dermal exposure led to Cr accumulation in the liver, kidneys, and spleen, causing tissue necrosis (Greef, 2018). Maternal exposure to Cr(VI) disrupts reproduction and transfers the metal to offspring via the placenta and lactation, leading to hormonal imbalances, oxidative stress, delayed sexual maturation, and ovarian tissue damage (Samuel et al., 2014). Birds also show Cr-induced carcinogenic, mutagenic, and teratogenic effects, with bioaccumulation observed in lung tissue (Kośla et al., 2019).

5.4 Aquatic Systems

Chromium contamination of aquatic environments leads to bioaccumulation in fish, invertebrates, and microalgae (Ghosh & Saha, 2022). Entry occurs via ingestion or gill absorption, after which Cr binds to DNA and proteins, causing cellular toxicity (Bakshi & Panigrahi, 2018). Cr(VI) biomagnifies through trophic levels, threatening biodiversity and ecosystem health. Sediment-bound Cr serves as a long-term contamination source, remobilizing under changing pH and redox conditions (Kośla et al., 2019). Fish exposed to Cr show behavioral abnormalities, physiological stress, immune suppression, hematological changes such as reduced hemoglobin and leukocyte counts, declined total lipid and protein contents, altered ROS and SOD levels, and carcinogenicity (Bakshi & Panigrahi, 2018; Biswas et al., 2023; Ghosh & Saha, 2022). In *Anabas testudineus*, Cr exposure resulted in eye deformation (e.g., scale loss, eyeball shrinkage, and eye redness), behavioral abnormalities (e.g., erratic swimming, hyperactivity, and frequent surfacing), and cellular damage (Biswas et al., 2023).

Microalgae such as *Mucidosphaerium pulchellum* and *Micractinium pusillum* also show suppressed growth, chlorophyll synthesis, and carbon assimilation under Cr(VI) stress. However, some increase lipid accumulation as a stress response (Bashir et al., 2021). Exposure to Cr adversely affected *Anacystis nidulans* and *Spirulina platensis*, significantly impairing their growth rate, biomass productivity, chlorophyll A synthesis, and ability to assimilate carbon (Azeez & Banerjee, 1988). These disruptions threaten aquatic primary productivity and oxygen cycling.

6 Risk Assessment and Monitoring

Effective Cr risk assessment and monitoring are essential for identifying human and ecological hazards, determining exposure levels, and informing regulatory decisions. These approaches combine chemical analysis, toxicological profiling, and epidemiological evidence to guide targeted interventions.

6.1 Risk Assessment Frameworks

Chromium risk assessment follows a four-step model: (1) hazard identification, (2) dose–response assessment, (3) exposure assessment, and (4) risk characterization (Ma et al., 2007). Cr(VI) is consistently identified as the primary toxic species due to its mutagenic, carcinogenic, and immunotoxic effects (Cieślak-Golonka, 1996). Dose–response models define thresholds for adverse effects by incorporating route-specific toxicokinetics (Thompson et al., 2018). Exposure assessments estimate intake through air, water, food, and dermal contact using environmental data. Risk characterization integrates these findings to calculate probabilistic health risks, forming the foundation for standards and public health guidance (Ray et al., 2014).

6.2 Biomonitoring Approaches

Biomonitoring offers direct insight into internal Cr exposure. Urinary Cr concentration, particularly in post-shift samples, serves as a key indicator of recent Cr(VI) intake due to rapid renal excretion (Wasowicz et al., 2025). Blood analysis provides additional information on total Cr levels and Cr speciation (Devoy et al., 2016). Hair and nail analyses can reflect long-term exposure but require stringent contamination control. Technological advances, particularly in inductively coupled plasma mass spectrometry (ICP-MS), have greatly improved detection sensitivity and specificity, making it feasible to monitor low-dose exposures across diverse populations (Taylor et al., 2017).

6.3 Environmental Monitoring Strategies

Environmental monitoring quantifies Cr levels in air, water, soil, sediments, and living organisms. Techniques such as atomic absorption spectrometry (AAS), graphite furnace AAS (GFAAS), and high-performance liquid chromatography (HPLC) with diode array or mass detection allow for accurate Cr(III) and Cr(VI) speciation (Sardans et al., 2011; Ščančar & Milačič, 2014). Regular surveillance of effluent discharge points, groundwater wells, and industrial zones is critical for detecting contamination early and ensuring compliance with environmental standards.

6.4 Current Challenges and Future Outlook

Major challenges include variability in environmental Cr levels across time and space, limited long-term biomonitoring data, and technical difficulties in accurately distinguishing Cr species in complex media. Promising innovations such as real-time monitoring devices, wearable exposure sensors, and predictive exposure modeling tools offer potential improvement (Lin et al., 2022; Mossotti et al., 2024; Zafar et al., 2022). Integrating environmental and biomonitoring data with Geographic Information Systems (GIS) can support spatiotemporal risk mapping and population vulnerability assessments (Rashid et al., 2023b). Going forward, standardizing methodologies, enhancing international data-sharing platforms, and promoting interdisciplinary collaboration will be critical to improving Cr exposure management.

7 Exposure Management and Mitigation Strategies

Chromium exposure management and mitigation are important to minimize toxicity. Certain strategies are adopted worldwide to reduce Cr effects on humans (Fig. 5). These strategies are explained in this section.

7.1 Community-Based Interventions

Community-level action is crucial in mitigating Cr exposure, particularly near industrial sites such as tanneries, electroplating units, and chromite mines (Coetzee et al., 2020). Participatory water monitoring empowers residents to detect Cr(VI) contamination and press for regulatory responses (Singh et al., 2023). In Hazaribagh, Bangladesh, grassroots advocacy led to the relocation of tanneries, stricter effluent standards, and the use of household filtration systems—resulting in measurable reductions in groundwater Cr(VI) levels (Khan et al., 2020). Additionally, zoning regulations and the creation of buffer zones between industrial and residential areas help reduce exposure through air and dermal contact (Caravanos et al., 2011). These community-led strategies often complement national environmental policies.

7.2 Education and Awareness Programs

Public education plays a vital role, especially for vulnerable groups such as industrial workers, pregnant women, and children. Awareness campaigns promote behaviors like avoiding contaminated water sources, maintaining hygiene after exposure, and understanding exposure risks (Bhattacharya et al., 2022). Workplace training focuses on proper use of personal protective equipment (PPE), safe material handling, and routine health screening (Santonen et al., 2019). In resource-limited regions, knowledge dissemination via radio, school programs, and multilingual community workshops helps strengthen local capacity and resilience.

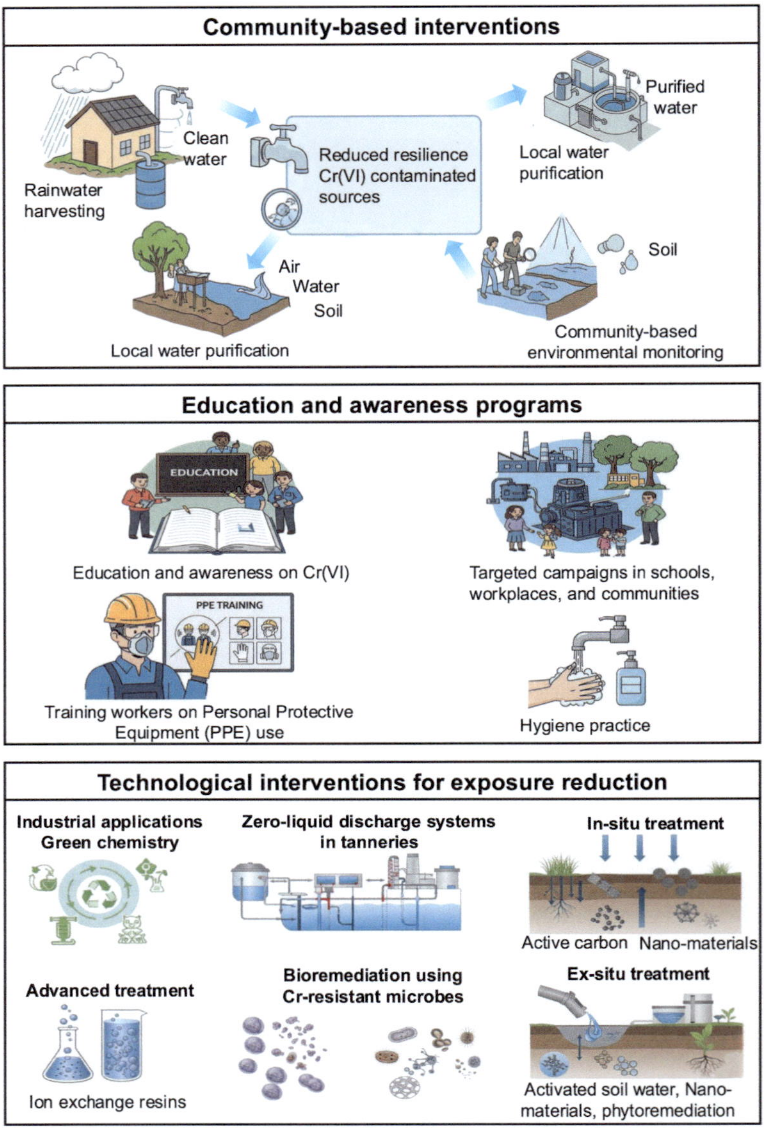

Fig. 5 Management and mitigation strategies of chromium exposure

7.3 Technological Interventions

Engineering and technological solutions are critical for removing Cr from contaminated environments. Effective water treatment methods include ion exchange, reverse osmosis, and chemical reduction using ferrous sulfate (Saleh et al., 2022). In workplaces, engineering controls such as local exhaust ventilation reduce inhalation risks

(Chen, 2024). Innovative approaches—such as zero-valent iron (ZVI) nanoparticles and biochar-based filtration systems—have achieved > 90% Cr(VI) removal in laboratory conditions via combined adsorption and reduction mechanisms (Qiu et al., 2020). However, further research is needed to assess the durability, scalability, and cost-effectiveness of these methods in real-world settings.

8 Success Stories in Chromium Exposure Reduction

While Cr exposure remains a significant environmental and public health concern globally, several initiatives have successfully mitigated its impact. Through a combination of strong regulations, technological innovation, and community engagement, these efforts have led to measurable improvements in environmental quality and human health protection.

Case Studies

8.1 Community-Driven Cleanup in Hinkley, USA

Hinkley, California, became a symbol of environmental justice after the Erin Brockovich case exposed groundwater contamination by Cr(VI) from Pacific Gas and Electric Company (PG&E) operation (Izbicki et al., 2023). In response, PG&E initiated a long-term remediation plan, using food-grade ethanol to chemically reduce Cr(VI) to the less toxic Cr(III) (Izbicki & Groover, 2018).

8.2 Zero Liquid Discharge (ZLD) Program, China

To curb Cr pollution from electroplating, China implemented a nationwide Zero Liquid Discharge (ZLD) policy (Huang et al., 2021). In provinces like Jiangsu and Guangdong, most industrial plating zones adopted membrane filtration, ion exchange, and evaporation technologies, cutting wastewater discharge to near zero (Li et al., 2022). Consequently, Cr(VI) concentrations in local water bodies dropped significantly compared to 2015 baselines.

8.3 California's Chromium (VI) Water Standard, USA

California became the first U.S. state to enforce a legal limit of 10 parts per billion (ppb) for Cr(VI) in public drinking water, following a detailed health risk assessment (Hausladen et al., 2018). Despite high treatment costs, the standard effectively reduced cancer risks and ensured safe water through enhanced removal technologies (Tan et al., 2020).

8.4 Glasgow's Green Remediation Model, United Kingdom

In the early 2000s, Glasgow faced widespread Cr contamination in soil from legacy dye and steel industries (Deoli et al., 2025). The city adopted phytoremediation techniques using willow and poplar trees to absorb Cr, reducing exposure risks, improving air quality, and converting polluted lands into green public spaces (Farraji et al., 2020).

8.5 The Kanpur Tannery Relocation, India

A landmark court ruling in India led to the shutdown of highly polluting tanneries along the Ganges River in Kanpur. These operations were moved to a centralized Common Effluent Treatment Plant (CETP) in an industrial zone (Simon & Joshi, 2022). The relocation significantly reduced Cr discharges into the river and improved worker safety through better waste management practices (Siddiqui et al., 2022).

9 Why is Environmental Chromium Exposure an Ongoing Challenge?

Despite some success stories being mentioned, Cr exposure remains a concern and is difficult to resolve quickly. Several factors contribute to the complete removal of this toxic element from the environment. The factors are described at this point.

9.1 Widespread Industrial Use and Legacy Contamination

Large volumes of Cr have traditionally been released into soils, sediments, and rivers by significant industries (such as tanneries, electroplating, the production of stainless steel and alloys, pigment and dye manufacturing, leather processing, certain mining, and metallurgy). Groundwater and surface water continue to receive Cr from many contaminated sites, which are "legacy" issues (contaminated sediments, tailings, sludge, and buried waste) long after the initial discharge. Source control is not consistent or comprehensive because new emissions also persist in areas with expanding industrial activity (Islam et al., 2023).

9.2 Chemical Properties and Environmental Behavior

In the environment, Cr exists in several oxidation states and continuously interconverts depending on mineral surface, pH, organic matter, and redox potential. The total Cr concentration in a medium does not necessarily reflect the toxicity potential, as Cr(VI) is the most toxic form and is converted to other non-toxic forms. Additionally, other non-toxic forms can be converted to the toxic Cr(VI) form. Cr(VI) is highly water-soluble and mobile, making it difficult to control. This dynamic chemistry makes monitoring (accurately measuring Cr(VI) in complex matrices) and designing robust, long-lasting remediation systems technically difficult. Many treatment decisions hinge on speciation, but field conditions often change in ways that undermine the assumptions made in the lab.

9.3 Remediation Cost and Durability

Until now, researchers have developed a wide range of remediation strategies, including chemical reduction/precipitation, adsorption (active carbon, ion exchange, biosorbents), electrochemical methods, phytoremediation, and bioremediation (microbial Cr(VI) reduction). However, each approach has limitations, e.g., they may be expensive at the municipal/industrial scale, sensitive to site chemistry, produce secondary wastes, or face time constraints (phytoremediation), or lose effectiveness if re-oxidation occurs. One of the biggest practical obstacles is still scaling promising lab/Bench techniques to diverse field locations while preserving long-term performance. While highlighting advancements, recent studies also stress the financial and technical barriers to extensive cleanup (Islam et al., 2023).

9.4 Regulatory Gaps and Uneven Enforcement

Different international organizations, including the WHO, the International Agency for Research on Cancer (IARC), the United Nations Environment Program (UNEP), and the Basal Conventional, regional, and national regulatory bodies, have set maximum permissible limits for environmental components and drinking water. However, many countries lack strict, enforceable standards or the resources to monitor and enforce them. Even in countries where standards exist, detection and remediation costs can be a barrier to compliance (particularly for small utilities and low-income regions). Regulatory variability and slow policy adoption mean communities remain at risk where debate, rulemaking, and resource allocation proceed. For example, recent advances in state-level regulation in the U.S. and the European Union illustrate both progress and the difficulty of uniform control.

9.5 Socioeconomic and Environmental Justice Dimensions

Workers in Cr-intensive industries, people living close to industrial clusters or unregulated disposal sites, and people living in low- and middle-income nations where industrial growth has outpaced pollution controls are frequently the communities most exposed, even though they also tend to have the least political clout and resources. Exposures persist across generations due to inadequate funding for remediation, limited access to safe water, and inadequate healthcare. Informal recycling/e-waste activities, occupational exposures, and inadequate PPE introduce other risk routes. Occupational dangers and their effects on health have been extensively documented.

9.6 Monitoring, Epidemiology, and Long-Term Health Uncertainties

Important uncertainties still surround the health effects of long-term low-level exposure, the combination effects with other pollutants, and the social and mental health burdens on impacted populations, despite the established acute and many chronic effects of Cr(VI), including carcinogenicity. Robust exposure monitoring and extensive, long-term epidemiological studies are costly and logistically challenging, but they are necessary for improved risk assessment and focused interventions.

9.7 Emerging and Compounding Pressure

Forecasts of exposure may become more challenging due to climate-driven changes that can remobilize deposited/precipitated Cr or modify its speciation (e.g., altered rainfall, flooding, groundwater dynamics, and redox shifts in soils). Legacy infrastructure failures, new industrial processes, and informal recycling can cause new contamination incidents. Even places that have been "remediated" in the past need constant monitoring due to these dynamic stresses.

10 Future Research and Recommendation

Future research should prioritize the development of cost-effective, eco-friendly, and scalable technologies to remove especially toxic Cr(VI) from contaminated environments, including advanced nanomaterials, biochar applications, and genetically engineered microorganisms. Additionally, it should focus on improving highly sensitive, rapid, and field-deployable detection systems for real-time monitoring. Long-term epidemiological studies are necessary to evaluate the chronic health effects of low-level exposure, particularly in vulnerable and occupationally exposed populations, alongside investigations into chromium speciation, transformation, and mobility under various environmental conditions. Policy-oriented work should integrate strict regulatory enforcement, industrial process redesign to minimize waste, and community-led monitoring programs, supported by targeted public awareness campaigns in high-risk areas. A coordinated, multidisciplinary approach, combining scientific innovation, public health research, regulatory action, and community engagement, is essential to shift from reactive mitigation toward preventive and sustainable chromium management.

11 Conclusion

Exposure to environmental Cr, particularly Cr(VI), remains an ongoing global challenge with profound ecological and public health implications due to its widespread use, persistence, and toxicological effects. Industrial activities, such as leather tanning, electroplating, and pigment production, continue to release Cr into the air, water, and soil, where it persists and often disproportionately affects vulnerable communities. Despite regulatory measures and remediation efforts, contamination persists, underscoring the need for improved detection methods, more effective remediation strategies, and stricter enforcement of environmental standards. Success requires strong coordination between working groups and the broader population, with a strong commitment to sustainable industrial practices and green technologies, to ensure a safer and more climate-resilient future.

References

Ali, S., Mir, R. A., Tyagi, A., Manzar, N., Kashyap, A. S., Mushtaq, M., Raina, A., Park, S., Sharma, S., Mir, Z. A., & Lone, S. A. (2023). Chromium toxicity in plants: Signaling, mitigation, and future perspectives. *Plants, 12*(7), 1502. https://doi.org/10.3390/plants12071502

Andleeb, S., Ahmad, Z., Mahmood, T., Bao, S., Saeed, A. A., & Jha, S. K. (2020). Evaluating toxicity impacts of environmental exposed chromium on small Indian mongoose (*Urva auropunctatus*) hematological, biochemical and histopathological functioning. *Chemosphere, 259*, Article 127485. https://doi.org/10.1016/j.chemosphere.2020.127485

Ao, M., Chen, X., Deng, T., Sun, S., Tang, Y., Morel, J. L., Qiu, R., & Wang, S. (2022). Chromium biogeochemical behaviour in soil-plant systems and remediation strategies: A critical review. *Journal of Hazardous Materials, 424*, Article 127233. https://doi.org/10.1016/j.jhazmat.2021.127233

ATSDR (Agency for Toxic Substances and Disease Registry). (2012). Toxicological profile for chromium. U.S. Department of Health and Human Services. https://www.atsdr.cdc.gov/toxprofiles/tp7.pdf. Accessed August 1, 2025.

Azeez, P. A., & Banerjee, D. K. (1988). Effect of chromium on cyanobacteria and its accumulation. *Toxicological & Environmental Chemistry, 16*(3), 229–240.

Bakshi, A., & Panigrahi, A. K. (2018). A comprehensive review on chromium induced alterations in freshwater fishes. *Toxicology Reports, 5*, 440–447.

Bashir, K. M. I., Lee, H. J., Mansoor, S., Jahn, A., & Cho, M. G. (2021). The effect of chromium on photosynthesis and lipid accumulation in two chlorophyte microalgae. *Energies, 14*(8), 2260. https://doi.org/10.3390/en14082260

Beaver, L. M., Stemmy, E. J., Schwartz, A. M., Damsker, J. M., Constant, S. L., Ceryak, S. M., & Patierno, S. R. (2009). Lung inflammation, injury, and proliferative response after repetitive particulate hexavalent chromium exposure. *Environmental Health Perspectives, 117*(12), 1896–1902.

Bhattacharya, A., Gupta, A., Kaur, A., & Malik, D. (2019). Alleviation of hexavalent chromium by using microorganisms: Insight into the strategies and complications. *Water Science and Technology, 79*(3), 411–424.

Bhattacharya, A., Sahu, S., Sahu, N., Das, A., Das, C., John, V., et al. (2022). Influence of mass-awareness campaign on community behavior pattern changes for safe drinking water availability in a groundwater arsenic-affected area of South Asia. *Groundwater for Sustainable Development, 18*, Article 100766. https://doi.org/10.1016/j.gsd.2022.100766

Biswas, S., Halder, P., Debnath, S., Ghosh, P., & Panigrahi, A. K. (2023). An assessment of Chromium (VI) induced toxicity in common freshwater fish *Anabas testudineus* (Bloch, 1792). *Biological Forum – An International Journal, 15*(5), 578–584.

Campillo-Cora, C., Rodríguez-Seijo, A., Pérez-Rodríguez, P., Fernández-Calvino, D., & Santás-Miguel, V. (2025). Effect of heavy metal pollution on soil microorganisms: Influence of soil physicochemical properties. A systematic review. *European Journal of Soil Biology, 124*, 103706. https://doi.org/10.1016/j.ejsobi.2024.103706

Caravanos, J., Clark, E., Fuller, R., & Lambertson, C. (2011). Assessing worker and environmental chemical exposure risks at an e-waste recycling and disposal site in Accra, Ghana. *Journal of Health and Pollution, 1*(1), 16–25.

Chakraborty, R., Renu, K., Eladl, M. A., El-Sherbiny, M., Elsherbini, D. M. A., Mirza, A. K., et al. (2022). Mechanism of chromium-induced toxicity in lungs, liver, and kidney and their ameliorative agents. *Biomedicine & Pharmacotherapy, 151*, Article 113119. https://doi.org/10.1016/j.biopha.2022.113119

Chen, K. (2024). Holistic understanding of ventilation rate in occupational health risk control. *Annals of the New York Academy of Sciences, 1531*(1), 3–11.

Cieślak-Golonka, M. (1996). Toxic and mutagenic effects of chromium(VI). A review. *Polyhedron, 15*(21), 3667–3689.

Coetzee, J. J., Bansal, N., & Chirwa, E. M. N. (2020). Chromium in environment, its toxic effect from chromite-mining and ferrochrome industries, and its possible bioremediation. *Exposure and Health, 12*, 51–62.

Costa, M., & Klein, C. B. (2006). Toxicity and carcinogenicity of chromium compounds in humans. *Critical Reviews in Toxicology, 36*(2), 155–163.

Del Real, A. P., Pérez-Sanz, A., García-Gonzalo, P., Castillo-Michel, H., Gismera, M. J., & Lobo, M. C. (2020). Evaluating Cr behaviour in two different polluted soils: Mechanisms and implications for soil functionality. *Journal of Environmental Management, 276*, Article 111073. https://doi.org/10.1016/j.jenvman.2020.111073

Deoli, V., Gupta, P. K., & Malik, A. (2025). Soil–water contamination at COPR sites and in adjoining agricultural lands: remediation strategies and practices. In D. Kumar, P. K. Gupta, B. Singh, S. Verma (Eds.), *Decontamination of subsurface water resources system using contemporary technologies* (pp. 145–152). Elsevier.

Devoy, J., Géhin, A., Müller, S., Melczer, M., Remy, A., Antoine, G., & Sponne, I. (2016). Evaluation of chromium in red blood cells as an indicator of exposure to hexavalent chromium: An in vitro study. *Toxicology Letters, 255*, 63–70.

Dhal, B., Thatoi, H. N., Das, N. N., & Pandey, B. D. (2013). Chemical and microbial remediation of hexavalent chromium from contaminated soil and mining/metallurgical solid waste: A review. *Journal of Hazardous Materials, 250*, 272–291.

Di Lisi, A., Dalla Valle, A., Menchetti, M., Crisafulli, C., & Fabbri, C. (2025). The potential benefits of chromium and selenium supplementation across psychiatric disorders and symptoms. *Journal of Clinical Psychopharmacology, 45*(4), 376–387.

Diene, E., Agrinier, N., Albessard, A., Cassadou, S., Schwoebel, V., & Lang, T. (2012). Relationships between impact on employment, working conditions, socio-occupational categories and symptoms of post-traumatic stress disorder after the industrial disaster in Toulouse, France. *Social Psychiatry and Psychiatric Epidemiology, 47*, 1309–1319.

Dintsi, B. S., Letsoalo, M. R., & Ambushe, A. A. (2023). Bioaccumulation and human health risk assessment of arsenic and chromium species in water–soil–vegetables system in Lephalale, Limpopo Province, South Africa. *Minerals, 13*(7), 930. https://doi.org/10.3390/min13070930

Dutta, S., Gorain, B., Choudhury, H., Roychoudhury, S., & Sengupta, P. (2022). Environmental and occupational exposure of metals and female reproductive health. *Environmental Science and Pollution Research, 29*(41), 62067–62092.

Farraji, H., Robinson, B., Mohajeri, P., & Abedi, T. (2020). Phytoremediation: Green technology for improving aquatic and terrestrial environments. *Nippon Journal of Environmental Science, 1*, 1–30.

Ferguson, A., Penney, R., & Solo-Gabriele, H. (2017). A review of the field on children's exposure to environmental contaminants: A risk assessment approach. *International Journal of Environmental Research and Public Health, 14*(3), 265. https://doi.org/10.3390/ijerph14030265

Fernandez-Luqueno, F., Lopez-Valdez, F., Gamero-Melo, P., Luna-Suarez, S., Aguilera-Gonzalez, E. N., Martínez, A. I., & Pérez-Velázquez, I. R. (2013). Heavy metal pollution in drinking water—a global risk for human health: A review. *African Journal of Environmental Science and Technology, 7*(7), 567–584.

Garg, V. K., Yadav, P., Mor, S., Singh, B., & Pulhani, V. (2014). Heavy metals bioconcentration from soil to vegetables and assessment of health risk caused by their ingestion. *Biological Trace Element Research, 157*(3), 256–265.

Gee, G. C., & Payne-Sturges, D. C. (2004). Environmental health disparities: A framework integrating psychosocial and environmental concepts. *Environmental Health Perspectives, 112*(17), 1645–1653.

Georgaki, M. N., & Charalambous, M. (2023). Toxic chromium in water and the effects on the human body: A systematic review. *Journal of Water and Health, 21*(2), 205–223.

Ghosh, D., & Saha, S. K. (2022). A comprehensive review on toxicity of chromium in freshwater fishes. *Applied Ecology and Environmental Sciences, 10*, 527–533.

Gill, R. A., Zhang, N., Ali, B., Farooq, M. A., Xu, J., & Gill, M. B. (2016). Role of exogenous salicylic acid in regulating physio-morphic and molecular changes under chromium toxicity in black- and yellow-seeded *Brassica napus* L. *Environmental Science and Pollution Research, 23*, 20483–20496.

Greef, D. (2018). The impact of heavy metal contamination on the morphology and physiology of small mammals. MS Thesis, University of the Witwatersrand, Johannesburg (South Africa). https://wiredspace.wits.ac.za/server/api/core/bitstreams/68885576-8c54-4ff5-8949-dc91f689bf56/content

Hausladen, D. M., Alexander-Ozinskas, A., McClain, C., & Fendorf, S. (2018). Hexavalent chromium sources and distribution in California groundwater. *Environmental Science & Technology, 52*(15), 8242–8251.

Huang, Z., Hwang, J. Y., Huang, C., & Shi, Y. (2021). Electroplating wastewater treatment in China. In B. Li et al. (Eds.), *Materials engineering—From ideas to practice: An EPD symposium in honor of Jiann-Yang Hwang. The minerals, metals & materials series* (pp. 225–232). Springer. https://doi.org/10.1007/978-3-030-65241-8_21

Islam, M. M., Mohana, A. A., Rahman, M. A., Rahman, M., Naidu, R., & Rahman, M. M. (2023). A comprehensive review of the current progress of chromium removal methods from aqueous solution. *Toxics, 11*(3), 252. https://doi.org/10.3390/toxics11030252

Islam, S., Kamila, S., & Chattopadhyay, A. (2022). Toxic and carcinogenic effects of hexavalent chromium in mammalian cells in vivo and in vitro: A recent update. *Journal of Environmental Science and Health, 40*(3–4), 282–315.

Iyer, M., Anand, U., Thiruvenkataswamy, S., Babu, H. W. S., Narayanasamy, A., Prajapati, V. K., et al. (2023). A review of chromium (Cr) epigenetic toxicity and health hazards. *Science of the Total Environment, 882*, Article 163483. https://doi.org/10.1016/j.scitotenv.2023.163483

Izbicki, J. A., & Groover, K. (2018). Natural and man-made hexavalent chromium, Cr (VI), in groundwater near a mapped plume, Hinkley, California—study progress as of May 2017, and a summative-scale approach to estimate background Cr (VI) concentrations (No. 2018–1045). *US Geological Survey*. https://doi.org/10.3133/ofr20181045

Izbicki, J. A., Warden, J. G., Groover, K. D., & Seymour, W. A. (2023). Evaluation of natural and anthropogenic (human-made) hexavalent chromium (No. 1885-G). *US Geological Survey*. https://doi.org/10.3133/pp1885G

James, B. R. (1994). Hexavalent chromium solubility and reduction in alkaline soils enriched with chromite ore processing residue. *Journal of Environmental Quality, 23*(2), 227–233.

Junaid, M., Hashmi, M. Z., Malik, R. N., & Pei, D. S. (2016). Toxicity and oxidative stress induced by chromium in workers exposed from different occupational settings around the globe: A review. *Environmental Science and Pollution Research, 23*, 20151–20167.

Khan, A., Michelsen, N., Marandi, A., Hossain, R., Hossain, M. A., Roehl, K. E., Zahid, A., Hassan, M. Q., & Schüth, C. (2020). Processes controlling the extent of groundwater pollution with chromium from tanneries in the Hazaribagh area, Dhaka, Bangladesh. *Science of the Total Environment, 710*, Article 136213. https://doi.org/10.1016/j.scitotenv.2019.136213

Khan, C., Malik, R. N., & Chen, J. (2024). Human exposure to chromite mining pollution, the toxicity mechanism and health impact. *Heliyon, 10*(21), Article e40083. https://doi.org/10.1016/j.heliyon.2024.e40083

Kośla, T., Lasocka, I., & Kołnierzak, M. (2019). Chromium, Cr. In E. Kalisińska (Ed.), *Mammals and birds as bioindicators of trace element contaminations in terrestrial environments* (pp. 57–124). Springer.

Laxmi, V., & Kaushik, G. (2020). Toxicity of hexavalent chromium in environment, health threats, and its bioremediation and detoxification from tannery wastewater for environmental safety. In G. Saxena, & R. Bharagava (Eds.), *Bioremediation of industrial waste for environmental safety* (pp. 223–243). Springer. https://doi.org/10.1007/978-981-13-1891-7_11

Li, Q. G., Liu, G. H., Qi, L., Wang, H. C., Ye, Z. F., & Zhao, Q. L. (2022). Heavy metal-contained wastewater in China: Discharge, management and treatment. *Science of the Total Environment, 808*, Article 152091. https://doi.org/10.1016/j.scitotenv.2021.152091

Lin, X., Luo, J., Liao, M., Su, Y., Lv, M., Li, Q., Xiao, S., & Xiang, J. (2022). Wearable sensor-based monitoring of environmental exposures and the associated health effects: A review. *Biosensors, 12*(12), 1131. https://doi.org/10.3390/bios12121131

Ma, G., & Garbers-Craig, A. (2006). A review on the characteristics, formation mechanisms and treatment processes of Cr (VI)-containing pyrometallurgical wastes. *Journal of the Southern African Institute of Mining and Metallurgy, 106*(11), 753–763.

Ma, H. W., Hung, M. L., & Chen, P. C. (2007). A systemic health risk assessment for the chromium cycle in Taiwan. *Environment International, 33*(2), 206–218.

McDermott, S., Salzberg, D. C., Anderson, A. P., Shaw, T., & Lead, J. (2015). Systematic review of chromium and nickel exposure during pregnancy and impact on child outcomes. *Journal of Toxicology and Environmental Health, 78*(21–22), 1348–1368.

Mishra, S., & Bharagava, R. N. (2016). Toxic and genotoxic effects of hexavalent chromium in environment and its bioremediation strategies. *Journal of Environmental Science and Health, 34*(1), 1–32.

Mossotti, G., Piscitelli, A., Catania, F., Aronne, M., Galfré, G., Lamberti, A., Ferrero, S., Scaltrito, L., & Bertana, V. (2024). Advances in water resource management: An in situ sensor solution for monitoring high concentrations of chromium in the electroplating industry. *Water, 16*(8), 1167. https://doi.org/10.3390/w16081167

Murthy, M. K., Khandayataray, P., Padhiary, S., & Samal, D. (2023). A review on chromium health hazards and molecular mechanism of chromium bioremediation. *Reviews on Environmental Health, 38*(3), 461–478.

National Toxicology Program. (2021). 15th Report on Carcinogens. Report on Carcinogens: Carcinogen Profiles 15:roc15.

Pokhrel, G. R., & Pokhre, G. (2022). The effect of chromium on human-health: A review. *BMC Journal of Scientific Research, 5*(1), 27–35.

Qin, C., Lian, H., Alqahtani, F. M., & Ahanger, M. A. (2024). Chromium mediated damaging effects on growth, nitrogen metabolism and chlorophyll synthesis in tomato can be alleviated by foliar application of melatonin and jasmonic acid priming. *Scientia Horticulturae, 323*, Article 112494. https://doi.org/10.1016/j.scienta.2023.112494

Qiu, Y., Zhang, Q., Gao, B., Li, M., Fan, Z., Sang, W., Hao, H., & Wei, X. (2020). Removal mechanisms of Cr(VI) and Cr(III) by biochar supported nanosized zero-valent iron: Synergy of adsorption, reduction and transformation. *Environmental Pollution, 265*, Article 115018. https://doi.org/10.1016/j.envpol.2020.115018

Rani, P., Arya, R. C., & Dwivedi, S. (2020). Chromium pollution: Impact on plants and its mitigation. In P. Mishra, R. R. Mishra, & C. O. Adetunji (Eds.), *Innovations in food technology* (pp. 323–340). Springer.

Rashid, A., Schutte, B. J., Ulery, A., Deyholos, M. K., Sanogo, S., Lehnhoff, E. A., & Beck, L. (2023a). Heavy metal contamination in agricultural soil: environmental pollutants affecting crop health. *Agronomy, 13*(6), 1521. https://doi.org/10.3390/agronomy13061521

Rashid, A., Ayub, M., Ullah, Z., Ali, A., Sardar, T., Iqbal, J., et al. (2023b). Groundwater quality, health risk assessment, and source distribution of heavy metals contamination around chromite mines. *International Journal of Environmental Research and Public Health, 20*(3), 2113. https://doi.org/10.3390/ijerph20032113

Ray, P. D., Yosim, A., & Fry, R. C. (2014). Incorporating epigenetic data into the risk assessment process for the toxic metals arsenic, cadmium, chromium, lead, and mercury: Strategies and challenges. *Frontiers in Genetics, 5*, 201. https://doi.org/10.3389/fgene.2014.00201

Rowbotham, A. L., Levy, L. S., & Shuker, L. K. (2000). Chromium in the environment: An evaluation of exposure of the UK general population and possible adverse health effects. *Journal of Toxicology and Environmental Health B Critical Reviews, 3*(3), 145–178.

Saha, R., Nandi, R., & Saha, B. (2011). Sources and toxicity of hexavalent chromium. *Journal of Coordination Chemistry, 64*(10), 1782–1806.

Saleh, T. A., Mustaqeem, M., & Khaled, M. (2022). Water treatment technologies in removing heavy metal ions from wastewater: A review. *Environmental Nanotechnology, Monitoring & Management, 17*, Article 100617. https://doi.org/10.1016/j.enmm.2021.100617

Samuel, J. B., Stanley, J. A., Sekar, P., Princess, R. A., Sebastian, M. S., & Aruldhas, M. M. (2014). Persistent hexavalent chromium exposure impaired the pubertal development and ovarian histoarchitecture in wistar rat offspring. *Environmental Toxicology, 29*(7), 814–828.

Santonen, T., Alimonti, A., Bocca, B., Duca, R. C., Galea, K. S., Godderis, L., et al. (2019). Setting up a collaborative European human biological monitoring study on occupational exposure to hexavalent chromium. *Environmental Research, 177*, Article 108583. https://doi.org/10.1016/j.envres.2019.108583

Sardans, J., Montes, F., & Peñuelas, J. (2011). Electrothermal atomic absorption spectrometry to determine As, Cd, Cr, Cu, Hg, and Pb in soils and sediments: A review and perspectives. *Soil and Sediment Contamination: An International Journal, 20*(4), 447–491.

Saud, S., Wang, D., Fahad, S., Javed, T., Jaremko, M., Abdelsalam, N. R., & Ghareeb, R. Y. (2022). The impact of chromium ion stress on plant growth, developmental physiology, and molecular regulation. *Frontiers in Plant Science, 13*, Article 994785. https://doi.org/10.3389/fpls.2022.994785

Sazakli, E. (2024). Human health effects of oral exposure to chromium: A systematic review of the epidemiological evidence. *International Journal of Environmental Research and Public Health, 21*(4), 406. https://doi.org/10.3390/ijerph21040406

Ščančar, J., & Milačič, R. (2014). A critical overview of Cr speciation analysis based on high performance liquid chromatography and spectrometric techniques. *Journal of Analytical Atomic Spectrometry, 29*(3), 427–443.

Shadreck, M., & Mugadza, T. (2013). Chromium, an essential nutrient and pollutant: A review. *African Journal of Pure and Applied Chemistry, 7*(9), 310–317.

Shanker, A. K. (2019). Chromium: Environmental pollution, health effects and mode of action. In J. Nriagu (Ed.), *Encyclopedia of Environmental Health* (pp. 624–633). Elsevier.

Sharma, A., Kapoor, D., Wang, J., Shahzad, B., Kumar, V., Bali, A. S., Jasrotia, S., Zheng, B., Yuan, H., & Yan, D. (2020). Chromium bioaccumulation and its impacts on plants: An overview. *Plants, 9*(1), 100. https://doi.org/10.3390/plants9010100

Sheehan, P. J., Meyer, D. M., Sauer, M. M., & Paustenbach, D. J. (1991). Assessment of the human health risks posed by exposure to chromium-contaminated soils. *Journal of Toxicology and Environmental Health, 32*(2), 161–201.

Shelnutt, S. R., Goad, P., & Belsito, D. V. (2007). Dermatological toxicity of hexavalent chromium. *Critical Reviews in Toxicology, 37*(5), 375–387.

Shukla, P., Srivastava, P., & Mishra, A. (2024). Routes of exposure (inhalation, ingestion, dermal contact) of heavy metals and their implications for human health. In V. Singh, A. Kumar, V. Mishra, & S. N. Rai (Eds.), *Heavy metal contamination in the environment* (pp. 114–123). CRC Press.

Siddiqui, N., Faiyaz, T., & Tari, V. S. (2022). Ganga and Yamuna Rivers: Through the Lens of the National Green Tribunal. *Nature Environment and Pollution Technology, 21*(4), 1657–1664.

Simon, M., & Joshi, H. (2022). Story of the Ganga River: Its pollution and rejuvenation. In A. Mukherjee (Eds.), *Riverine systems* (pp. 21–55). Springer. https://doi.org/10.1007/978-3-030-87067-6_2

Singh, A., Tejavath, R., Bandyopadhyay, T., Bhojwani, V., Bhattacharya, M., Prakash, M., & Sharma, L. M. (2023). Enabling marginalized communities to monitor and treat chromium-polluted groundwater with decentralized and affordable technologies. *Environmental Engineering Science, 40*(11), 584–595.

Singh, S., Naik, T. S. K., Chauhan, V., Shehata, N., Kaur, H., Dhanjal, D. S., et al. (2022a). Ecological effects, remediation, distribution, and sensing techniques of chromium. *Chemosphere, 307*, Article 135804. https://doi.org/10.1016/j.chemosphere.2022.135804

Singh, V., Singh, N., Verma, M., Kamal, R., Tiwari, R., Sanjay Chivate, M., et al. (2022b). Hexavalent-chromium-induced oxidative stress and the protective role of antioxidants against cellular toxicity. *Antioxidants, 11*(12), 2375. https://doi.org/10.3390/antiox11122375

Srivastava, D., Tiwari, M., Dutta, P., Singh, P., Chawda, K., Kumari, M., et al. (2021). Chromium stress in plants: Toxicity, tolerance and phytoremediation. *Sustainability, 13*, 4629. https://doi.org/10.3390/su13094629

Suljević, D., Fočak, M., & Alijagic, A. (2024). Assessing chromium toxicity across aquatic and terrestrial environments: A cross-species review. *Drug and Chemical Toxicology, 47*(6), 1312–1324.

Tan, C., Avasarala, S., & Liu, H. (2020). Hexavalent chromium release in drinking water distribution systems: New insights into zerovalent chromium in iron corrosion scales. *Environmental Science & Technology, 54*, 13036–13045.

Taylor, A., Barlow, N., Day, M. P., Hill, S., Patriarca, M., & White, M. (2017). Atomic spectrometry update: Review of advances in the analysis of clinical and biological materials, foods and beverages. *Journal of Analytical Atomic Spectrometry, 32*(3), 432–476.

Teklay, A. (2016). Physiological effect of chromium exposure: A review. *International Journal of Food Science Nutrition Diet, S7*(001), 1–11.

Thompson, C. M., Kirman, C. R., Hays, S. M., Suh, M., Harvey, S. E., Proctor, D. M., & Harris, M. A. (2018). Integration of mechanistic and pharmacokinetic information to derive oral reference dose and margin-of-exposure values for hexavalent chromium. *Journal of Applied Toxicology, 38*(3), 351–365.

Tumolo, M., Ancona, V., De Paola, D., Losacco, D., Campanale, C., Massarelli, C., & Uricchio, V. F. (2020). Chromium pollution in European water, sources, health risk, and remediation strategies: An overview. *International Journal of Environmental Research and Public Health, 17*(15), 5438. https://doi.org/10.3390/ijerph17155438

Uddin, S., Afroz, H., Hossain, M., Briffa, J., Blundell, R., & Islam, M. R. (2023). Heavy metals/metalloids in food crops and their implications for human health. *Heavy metal toxicity and tolerance in plants: A biological, omics, and genetic engineering approach* (pp. 59–86). https://doi.org/10.1002/9781119906506.ch3

Uddin, S., Khanom, S., Islam, M. R., & Hossain, M. (2024). Toxicity of rhizospheric cadmium-contaminated soil and its remediation. In N. Kumar (Eds.) *Cadmium toxicity: Challenges and solutions* (pp. 321–377). Springer. https://doi.org/10.1007/978-3-031-65611-8_14

Wasowicz, W., Janasik, B., Reszka, E., Kasperczyk, E., Chrzanowski, J., & Fendler, W. (2025). Metals (Cr, Mn Co, Ni) concentration in the blood plasma and urine of Polish welders and telomere length as a potential indicator of toxicity of metals welding fumes exposure. *International Journal of Occupational Medicine and Environmental Health, 38*(1), 70–90.

WHO (World Health Organization). (2020). Preventing disease through healthy environments: exposure to chromium and health effects. https://iris.who.int/bitstream/handle/10665/204585/9789241565196_eng.pdf. Accessed August 1, 2025.

Xing, Y., Zheng, Y., & Wang, X. (2025). Integrated strategies for effective remediation of chromium-contaminated soils: advancements, challenges, and sustainability implications. *Environmental Advances*, 100614. https://doi.org/10.1016/j.envadv.2025.100614

Yasmeen, R., & Hafeez, F. (2023). Occupational exposure of the emissions on the respiratory system of the workers belonging to different industries. *Archives of Respiratory Research, 2*(1).

Zafar, M., Aggarwal, A., Rene, E. R., Barbusiński, K., Mahanty, B., & Behera, S. K. (2022). Data-driven machine learning intelligent tools for predicting chromium removal in an adsorption system. *Processes, 10*(3), 447. https://doi.org/10.3390/pr10030447

Zhang, K., Li, N., Liao, P., Jin, Y., Li, Q., Gan, M., Chen, Y., He, P., Chen, F., Peng, M., & Zhu, J. (2021). Conductive property of secondary minerals triggered Cr(VI) bioreduction by dissimilatory iron reducing bacteria. *Environmental Pollution, 286*, Article 117227. https://doi.org/10.1016/j.envpol.2021.117227

Zhitkovich, A. (2011). Chromium in drinking water: Sources, metabolism, and cancer risks. *Chemical Research in Toxicology, 24*, 1617–1629.

Zhu, S., Chen, J., Zhu, Y., Zhang, B., Jia, J., Pan, M., Yang, Z., Cao, J., & Shen, Y. (2025). Influencing factors of hexavalent chromium speciation transformation in soil from a northern China chromium slag site. *Molecules, 30*(15), 3076. https://doi.org/10.3390/molecules30153076

Zulfiqar, U., Haider, F. U., Ahmad, M., Hussain, S., Maqsood, M. F., Ishfaq, M., Shahzad, B., Waqas, M. M., Ali, B., Tayyab, M. N., Ahmad, S. A., Khan, I., & Eldin, S. M. (2023). Chromium toxicity, speciation, and remediation strategies in soil-plant interface: A critical review. *Frontiers in Plant Science, 13*, 1081624. https://doi.org/10.3389/fpls.2022.1081624

Health Risks Linked to Chromium Toxicity

Effects of Chromium Exposure on Human Health

Arti Chamoli, Santosh Kumar Karn, Neetu Pandey, Vivek Raj, and Elayaraja Sivaramasamy

Abstract Chromium (Cr) is a heavy metal that is naturally present in the environment, posing significant ecological and public health risks, specifically in its hexavalent form (Cr(VI)). This review synthesizes current knowledge on chromium's sources, exposure pathways, toxic kinetics, and health effects. Industrial processes like alloy manufacturing, tanning, and electroplating are major contributors to environmental Cr(VI) contamination. Humans can be exposed to Cr(VI) via aspiration, swallowing, or skin absorption, particularly in places where exposure risks are higher. Cr(VI) is highly toxic and carcinogenic, inducing oxidative stress, DNA damage, and epigenetic alterations. Acute exposure can lead to respiratory distress, gastrointestinal injury, and skin ulceration, while chronic exposure is linked to lung cancer, nephrotoxicity, hepatotoxicity, reproductive dysfunction, and neurotoxicity. In contrast, trivalent chromium (Cr(III)) is a necessary micronutrient with relatively low toxicity. Biomonitoring studies identify biomarkers such as urinary 8-OHdG, MDA, and exhaled breath condensate chromium levels as indicators of exposure and oxidative damage. Preventive strategies include industrial hygiene, personal protective equipment, environmental remediation, and nutritional interventions. This review underscores the compelling requirement for regulatory oversight and public health initiatives to alleviate chromium-related risks.

A. Chamoli · S. K. Karn (✉)
Department of Biochemistry and Biotechnology, Sardar Bhagwan Singh University, Balawala, Dehradun, India
e-mail: santoshkarn.sbsu@gmail.com

N. Pandey
Department of Applied Chemistry and Basic Science, Sardar Bhagwan Singh University, Balawala, Dehradun, India

V. Raj
Department of Biotechnology, Patna Women's College Patna (Autonomous) Patna University, Patna, India

E. Sivaramasamy
Faculty of Fisheries and Protection of Waters, Institute of Aquaculture and Protection of Waters, Laboratory of Controlled Reproduction and Intensive Aquaculture, University of South Bohemia in České Budějovice, České Budějovice, Czech Republic

N. Kumar (ed.), *Chromium Toxicity*, https://doi.org/10.1007/978-3-032-14323-5_5

Keywords Hexavalent chromium · Toxicokinetics · Oxidative stress · Carcinogenicity · Genotoxicity

1 Introduction

In the contemporary world, pollution arising from heavy metals is a critical environmental problem that adversely affects all life forms on the planet. Heavy metals are now regarded as the most significant ecological pollutants rising from industries, impairing and contaminating animals and humans (Hossini et al., 2022). Chromium is naturally present in diverse sources, including living organisms (animals and plants), soil, rocks, volcanic ash, and atmospheric gases. Chromium (Cr) is a silvery-white transition metal with a relative atomic mass of 52.996 g/mol and an atomic number of 24. The Cr content in commercial alloys is generally restricted to below 30–40% by weight due to inherent brittleness. Chromium is frequently added to alloys used in dental and medical applications as instruments, implants, prosthetic bridges, and various other metal-based devices. Electroplating, the catalyst industry, leather tanning, water cooling systems, pulp manufacturing, pigments and dyes, wood treatment, and alloy manufacturing are among the most common and significant sources of chromium exposure. Chromium salts were used in photographic processing. The refining of petroleum and the processing of chromite ore released chromium contaminants into the environment, affecting the soil, atmosphere, and water sources.

Hexavalent and trivalent chromium baths are the two main types used in plating. The use of hexavalent chromium baths is widespread (Saha et al., 2022). Adequately excessive levels of chromium exposure to humans could likely cause damage because of its genotoxic and cancer-inducing effects. The WHO has recognized chromium as a carcinogenic substance for humans, and according to data published by the Agency for Toxic Substances and Disease Registry (ATSDR), Cr is listed among one of eight heavy metals in the top 50 most dangerous compounds worldwide (Achmad & Auerkari et al., 2017). There are six variants of chromium, out of which trivalent (III) and hexavalent (VI) chromium are the most toxic forms that affect the environment (Mohanty et al., 2023). Chromium (III), trivalent, is the naturally occurring stable valence state, although elemental chromium (0) does not occur naturally. The second most stable form is the hexavalent (VI) form. However, chromium VI is mostly produced by human activities and is rarely found in nature (Shekhawat et al., 2015). Figure 1 illustrates the major sources, toxic forms, and health effects of Chromium (Cr) in the environmental and biological system.

Human's unmediated contact with Cr (VI) causes chromium toxicity and even cancer. Additionally, Cr (VI) enters the human body via the food chain when it is accumulated in the soil and reaches plants and is absorbed. It also affects soil health, including soil pH, photosynthesis, and enzymatic activities, causing oxidative damage, infertility, and death of microorganisms present in the soil as a result of Cr toxicity, leading to lower crop yield. Animals consuming the Cr affected soil's

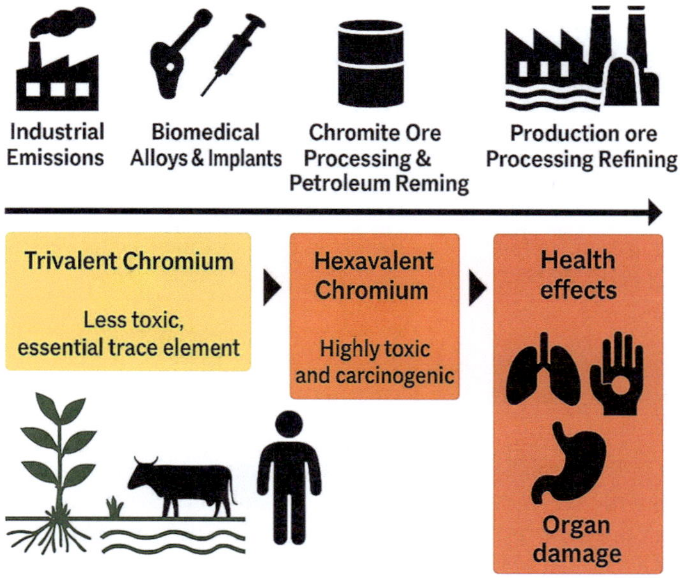

Fig. 1 Major sources, toxic forms, and health effects of Chromium (Cr) in the environmental and biological system

agricultural waste end up with organ damage as well as acute or chronic toxicity. Lung cancer, respiratory conditions, skin allergies, kidney and liver damage, and reproductive toxicity are among the illnesses linked to chromium exposure in humans (Kapoor et al., 2022; Sawicka et al., 2021; Yan et al., 2023). Ingestion is the chief non-occupational exposure pathway to Cr. However, occupational exposure happens through aspiration or skin absorption. (Wilbur et al., 2012). Bronchitis (Khan et al., 2013), Asthma, ulceration and rupture of the nasal septum, and respiratory tract inflammation occur as a result of detrimental respiratory and pulmonary health (Bright et al., 1997). Long exposure to chromate and dichromate leads to contact dermatitis (Leijding et al., 2018), which can include more critical skin reactions like blisters, burns, and skin ulcers (Wilbur et al., 2012). Gastritis, chronic dyspepsia, and gastric ulcers are the gastrointestinal complications resulting from occupational exposure. (Chen et al., 2019; Khan et al., 2013). Chromium exists in several forms, but when it comes to health, Chromium (VI) and Chromium (III) are worlds apart. Chromium (VI) is significantly more noxious, both in the short term and over prolonged exposure. When inhaled, it primarily targets the respiratory system, causing symptoms like wheezing, coughing, and shortness of breath, especially at high concentrations such as those found in chromium trioxide. In severe cases, it can also affect the digestive and nervous systems, while dermal contact can cause painful burns. Swallowing large amounts of Chromium (VI), whether by accident or occupational exposure, can lead to serious gastrointestinal distress, think abdominal pain, vomiting, and even internal bleeding. Animal studies back this up, showing that Chromium (VI) is extremely toxic when inhaled or ingested.

On the other hand, Chromium (III) plays a very different role in the body. It's a trace nutrient, with adults needing about 50–200 μg per day to help regulate blood sugar and lipid metabolism. While it's not entirely harmless, animal studies suggest moderate toxicity at high doses, it's far less dangerous than its hexavalent counterpart. In summary, Chromium (III) helps; Chromium (VI) harms. Understanding the difference is crucial for both public health and occupational safety (DEFRA & EA, 2002; Hessel et al., 2021).

2 Types and Sources of Chromium Exposure

Cr is found in the Earth's crust in copious amounts (100 μg/g), but the degree of its morbidity depends on its chemical state (Hossini et al., 2022). Trivalent (III) and hexavalent (VI) Cr are its two most prevalent oxidation forms; Cr(III) is generally non-toxic, whereas Cr(VI) is not only toxic butalso cancer-causing to living organisms. Chromate, derived from trivalent and hexavalent Cr compounds, produces dust during manufacturing procedures, which has noxious effects (Fontaine et al., 2019). Different sources and forms of Chromium are presented in Fig. 2.

Trivalent Cr occurs as a trace element found naturally in many foods and is accessible as a nutritional supplement. (Anderson & Cefalu, 2010). Additionally, it is present in many multivitamins, nutritional supplements, and meals. Supplementation with chromium often includes picolinate, histidinate, dinicocysteinate, and niacin-bound chromium. Chromium in the hexavalent form is a hazardous byproduct of stainless steel and other production processes.

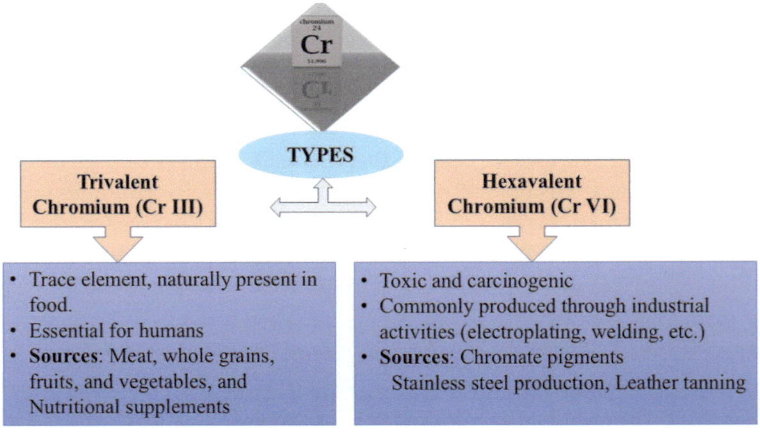

Fig. 2 Sources and forms of chromium

3 Routes of Human Exposure

Heavy metals are pervasive; therefore, human exposure may result via aspiration, ingestion, or skin contact. Due to their lethal activity, toxicological studies on numerous heavy metals have been intensively searched for, and the metals are handled through regulation. Heavy metal exposure also causes many types of cancer. The International Agency for Research on Cancer (IARC) has categorized arsenic, cadmium, chromium, and nickel as Group 1 carcinogens. (Forte et al., 2020; Shin et al., 2023). The National Institute for Occupational Safety and Health (NIOSH) stated that the health risks linked to chromium exposure depend on how the substance enters the body, whether through skin contact, breathing it in, or swallowing it. Each route can lead to different types and severities of adverse effects (Santonen et al., 2022).

Chromium occurs in the environment at various concentrations. Cr (VI) is the most toxic form brought into the ecosystem because of human activities, while the least toxic form, Cr (III), is present naturally in the environment. The industrial use of Cr (VI) as an anticorrosive agent in combustion and cooling systems results in its exposure to the environment (Slooff et al., 1990). Because of the omnipresent property of heavy metals, humans can be exposed to them by ingestion, inhalation, and dermal routes. Due to their high toxicity, toxicological research of numerous heavy metals has been carried out, and these metals are regulated by regulatory measures. Various types of cancer are caused by exposure to heavy metals. The International Agency for Research on Cancer (IARC) classified arsenic, cadmium, chromium, and nickel as Group 1 carcinogens (Shin et al., 2023). Based on the duration, chromium exposure is classified into three categories: acute exposure of about 14 days, intermediate exposure lasting up to 75 to 364 days, and chronic exposure extending for 365 days or more (Guertin, 2004). Water is the major source of oral exposure to chromium in humans. Soil type and industrial sources influence the chromium concentration in drinking water. However, water sources are the main concern, contaminated with industrial waste. Chromium poisoning has been reported to cause kidney failure, mouth ulcers, acute tubular necrosis, vomiting, abdominal pain, indigestion, and in severe cases, it may lead to death (Beaumont et al., 2008). Among the general public, the dermal exposure to chromium takes place through skin contact with soil and certain consumer products containing chromium. Chromium compounds could be found in leather and cosmetic products in the form of both Cr (VI) and Cr (III) (Zhao et al., 2020). Chromium compounds are capable of penetrating into human skin to a certain degree, especially when the skin is injured. Dermal toxicity is caused by penetrated chromium when bound to the immune cells of the skin. Sensitization and induction of contact dermatitis can be caused by chromium exposure through the formation of Cr-protein conjugates. Additionally, skin irritation can also be caused by chromium exposure that might lead to ulceration (Hagvall et al., 2021). Metastasis, immune system disorders and DNA damage are mainly caused by the exposure of chromium and chromium compounds by dermal contact. The researchers demonstrated that the chromium exposure induced DNA damage in human dermal fibroblasts. The dermal exposure to chromium leads to skin diseases like allergic reactions, hypersensitivity, skin damage, and ulceration (Rudolf & Červinka, 2006). Industrial workers are mainly affected through respiratory exposure to chromium. Various studies on occupational epidemiology revealed that a higher chance of lung cancer is linked with the inhalation of Cr (VI). Nasal irritation and perforation were observed among workers at increased risk of lung cancer (Alvarez et al., 2021). The relationship within inhaled chromium and lung cancer would be supported by these observations, while the mechanism of action for hexavalent chromium Cr (VI), which induces lung tumors, is still unknown. Chromium and its compounds are exposed during inhalation, leading to cell cycle progression, DNA damage, tumor development and metastasis, and different respiratory diseases like lung cancer, pneumonia, lung neoplasm, and pulmonary fibrosis. Lung tumor formation was promoted by the inhalation of hexavalent chromium through the non-thresholded genotoxic mechanism, including DNA adduct formation (Seidler et al., 2013). In comparison, the orally ingested chromium toxicity has limited evidence. The study reported that in the Chinese population, exposure to chromium increased the risk of cancer through drinking water. Food ingestion is the main exposure pathway for non-occupational workers. A community health goal for the concentration of chromium in potable water was established by the California Environmental Protection Agency (CalEPA) based on its environmental health hazard assessment. Tumorigenic cell processes, metastasis, and DNA damage are primarily caused by exposure to chromium and its compounds. Ingestion of chromium was also linked with male infertility, kidney disease, and gastrointestinal tract diseases like squamous cell carcinoma and small intestine neoplasms. Oxidative stress was induced by ingested chromium through various pathways. Chromium-induced oxidative stress is based on the levels of production of reactive oxygen species, which may cause the formation of tumors and cell death (Beaumont et al., 2008).

4 Toxicokinetics of Chromium

The mechanism(s) that underlie the carcinogenicity of Cr(VI)are unclear. The intracellular reduction of Cr(VI) to Cr(III) is accompanied by oxidative damage to key cellular structures. such as DNA damage and the generation of free radicals, may result in chromium toxicity within the cell (ATSDR, 2000). The biological significance of nonoxidative pathways in Cr(VI) carcinogenesis has been revealed by recent research (Zhitkovich et al., 2001).

Cr(VI), in the form of chromate oxyanions, imitates essential anions like sulfate and phosphate, and enters the cell through active transport via respective channels. Some Cr(VI) could be minimized outside the cell to Cr(III), which binds to extracellular ligands. However, Cr(III) has poor membrane permeability and diffuses passively at a very low rate, making Cr(VI) the primary species that infiltrates the cell. Once inside, Cr(VI) undergoes stepwise reduction to Cr(V), Cr(IV), and finally Cr(III). This redox cycling generates reactive oxygen species (ROS), counting superoxide and hydroxyl radicals, which contribute to oxidative stress. Chromium also interacts with cellular molecules, such as ascorbate, cysteine, and glutathione, forming bulky conjugates. Some chromium intermediates are detoxified and excreted, but others migrate toward the nucleus. In the nucleus, chromium species form DNA adducts and cross-links with nuclear proteins, leading to DNA strand breaks, genotoxicity, and epigenetic alterations. These changes can disrupt gene expression and DNA repair mechanisms, increasing the risk of mutations and carcinogenesis. Chromium exposure has also been shown to disrupt the epigenetic landscape of cells, affecting both DNA methylation patterns and histone modifications. These changes can interfere with how genes are regulated and expressed, potentially contributing to long-term cellular dysfunction and disease development (Rager et al., 2019).

5 Health Effects of Chromium Exposure

5.1 Acute Effects

Hexavalent chromium (Cr(VI)) is primarily linked to acute toxic effects, with exposure occurring mainly through inhalation or ingestion, or dermal contact, especially in occupational or accidental settings (Chatterjee, 2015). Acute exposure to Cr(VI) causes severe health concerns, primarily because of its high solubility and ability to penetrate biological membranes. Inhalation of Cr(VI) compounds common in occupational settings such as welding, chrome plating, or pigment manufacturing can lead to immediate respiratory symptoms, including coughing, wheezing, and bronchospasm. Severe cases may progress to pulmonary edema or acute respiratory distress syndrome (ARDS), with nasal septum ulceration or perforation observed in high-exposure scenarios. Oral ingestion, though less common, is highly toxic and can result in corrosive damage to the gastrointestinal tract, hemolysis, acute renal failure, hepatic injury, and potentially fatal multi-organ dysfunction (Zhigalenok et al., 2025). Dermal contact, especially with compromised skin, may cause chrome ulcers, characteristic necrotic lesions, and allergic contact dermatitis, which can be persistent and debilitating (Fig. 3). Systemic toxicity may occur following extensive dermal exposure, manifesting as fever, muscle cramps, and circulatory collapse (Sharma et al., 2022).

The underlying mechanism involves intracellular reduction of Cr(VI) to Cr(III), generating reactive oxygen species (ROS) that damage DNA, proteins, and lipids, contributing to cytotoxicity, nephrotoxicity, and genotoxicity (DesMarias & Costa, 2019). These acute effects are dose-dependent and influenced by the route of exposure, duration, and individual susceptibility. Prompt identification and removal from exposure, along with supportive medical care, are critical to mitigating outcomes. Due to its high toxicity and strong carcinogenic potential, hexavalent chromium (Cr(VI)) has been designated a Group 1 carcinogen by the International Agency for

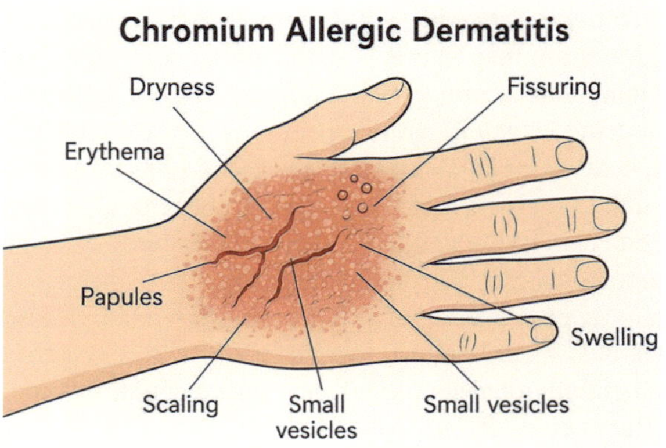

Fig. 3 Clinical presentation of chromium allergic dermatitis: characteristic symptoms include erythema, dryness, fissuring, papules, scaling, vesiculation, and swelling, typically resulting from prolonged dermal exposure to hexavalent chromium compounds

Research on Cancer (IARC), and stringent occupational safety measures are essential to prevent acute incidents. Understanding these acute manifestations is vital for clinicians, toxicologists, and occupational health professionals to ensure timely diagnosis and intervention, especially in industrial environments where chromium compounds are routinely handled (OSHA, 2006).

5.2 Chronic Effects

As a heavy metal, chromium poses serious health risks when it accumulates in the body over time. Prolonged exposure can lead to toxic effects and trigger a range of physiological disorders, including hypersensitivity and ulcers, allergic reactions, anemia, and even impairment of the male reproductive system and disturbances of other biological systems (Zhitkovich, 2011). Chronic exposure to hexavalent chromium [Cr(VI)] via inhalation in humans can severely affect the respiratory tract, causing nasal septum perforations, bronchitis, asthma, and reduced lung function. It may also impact the liver, kidneys, gastrointestinal and immune systems, and potentially the blood. Dermal contact can lead to dermatitis and skin ulceration. Animal studies show Cr(VI) accumulates primarily in the lungs and kidneys. The U.S. EPA has established Reference Concentrations (RfCs) for Cr(VI) as 0.0001 mg/m^3 for particulates (based on rat respiratory effects, with medium confidence) and 0.000008 mg/m^3 for chromic acid mists and dissolved aerosols (based on human data, with low confidence due to exposure uncertainties). The Reference Dose (RfD) for oral Cr(VI) exposure is 0.003 mg/kg/day, derived from rat studies with no observed effects, though EPA assigns low confidence due to limited data and study design flaws. For trivalent chromium [Cr(III)], the EPA has not set an RfC, as inhalation effects are not considered analogous to Cr(VI). The RfD for Cr(III) is 1.5 mg/kg/day, based on dietary exposure in rats, but confidence remains low due to insufficient study detail and lack of high-dose data. Overall, regulatory values reflect significant uncertainty in chromium toxicity data.

5.3 *Carcinogenicity Linked to Lung Cancer*

The hexavalent chromium [Cr(VI)] exposure has been intensely associated with lung cancer and was first documented in the U.S. by Machle and Gregorius in 1948. Since then, a substantial body of research has consistently reported elevated lung cancer rates amid workers exposed to chromium compounds (Alderson et al., 1981; Langard, 1990; Gibb et al., 2000; Luippold et al., 2003; Birk et al., 2006). Epidemiological evidence further indicates that water-insoluble particulate forms of Cr(VI) pose a significantly higher carcinogenic risk compared to their water-soluble counterparts. This distinction has been supported by epidemiological findings from various studies (Hayes et al., 1979; Alderson et al., 1981; OSHA, 2006). Squamous cell carcinoma is the most common form of lung cancer among chromate workers (Ishikawa et al., 1994). Curiously, exposure to Cr(VI) was linked with the emergence of many cancers in multiple participants, indicating the potential for discrete susceptibility factors in Cr-induced carcinogenesis. Within a year, dysplastic lesions typically give rise to squamous cell carcinoma (Satoh et al., 1997). Where Cr builds up is where lung cancers are found, and in former employees exposed to soluble chromates, the highest Cr levels are found in bronchial bifurcations.

Genotoxicity occurs when any toxic chemical interacts with DNA, leading to mutations, chromosomal damage, and carcinogenesis. Chromium trivalent (III) and hexavalent (VI) are the pervasive chemicals in the environment that are thought to be genotoxic (Sawicka et al., 2021). According to legislative organizations, chromium(VI) has been a potential human lung carcinogen (Buters & Biedermann, 2017).

5.4 *Nephrotoxicity*

Hexavalent chromium is a potent nephrotoxic heavy metal that damages the kidneys, including tubular damage and glomerular impairment, primarily driven by inflammatory responses, redox imbalance, and programmed cell death (Mishra et al., 2022).

5.5 *Hepatotoxicity*

Chromium (Cr), especially the hexavalent form (Cr(VI)), is an evidenced ecological and occupational toxicant. While trivalent chromium (Cr(III)) is an important micronutrient required in glucose and lipid metabolism, Cr(VI) exhibits high toxicity as it can readily cross the cell membranes via anion transporters and undergo intracellular reduction, generating reactive intermediates and reactive oxygen species (ROS) (Hassan et al., 2019; Li et al., 2023a, 2023b). Cr(VI)-induced hepatotoxicity is primarily mediated through oxidative stress. Upon cellular entry, Cr(VI) is reduced to Cr(III), creating ROS such as superoxide anions, hydrogen peroxide, and hydroxyl radicals. These species disrupt mitochondrial function, impair antioxidant defenses (e.g., glutathione, superoxide dismutase), and initiate lipid peroxidation, protein oxidation, and DNA destruction (Hassan et al., 2019). Histopathological studies in animal models reveal hepatic sinusoidal dilation, hepatocyte necrosis, vacuolization, and inflammatory infiltration following Cr(VI) exposure. Increased hepatic serum enzymes, alanine aminotransferase (ALT) and aspartate aminotransferase (AST) further confirm hepatocellular injury (Eman & Farag, 2020). Recent transcriptomic analyses have identified key molecular pathways involved in Cr(VI)-induced liver damage. Differentially expressed genes (DEGs) are enriched in pathways such as p53, PI3K-Akt, MAPK, AMPK, and PPAR signaling, implicating apoptosis, cell cycle dysregulation, and metabolic disturbances. Notably, Cr(VI) down regulates SIRT1, impairing deacetylation of transcription factors like NF-κB and FOXO3, thereby exacerbating inflammation and apoptosis (Tian et al., 2018). A comprehensive evaluation of human, animal, and mechanistic data concluded that Cr(VI) exposure leads to hepatic lipid accumulation, chronic inflammation, and elevated ALT/AST levels. The proposed mode of action involves oxidative stress, mitochondrial dysfunction, and apoptosis (Yu et al., 2021).

5.6 *Effect on Fertility*

Over the past few decades, declining human fecundity has been increasingly linked to ecological contaminants, with hexavalent chromium [Cr(VI)] emerging as a significant threat to male reproductive health. Industrial activities and improper disposal of chromium-contaminated waste have resulted in extensive environmental Cr(VI) exposure via air, water, and food. Epidemiological and experimental studies confirm that excess Cr(VI) disrupts key testicular functions, impairing spermatogenesis, steroidogenesis, and sperm maturation. Cr(VI) induces multiple sperm anomalies, including reduced count, motility, and viability, along with acrosomal damage and compromised blood-testis barrier integrity in Sertoli cells—particularly affecting spermatogenesis at the round spermatid stage. Leydig cells, essential for androgen production, are highly susceptible to Cr(VI), resulting in impaired steroidogenic activity and hypoandrogenism, which further disrupts sperm development and maturation. The review proposes a mechanistic hypothesis: exposure to Cr(VI) during critical developmental windows interferes with the hypothalamo-hypophyseal-testicular axis by triggering oxidative stress, ultimately compromising male fecundity. These findings underscore the urgent need for regulatory control of chromium emissions and further research into its reproductive toxicity, especially during sensitive periods of testicular differentiation and maturation. (Navin & Aruldhas, 2021; Li et al., 2023a, 2023b). Hexavalent Cr has been shown to adversely affect female fertility through multiple mechanisms, including ovarian dysfunction, endocrine disruption, oxidative stress, and direct cytotoxicity to reproductive tissues. Marouani et al. evaluated the impacts of Cr(VI) on female reproductive function in Wistar rats. Rats received daily intraperitoneal injections of potassium dichromate (1 or 2 mg/kg) for 15 days. CrVI exposure led to reduced ovarian and uterine weights, decreased follicle numbers, and prolonged estrous cycles. Serum levels of LH and FSH increased dose-dependently, alongside elevated chromium concentrations in blood and ovaries. Histological analysis revealed atretic follicles, arrested follicle growth, and uterine epithelial atrophy with thinning of the myometrium. These findings suggest that subacute CrVI exposure induces significant reproductive toxicity, impairing ovarian and uterine structure and function in female rats (Marouani et al., 2015).

5.7 *Neurotoxicity*

Cr(VI), Hexavalent chromium is a common environmental toxicant, exerts severe neurotoxic effects that remain underexplored at the primary neuron level. Using potassium dichromate ($K_2Cr_2O_7$), this study investigated Cr(VI)-induced neurotoxicity in rat primary hippocampal neurons. Cr(VI) dose- and time-dependently reduced neuronal viability, as shown by MTT assay. Mitochondrial dysfunction was confirmed via JC-1 and Mito-Tracker probes, while DCFH-DA and Mito-SOX Red revealed elevated oxidative stress. Western blotting indicated altered Bcl-2 family expression and MAPKs activation, implicating mitochondrial damage and ROS-mediated signaling in the toxic cascade. Cr(VI) also induced DNA damage. Notably, glutathione (GSH) treatment mitigated ROS levels, restored Bcl-2 family balance, suppressed MAPKs activation, and improved neuronal survival. The observations indicated that Cr(VI) neurotoxicity is driven by mitochondrial impairment, oxidative stress, and MAPK pathway activation, and that antioxidant intervention may offer neuroprotection (Zhang et al., 2024a, 2024b). (Cr(VI)) can penetrate the hypothalamus (HY) because of the relatively weak blood–brain barrier (BBB), leading to time-dependent neurotoxicity in mice. Intraperitoneal Cr(VI) exposure for 14 and 28 days resulted in Cr accumulation in HY, with more severe neuropathologies observed at 28 days. These included gliosis, synaptic degeneration, neuronal loss, and BBB disruption. Mechanistically, Cr(VI) impaired Nrf2-mediated antioxidant defense and activated NF-κB-driven inflammation. Findings highlight HY as a vulnerable CNS region to Cr(VI) exposure and suggest that targeting oxidative stress and inflammation might suggest therapeutic strategies for Cr(VI)-induced neurotoxicity in occupational settings (Ding et al., 2024). Intranasal exposure to Cr(VI) via in potassium dichromate (PDC) induces brain toxicity in rats in a dose- and time-dependent manner, as evidenced by biochemical, immunohistochemical, and histopathological analyses. Behavioral assessments revealed pronounced impairments in locomotor activity and cognitive function, while motor coordination was only mildly affected. Although short-term exposure at the tested doses did not result in significant brain injury, the progressive nature of observed neurotoxic changes over time suggests a potential risk of more severe neurological damage with prolonged exposure. These findings underscore the importance of monitoring duration-dependent effects in Cr(VI) neurotoxicity (Hegazy et al., 2021).

6 Chromium and DNA Damage

Cr(VI) is generally more soluble and reactive than Cr(III), as it can enter the cellular compartment by easily crossing the plasma membrane via various receptors present on the cell surface, anion transporters, and phosphate transporters (Mishra & Bharagav, 2016). Due to its greater solubility and higher oxidative state, Cr (VI) is much toxic, causing damage to the kidneys, lungs, and liver, among other organs (Chakraborty et al., 2022). Chromium chemistry has a significant impact on its cellular entrance and hazardous properties. The chromate oxyanion (CrO_4) is the most abundant form of hexavalent chromium in the environment. The chromate oxyanion, which exhibits an analogous structure to the sulfate oxyanion (SO4), enters the cell via general sulfate transporters on the cell surface (Salnikow & Zhitkovich, 2008). After chromium exposure, within the cells, chromium generates free radicals in both the cytoplasm and nucleus. These reactive molecules can latch onto DNA, disrupting its integrity. In studies using human fibroblasts, chromium exposure triggered DNA destruction by causing cells to accumulate in the sub-G1 phase and halting their cycle at the G2/M checkpoint, both signs of cellular distress. When inhaled, chromium doesn't just irritate lung tissue; it sparks inflammation, damages cells, and can ultimately lead to lung cancer (Shin et al., 2023). Even when consumed through drinking water, chromium can bind directly to DNA, forming harmful adducts that compromise genetic stability. In short, chromium doesn't just pass through the body; it leaves a molecular fingerprint that can interfere with how cells grow, divide, and repair themselves. That's why understanding its toxic mechanisms is so critical, especially for environmental and occupational health (Lu et al., 2024).

7 Biomarkers of Exposure and Effect

A study by Pan et al. analyzed oxidative DNA damage and lipid peroxidation among 230 non-smoking male workers, including 105 workers in the electroplating section exposed to hexavalent chromium (Cr(VI)) and 125 office-based controls. Researchers reportedly used biological samples and questionnaires to assess Cr(VI) exposure and its effects. Findings indicated significantly higher Cr levels in air, hair, urine, and nails among exposed workers. Biomarkers such as urinary 8-OHdG and MDA were elevated in this group. Statistical analysis suggested strong correlations between Cr exposure and oxidative stress markers, with ten-fold increases in urinary Cr linked to 1.73-fold and 1.45-fold rises in 8-OHdG and MDA, respectively. Another study was conducted to examine chromium levels in exhaled breath condensate (EBC) of chrome-plating workers and their association with airway oxidative stress biomarkers. It was reported that EBC samples taken before and after a Friday working hour and again on Monday showed elevated Cr, hydrogen peroxide (H_2O_2), and malondialdehyde (MDA) levels post-shift, with reductions by Monday. Cr-EBC levels were found to correlate significantly with H_2O_2-EBC, MDA-EBC, and urinary Cr levels. These findings suggested that EBC could serve as a reliable non-invasive matrix for monitoring chromium exposure and oxidative stress in the respiratory tract of workers exposed to Cr(VI) (Caglieri et al., 2006). Club cell protein (CC16) and surfactant-associated protein D (SP-D) could serve as biomarkers for lung injury in workers exposed to hexavalent chromium (Cr(VI)). Researchers reportedly measured chromium concentrations in air (CrA) and blood (CrB), along with lung function parameters and levels of CC16, SP-D, TNF-α, and IL-6 in 91 exposed workers and 38 controls. It was noted that Cr(VI) primarily enters the body via inhalation and is collected within organs such as the lungs, liver, kidneys, and endocrine glands. The findings suggested that occupational exposure to Cr(VI) led to epithelial damage in the lungs, as evidenced by significantly reduced values of FEV1, FEV1/FVC, MEF, and MVV among the exposed group relative to controls. These reductions in pulmonary function were interpreted as signs of chromium-induced respiratory impairment. The study also indicated that CC16 and SP-D levels may reflect such injury, supporting their potential use as non-invasive biomarkers for chromium-related lung damage. Overall, the results highlighted the respiratory risks linked with occupational Cr(VI) exposure and underscored the importance of monitoring both exposure levels and biological indicators of lung epithelial injury in affected workers (Li et al., 2015).

8 Preventive Measures and Management

Effective prevention and management of chromium exposure, particularly hexavalent chromium, requires a multifaceted approach combining industrial hygiene, personal protection, environmental remediation, and nutritional support. Industrial hygiene practices are foundational, involving source control through closed systems, local exhaust ventilation, and regular air and surface monitoring to limit occupational exposure. Worker education and process substitution, replacing Cr^{6+} compounds with less toxic alternatives, further reduces risk (Yan et al., 2023). Personal protective equipment (PPE) plays a critical role, with chemical-resistant gloves, respirators, and protective clothing forming barriers against dermal and inhalation exposure. Decontamination facilities and proper PPE disposal protocols enhance safety (ATSDR, 2000). Environmental cleanup strategies target chromium-contaminated soil and water using chemical reduction (e.g., ferrous sulfate), filtration technologies, and bioremediation with chromium-resistant microbes, such as Pseudomonas spp., to restore ecological balance (Narayani & Shetty, 2013). Nutritional interventions complement these efforts by mitigating oxidative stress and supporting the body's detoxification processes. Antioxidants such as vitamins C and E, selenium, and glutathione precursors help neutralize reactive chromium species, while chelation-supportive nutrients like alpha-lipoic acid and N-acetylcysteine enhance excretion (Flora, 2009). Dietary zinc and magnesium may competitively inhibit chromium uptake, and adequate hydration and fiber intake promote renal clearance. Together, these strategies form a comprehensive framework for minimizing chromium toxicity and safeguarding human health in occupational and environmental contexts.

9 Case Studies and Epidemiological Evidence

Chromium exposure generally occurs through inhaling aerosols and dermal contact. Chromium bioaccumulation can cause toxicity and damage to numerous ecological systems. Although there have been very few reports of any mortality due to chromium oral ingestion. One such case was reported where a female worker accidentally ingested chromium containing electroplating water, resulting in multiple organ failure and eventually death. Nausea and abdominal pain were the early symptoms that are common with chromium toxicity. Progressing to successive abnormalities of multiple organ systems, initially damaging the kidneys, then expanding to the hepatic, digestive system, cardiovascular system, and ultimately the immune system, indicating the extensive noxious impacts of chromium on the human body (Zhang et al., 2024a, 2024b).

Acknowledgements Authors are thankful to Department of Biochemistry and Biotechnology, Sardar Bhagwan Singh University, Balawala, Dehradun.

References

Achmad, R. T., & Auerkari, E. I. (2017). Effects of chromium on human body. *Annual Research and Review in Biology, 13*(2), ARRB-33462.

Alderson, M. R., Rattan, N. S., & Bidstrup, L. (1981). Health of workmen in the chromate-producing industry in Britain. *Occupational and Environmental Medicine, 38*(2), 117–124.

Alvarez, C. C., Gómez, M. E. B., & Zavala, A. H. (2021). Hexavalent chromium: Regulation and health effects. *Journal of Trace Elements in Medicine and Biology, 65*, Article 126729.

Anderson, R. A. & Cefalu, W. T. (2010). Chromium. In P. M. Coates, J. M. Betz, M. R. Blackman et al., (Eds.), *Encyclopedia of dietary supplements* (2nd ed.). Informa Healthcare.

ATSDR, T. (2000). ATSDR (Agency for toxic substances and disease registry). *Prepared by Clement International Corp., under contract* (Vol. 205, pp. 88–0608).

Beaumont, J. J., Sedman, R. M., Reynolds, S. D., Sherman, C. D., Li, L. H., Howd, R. A., Sandy, M. S., Zeise, L., & Alexeeff, G. V. (2008). Cancer mortality in a Chinese population exposed to hexavalent chromium in drinking water. *Epidemiology, 19*(1), 12–23.

Birk, T., Mundt, K. A., Dell, L. D., Luippold, R. S., Miksche, L., Steinmann-Steiner-Haldenstaett, W., & Mundt, D. J. (2006). Lung cancer mortality in the German chromate industry, 1958 to 1998. *Journal of Occupational and Environmental Medicine, 48*(4), 426–433.

Bright, P., Burge, P. S., O'Hickey, S. P., Gannon, P. F., Robertson, A. S., & Boran, A. (1997). Occupational asthma due to chrome and nickel electroplating. *Thorax, 52*(1), 28–32.

Buters, J., & Biedermann, T. (2017). Chromium (VI) contact dermatitis: Getting closer to understanding the underlying mechanisms of toxicity and sensitization! *Journal of Investigative Dermatology, 137*(2), 274–277.

Caglieri, A., Goldoni, M., Acampa, O., Andreoli, R., Vettori, M. V., Corradi, M., Apostoli, P., & Mutti, A. (2006). The effect of inhaled chromium on different exhaled breath condensate biomarkers among chrome-plating workers. *Environmental Health Perspectives, 114*(4), 542–546.

Chakraborty, R., Renu, K., Eladl, M. A., El-Sherbiny, M., Elsherbini, D. M. A., Mirza, A. K., Vellingiri, B., Iyer, M., Dey, A., & Gopalakrishnan, A. V. (2022). Mechanism of chromium-induced toxicity in lungs, liver, and kidney and their ameliorative agents. *Biomedicine and Pharmacotherapy, 151*, Article 113119.

Chatterjee, S. (2015). Chromium toxicity and its health hazards. *International Journal of Advanced Research, 3*(7), 167–172.

Chen, Q. Y., Murphy, A., Sun, H., & Costa, M. (2019). Molecular and epigenetic mechanisms of Cr (VI)-induced carcinogenesis. *Toxicology and Applied Pharmacology, 377*, Article 114636.

Department for Environment, Food and Rural Affairs (DEFRA) and Environment Agency (EA). (2002). Contaminants in Soil: Collation of Toxicological Data and Intake Values for Humans. Chromium. R&D Publications TOX 4. Environment Agency.

DesMarias, T. L., & Costa, M. (2019). Mechanisms of chromium-induced toxicity. *Current Opinion in Toxicology, 14*, 1–7.

Ding, J., Sun, B., Gao, Y., Zheng, J., Liu, C., Huang, J., Jia, N., Pei, X., Jiang, X., Hu, S., & Xia, B. (2024). Evidence for chromium crosses blood brain barrier from the hypothalamus in chromium mice model. *Ecotoxicology and Environmental Safety, 273*, Article 116179.

Eman, S., & Farag, A. I. (2020). Chromium-induced hepatotoxicity and potential protective effect of selenium in adult male albino rat: a histological, immuno-histochemical and molecular study. *The Medical Journal of Cairo University, 88*, 187–196.

Epa, U.S. (1999). Integrated Risk Information System (IRIS) on Baygon. *National Center for Environmental Assessment, Office of Research and Development: Washington, DC, USA.*

Flora, S. J. (2009). Structural, chemical and biological aspects of antioxidants for strategies against metal and metalloid exposure. *Oxidative Medicine and Cellular Longevity, 2*(4), 191–206.

Fontaine, M., Clement, Y., Blanc, N., & Demesmay, C. (2019). Hexavalent chromium release from leather over time natural ageing versus accelerated ageing according to a multivariate approach. *Journal of Hazardous Materials, 368*, 811–818.

Forte, I. M., Indovina, P., Costa, A., Iannuzzi, C. A., Costanzo, L., Marfella, A., Montagnaro, S., Botti, G., Bucci, E., & Giordano, A. (2020). Blood screening for heavy metals and organic pollutants in cancer patients exposed to toxic waste in southern Italy: A pilot study. *Journal of Cellular Physiology, 235*(6), 5213–5222.

Gibb, H. J., Lees, P. S., Pinsky, P. F., & Rooney, B. C. (2000). Lung cancer among workers in chromium chemical production. *American Journal of Industrial Medicine, 38*(2), 115–126.

Guertin, J. (2004). Toxicity and health effects of chromium (all oxidation states). *Chromium (VI) handbook* (pp. 215–234).

Hagvall, L., Pour, M. D., Feng, J., Karma, M., Hedberg, Y., & Malmberg, P. (2021). Skin permeation of nickel, cobalt and chromium salts in ex vivo human skin, visualized using mass spectrometry imaging. *Toxicology in Vitro, 76*, Article 105232.

Hassan, M., Abd-Elwahab, W., Megahed, R., & Mohammed, A. (2019). An evaluation of hepatotoxicity, nephrotoxicity, and genotoxicity induced by acute toxicity of hexavalent chromium and comparison of the possible protective role of selenium and vitamin E on these effects. *Ain Shams Journal of Forensic Medicine and Clinical Toxicology, 33*(2), 48–58.

Hayes, R. B., Lilienfeld, A. M., & Snell, L.M. (1979). Mortality in chromium chemical production workers: a prospective study. *International Journal of Epidemiology, 8*(4), 365–374.

Hegazy, R., Mansour, D., Salama, A., Hassan, A., & Saleh, D. (2021). Exposure to intranasal chromium triggers dose and time-dependent behavioral and neurotoxicological defects in rats. *Ecotoxicology and Environmental Safety, 216*, Article 112220.

Hessel, E. V., Staal, Y. C., Piersma, A. H., den Braver-Sewradj, S. P., & Ezendam, J. (2021). Occupational exposure to hexavalent chromium. Part I. Hazard assessment of non-cancer health effects. *Regulatory Toxicology and Pharmacology, 126*, 105048.

Hossini, H., Shafie, B., Niri, A. D., Nazari, M., Esfahlan, A. J., Ahmadpour, M., Nazmara, Z., Ahmadimanesh, M., Makhdoumi, P., Mirzaei, N., & Hoseinzadeh, E. (2022). A comprehensive review on human health effects of chromium: Insights on induced toxicity. *Environmental Science and Pollution Research, 29*(47), 70686–70705.

Ishikawa, Y., Nakagawa, K., Satoh, Y., Kitagawa, T., Sugano, H., Hirano, T., & Tsuchiya, E. (1994). Characteristics of chromate workers' cancers, chromium lung deposition and precancerous bronchial lesions: An autopsy study. *British Journal of Cancer, 70*(1), 160–166.

Kapoor, R. T., Mfarrej, M. F. B., Alam, P., Rinklebe, J., & Ahmad, P. (2022). Accumulation of chromium in plants and its repercussions in animals and humans. *Environmental Pollution, 301*, Article 119044.

Khan, D. A., Mushtaq, S., Khan, F. A., & Khan, M. Q. A. (2013). Toxic effects of chromium on tannery workers at Sialkot (Pakistan). *Toxicology and Industrial Health, 29*(2), 209–215.

Langrrd, S. (1990). One hundred years of chromium and cancer: A review of epidemiological evidence and selected case reports. *American Journal of Industrial Medicine, 17*(2), 189–214.

Lejding, T., Mowitz, M., Isaksson, M., Bruze, M., Pontén, A., Svedman, C., Zimerson, E., & Engfeldt, M. (2018). A retrospective investigation of hexavalent chromium allergy in southern Sweden. *Contact Dermatitis, 78*(6), 386–392.

Li, P., Li, Y., Zhang, J., Yu, S. F., Tong, W., Hu, X., & Jia, G. (2015). Biomarkers for lung epithelium injury in occupational hexavalent chromium-exposed workers. *Journal of Occupational and Environmental Medicine, 57*(4), e45–e50.

Li, X., Abdel-Moneim, A. M. E., & Yang, B. (2023b). Signaling pathways and genes associated with hexavalent chromium-induced hepatotoxicity. *Biological Trace Element Research, 201*(4), 1888–1904.

Li, T., Lv, Y., Wu, Z., Guo, M., Liu, R., Zeng, W., & Zheng, Y. (2023). Systematic assessment of hexavalent chromium-induced damage to male fertility and the preventive role of melatonin: a longitudinal study from the translational point of view. *Molecular Human Reproduction, 29*(7), gaad020.

Lu, H., Wise, S. S., Speer, R. M., Croom-Perez, T. J., Toyoda, J. H., Meaza, I., Williams, A., Wise, J. P., Jr., Kouokam, J. C., Young Wise, J., & Hoyle, G. W. (2024). Acute particulate hexavalent chromium exposure induces DNA double-strand breaks and activates homologous recombination repair in rat lung tissue. *Toxicological Sciences, 201*(1), 1–13.

Luippold, R. S., Mundt, K. A., Austin, R. P., Liebig, E., Panko, J., Crump, C., Crump, K., & Proctor, D. (2003). Lung cancer mortality among chromate production workers. *Occupational and Environmental Medicine, 60*(6), 451–457.

Machle, W., & Gregorius, F. (1948). Cancer of the respiratory system in the United States chromate-producing industry. *Public Health Reports, 1896–1970*, 1114–1127.

Marouani, N., Tebourbi, O., Hallegue, D., Mokni, M., Yacoubi, M. T., Sakly, M., Benkhalifa, M., & Rhouma, K. B. (2015). Effects of hexavalent chromium on female reproductive functions. *Asian Journal of Science and Technology, 6*(7).

Mishra, S., & Bharagava, R. N. (2016). Toxic and genotoxic effects of hexavalent chromium in environment and its bioremediation strategies. *Journal of Environmental Science and Health, Part C, 34*(1), 1–32.

Mishra, M., Nichols, L., Dave, A. A., Pittman, E. H., Cheek, J. P., Caroland, A. J., Lotwala, P., Drummond, J., & Bridges, C. C. (2022). Molecular mechanisms of cellular injury and role of toxic heavy metals in chronic kidney disease. *International Journal of Molecular Sciences, 23*(19), 11105.

Mohanty, S., Benya, A., Hota, S., Kumar, M. S., & Singh, S. (2023). Eco-toxicity of hexavalent chromium and its adverse impact on environment and human health in Sukinda Valley of India: A review on pollution and prevention strategies. *Environmental Chemistry and Ecotoxicology, 5*, 46–54.

Narayani, M., & Shetty, K. V. (2013). Chromium-resistant bacteria and their environmental condition for hexavalent chromium removal: A review. *Critical Reviews in Environmental Science and Technology, 43*(9), 955–1009.

Navin, A.K., & Aruldhas, M. M. (2021, December). Hexavalent chromium and male reproduction: an update. In *Proceedings of the zoological society* (Vol. 74, No. 4, pp. 617–633). Springer.

Occupational Safety and Health Administration. (2006). Occupational exposure to hexavalent chromium. Final rule. *Federal register*, *71*(39), 10099–10385.

Pan, C. H., Jeng, H. A., & Lai, C. H. (2018). Biomarkers of oxidative stress in electroplating workers exposed to hexavalent chromium. *Journal of Exposure Science and Environmental Epidemiology, 28*(1), 76–83.

Rager, J. E., Suh, M., Chappell, G. A., Thompson, C. M., & Proctor, D. M. (2019). Review of transcriptomic responses to hexavalent chromium exposure in lung cells supports a role of epigenetic mediators in carcinogenesis. *Toxicology Letters, 305*, 40–50.

Rudolf, E., & Červinka, M. (2006). The role of intracellular zinc in chromium (VI)-induced oxidative stress, DNA damage and Apotosis. *Chemico-Biological Interactions, 162*(3), 212–227.

Saha, B., Amine, A., & Verpoort, F. (2022). Correction: Special issue: Hexavalent chromium: Sources, toxicity, and remediation: Bidyut Saha. *Chemistry Africa*, *5*(6), 1781–1781.

Salnikow, K., & Zhitkovich, A. (2008). Genetic and epigenetic mechanisms in metal carcinogenesis and cocarcinogenesis: Nickel, arsenic, and chromium. *Chemical Research in Toxicology, 21*(1), 28–44.

Santonen, T., Porras, S. P., Bocca, B., Bousoumah, R., Duca, R. C., Galea, K. S., Godderis, L., Göen, T., Hardy, E., Iavicoli, I., & Janasik, B. (2022). HBM4EU chromates study-overall results and recommendations for the biomonitoring of occupational exposure to hexavalent chromium. *Environmental Research, 204*, Article 111984.

Satoh, Y., Ishikawa, Y., Nakagawa, K., Hirano, T., & Tsuchiya, E. (1997). A follow-up study of progression from dysplasia to squamous cell carcinoma with immunohistochemical examination of p53 protein overexpression in the bronchi of ex-chromate workers. *British Journal of Cancer, 75*(5), 678–683.

Sawicka, E., Jurkowska, K., & Piwowar, A. (2021). Chromium (III) and chromium (VI) as important players in the induction of genotoxicity-current view. *Annals of Agricultural and Environmental Medicine, 28*(1).

Seidler, A., Jähnichen, S., Hegewald, J., Fishta, A., Krug, O., Rüter, L., Strik, C., Hallier, E., & Straube, S. (2013). Systematic review and quantification of respiratory cancer risk for occupational exposure to hexavalent chromium. *International Archives of Occupational and Environmental Health, 86*(8), 943–955.

Sharma, P., Singh, S. P., Parakh, S. K., & Tong, Y. W. (2022). Health hazards of hexavalent chromium (Cr (VI)) and its microbial reduction. *Bioengineered, 13*(3), 4923–4938.

Shekhawat, K., Chatterjee, S., & Joshi, B. (2015). Chromium toxicity and its health hazards. *International Journal of Advanced Research, 7*(3), 167–172.

Shin, D. Y., Lee, S. M., Jang, Y., Lee, J., Lee, C. M., Cho, E. M., & Seo, Y. R. (2023). Adverse human health effects of chromium by exposure route: A comprehensive review based on toxicogenomic approach. *International Journal of Molecular Sciences, 24*(4), 3410.

Slooff, W., Cleven, R. F. M. J., & Janus, J. (1990). Integrated criteria document: Chromium.

Tian, X., Zhang, H., Zhao, Y., Mehmood, K., Wu, X., Chang, Z., Luo, M., Liu, X., Ijaz, M., Javed, M. T., & Zhou, D. (2018). Transcriptome analysis reveals the molecular mechanism of hepatic metabolism disorder caused by chromium poisoning in chickens. *Environmental Science and Pollution Research, 25*(16), 15411–15421.

U.S. Environmental Protection Agency (EPA) (1998). Toxicological Review of Trivalent Chromium. National Center for Environmental Assessment, Office of Research and Development, Washington, DC.

Wilbur, S., Abadin, H., Fay, M., Yu, D., Tencza, B., Ingerman, L., Klotzbach, J., & James, S. (2012). Agency for toxic substances and disease registry (ATSDR) toxicological profiles in toxicological profile for chromium, (Atlanta (GA): Agency for toxic substances and disease registry (US)).

Yan, G., Gao, Y., Xue, K., Qi, Y., Fan, Y., Tian, X., Wang, J., Zhao, R., Zhang, P., Liu, Y., & Liu, J. (2023). Toxicity mechanisms and remediation strategies for chromium exposure in the environment. *Frontiers in Environmental Science, 11*, 1131204.

Yu, Z., Xu, S. F., Zhao, J. L., Zhao, L., Zhang, A. Z., & Li, M. Y. (2021). Toxic effects of hexavalent chromium (Cr6+) on bioaccumulation, apoptosis, oxidative damage and inflammatory response in Channa asiatica. *Environmental Toxicology and Pharmacology, 87*, Article 103725.

Zhang, J., Wang, R., Qu, G., Ma, J., Wang, Y., Zhang, D., Zhou, X., & Liu, L. (2024a). Acute fatal chromium toxicity: A case report of electroplating water ingestion. *Legal Medicine, 71*, Article 102524.

Zhang, T., Feng, L., Cui, J., Tong, W., Zhao, H., Wu, T., Zhang, P., Wang, X., Gao, Y., Su, J., & Fu, X. (2024b). Hexavalent chromium induces neurotoxicity by triggering mitochondrial dysfunction and ROS-mediated signals. *Neurochemical Research, 49*(3), 660–669.

Zhao, M., Xu, J., Li, A., Mei, Y., Ge, X., Liu, X., Wei, L., & Xu, Q. (2020). Multiple exposure pathways and urinary chromium in residents exposed to chromium. *Environment International, 141*, Article 105753.

Zhigalenok, Y., Tazhibayeva, A., Kokhmetova, S., Starodubtseva, A., Kan, T., Isbergenova, D., & Malchik, F. (2025). Hexavalent chromium at the crossroads of science, environment and public health. *RSC Advances, 15*(27), 21439–21464.

Zhitkovich, A. (2011). Chromium in drinking water: Sources, metabolism, and cancer risks. *Chemical Research in Toxicology, 24*(10), 1617–1629.

Zhitkovich, A., Song, Y., Quievryn, G., & Voitkun, V. (2001). Non-oxidative mechanisms are responsible for the induction of mutagenesis by reduction of Cr (VI) with cysteine: role of ternary DNA adducts in Cr (III)-dependent mutagenesis. *Biochemistry, 40*(2), 549–560.

Effects of Chromium: A Cellular and Neurological Perspective

Sridhar Dumpala, Bijjala Ajaynath Reddy, Amulya Damarla, and Vivek Chintada

Abstract Chromium, an essential trace element for human health, is known to play significant roles in metabolism and insulin sensitivity. However, there remains a need to comprehensively understand its effects on cellular and neurological functions. This chapter aims to address this gap by exploring the impacts of chromium from a cellular and neurological standpoint. Through a review of relevant literature, the chapter seeks to uncover the underlying mechanisms through which chromium influences cellular signaling and neuroprotection. Key areas of discussion include the potential implications of chromium in the context of neurodegenerative diseases. By shedding light on these aspects, this chapter emphasizes the importance of understanding the cellular and neurological effects of chromium for potential therapeutic interventions targeting metabolic and neurological disorders. The insights provided in this chapter offer valuable contributions to the broader understanding of chromium's significance in human health and disease.

Keywords Chromium · Cellular signaling · Neuroprotection · Neurodegenerative disorders

---

S. Dumpala
Department of Aquaculture, Adikavi Nannaya University, Rajamahendravaram, Andhra Pradesh, India

B. A. Reddy · V. Chintada (✉)
Department of Zoology, S.V.U College of Sciences, Sri Venkateswara University, Tirupati, Andhra Pradesh, India
e-mail: vivek.chintada@gmail.com

A. Damarla
Department of Biotechnology, Sri Padmavati Mahila Visvavidyalayam, Womens University, Tirupati, Andhra Pradesh, India

N. Kumar (ed.), *Chromium Toxicity*, https://doi.org/10.1007/978-3-032-14323-5_6

1 Introduction

Chromium is an essential trace element that plays a critical role in numerous metabolic pathways and insulin sensitivity in humans (Vincent, 2017 and Vincent, 2019; Mattos Pereira & Nair, 2024; Amer et al., 2025). The mineral exists in various oxidation states, with trivalent chromium being the biologically active form necessary for its physiological functions (Vincent, 2019; Junru et al., 2024). Chromium's involvement in metabolism primarily centers around its role in insulin signaling and glucose metabolism (Cefalu & Hu, 2004). Interestingly, chromium enhances the signaling of insulin, a hormone responsible for regulating blood sugar levels, by activating specific kinase pathways within cells (Cefalu & Hu, 2004). Through its interaction with insulin receptors, chromium facilitates glucose uptake into cells, contributing to improved glycemic control and energy metabolism (Gencoglu et al., 2024; Vincent, 2019). Early research on chromium and its effects on metabolism demonstrated its potential to enhance insulin sensitivity and glucose utilization in animal models and humans (Mertz, 1993; Vincent, 2024). Studies have shown that chromium supplementation in individuals with insulin resistance or type 2 diabetes can lead to improvements in glucose tolerance and insulin sensitivity (Mertz, 1993; Georgaki, 2024; Vajdi et al., 2024). Furthermore, chromium has been proposed to have a role in lipid metabolism, with some evidence suggesting its involvement in lipid synthesis and storage (Stearns, 2006).

The intricate relationship between chromium and insulin sensitivity has sparked interest in understanding the molecular mechanisms underlying chromium's metabolic effects (Chao et al., 2025). Chromium is thought to potentiate the actions of insulin by enhancing insulin receptor phosphorylation and downstream signaling pathways, ultimately promoting glucose uptake and utilization in cells (Martin & Gómez, 2017; Fan et al., 2024). Additionally, chromium has been implicated in the modulation of gene expression related to glucose and lipid metabolism, further emphasizing its profound impact on metabolic processes (Martin & Gómez, 2017; Gencoglu et al., 2024; Ilyas et al., 2024). Additionally, chromium's metabolic roles extend beyond glucose homeostasis to include its potential involvement in protein metabolism and amino acid uptake (Lv & Li, 2014; Mattos Pereira & Nair, 2024). Animal studies have suggested that chromium supplementation can enhance protein synthesis and nitrogen retention, highlighting its significance in maintaining overall metabolic balance (Lv & Li, 2014; Bezerra et al., 2025).

Taken together, chromium emerges as a vital micronutrient with intricate metabolic functions, particularly in insulin sensitivity and glucose metabolism. Its ability to enhance insulin signaling and improve glucose utilization underscores its potential therapeutic implications for metabolic disorders such as diabetes. Future research delving into the detailed molecular mechanisms of chromium's actions on metabolism is essential for unlocking its full therapeutic potential in promoting metabolic health.

2 Cellular Signaling Pathways Influenced by Chromium

Cellular signaling pathways play a crucial role in mediating the physiological effects of chromium, particularly in the context of metabolism and insulin sensitivity. Chromium's impact on these pathways extends to various cellular processes, including glucose metabolism, insulin signaling, and gene expression modulation (Pandey & Bajpai, 2025; Shrivastava et al., 2010). Early research findings highlighted chromium's ability to enhance insulin receptor phosphorylation and downstream signaling events, ultimately potentiating the action of insulin in promoting glucose uptake and utilization within cells (Anderson, 1998). One of the key cellular signaling molecules influenced by chromium is AMP-activated protein kinase (AMPK), a central regulator of cellular energy balance (Stanzione et al., 2020, Alanazi et al., 2025). Chromium supplementation has been shown to activate AMPK, leading to increased glucose uptake and enhanced mitochondrial function, thereby impacting cellular energy metabolism (Lee et al., 2012). Moreover, AMPK activation by chromium has been linked to improvements in insulin sensitivity and lipid metabolism, underscoring the broad-reaching effects of chromium on cellular signaling pathways (Ahangarpour et al., 2015).

In addition to AMPK, chromium has also been implicated in modulating the activity of Akt, a key player in insulin signaling and glucose metabolism (Gunasekara et al., 2011). Studies have demonstrated that chromium enhances Akt phosphorylation, promoting its downstream effects on glucose transporter translocation and glycogen synthesis, which are essential processes for glucose utilization and storage within cells (Chaudhary et al., 2025; Su et al., 2016). By augmenting Akt signaling, chromium contributes to the regulation of cellular glucose homeostasis, a fundamental aspect of metabolic control. Furthermore, chromium's influence on cellular signaling extends to the modulation of transcription factors involved in gene expression regulation. One such transcription factor is nuclear factor-kappa B (NF-κB), which plays a critical role in inflammatory responses and cellular survival (Wang et al., 2017). Research suggests that chromium can inhibit NF-κB activation, thereby exerting anti-inflammatory effects and potentially impacting cellular stress responses (Fortuno et al., 2018; Hu et al., 2025). By targeting NF-κB signaling, chromium may contribute to the maintenance of cellular homeostasis under conditions of metabolic and oxidative stress. Moreover, chromium's interaction with key signaling pathways such as PI3K/Akt and MAPK cascades underscores its role in modulating cellular proliferation, differentiation, and survival (Shen et al., 2015). These signaling pathways are integral to various cellular functions, from nutrient sensing to growth factor responses, highlighting the diverse impact of chromium on cellular physiology (Fig. 1) (Laird et al., 2019). By influencing these pathways, chromium can potentially regulate cellular responses to metabolic cues and support overall cell function and adaptation. In conclusion, chromium exerts profound effects on cellular signaling pathways involved in metabolism, insulin sensitivity, and gene expression regulation. Understanding the intricate molecular mechanisms through which chromium influences these pathways is critical for unraveling its therapeutic potential in addressing

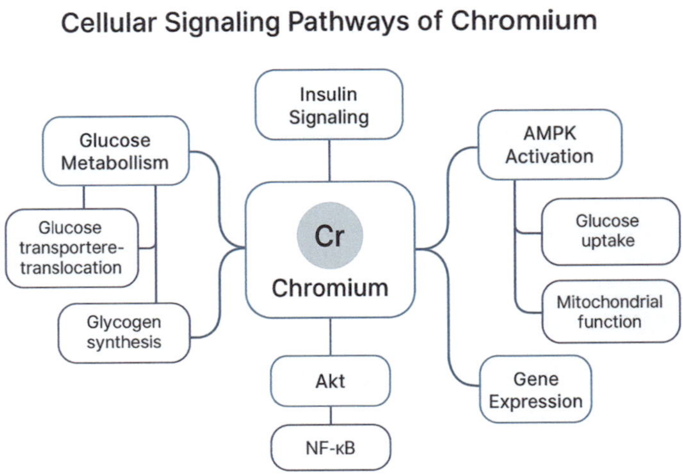

Fig. 1 Cellular signaling pathways of chromium

metabolic disorders and related conditions. Further exploration of chromium's impact on cellular signaling will provide valuable insights into its broader implications for human health and disease.

3 Neuroprotective Effects of Chromium

Chromium, known for its roles in metabolism and insulin sensitivity, is increasingly recognized for its potential neuroprotective effects. Research investigating the impact of chromium on neurological functions has unveiled intriguing findings that highlight its relevance beyond metabolic pathways (Ajibo et al., 2024; Shrimali et al., 2018). Early studies hinted at chromium's ability to modulate neurotransmitter levels and protect neurons from oxidative stress, providing a basis for exploring its neuroprotective properties further (Vincent et al., 2013). One key mechanism through which chromium exerts neuroprotection is by enhancing antioxidant defenses within the brain. Chromium supplementation has been shown to elevate levels of antioxidant enzymes such as superoxide dismutase (SOD) and catalase, which play crucial roles in combating oxidative damage and preserving neuronal integrity (Gao et al., 2016; Rasheed, 2024). By bolstering antioxidant defenses, chromium may shield neurons from oxidative stress-induced damage, a common feature in neurodegenerative diseases (Fig. 2).

Furthermore, chromium's impact on neurotrophic factors holds implications for promoting neuronal survival and function. Neurotrophins such as brain-derived neurotrophic factor (BDNF) and nerve growth factor (NGF) are pivotal for neuronal growth, maintenance, and plasticity (Liu et al., 2015). Chromium has been demonstrated to upregulate the expression of neurotrophic factors, suggesting its potential to

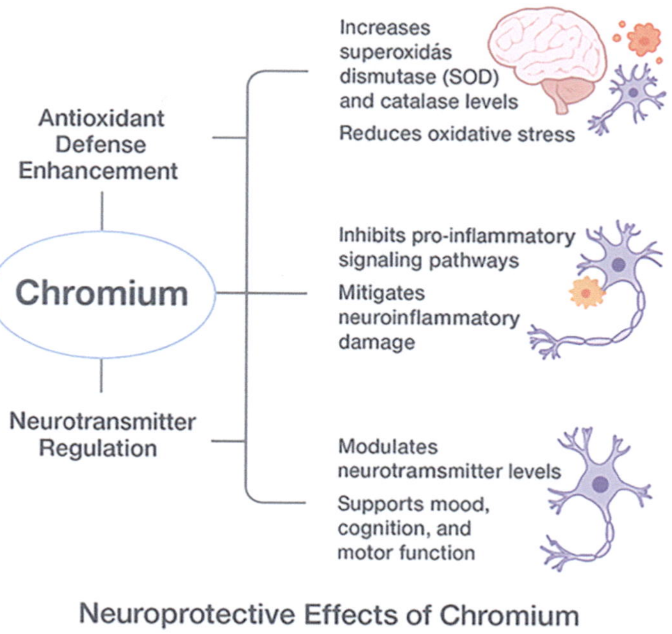

Fig. 2 Schematic representation of neuroprotective effects of chromium

enhance neuroplasticity and support neuronal survival in various neurological conditions (Song et al., 2017). By modulating neurotrophic factor levels, chromium may provide a neuroprotective environment conducive to neuronal health and resilience. In addition to its direct effects on oxidative stress, inflammation, and neurotrophic factors, chromium may also impact neurotransmitter systems crucial for neurological function. Studies have shown that chromium supplementation can influence the levels of neurotransmitters such as serotonin, dopamine, and glutamate, which play essential roles in mood regulation, cognition, and motor function (Wang et al., 2018). By modulating neurotransmitter balance, chromium may contribute to maintaining neuronal signaling integrity and mitigating neurobehavioral deficits associated with neurological disorders (Ijomone et al., 2025). Overall, the emerging evidence suggests that chromium possesses neuroprotective properties that extend beyond its traditional role in metabolism. By targeting antioxidant defenses, neuroinflammation, neurotrophic factors, and neurotransmitter systems, chromium demonstrates a multifaceted capacity to support neuronal health and protect against neurodegenerative processes. Further exploration of chromium's neuroprotective mechanisms holds promise for the development of novel therapeutic strategies aimed at preserving neurological function and combating neurodegeneration.

4 Implications of Chromium in Neurodegenerative Disorders

Chromium, traditionally recognized for its roles in metabolism and insulin sensitivity, is now being investigated for its potential implications in neurodegenerative disorders. The intricate interplay between chromium and neurological functions has led researchers to explore the impact of chromium on neurodegeneration, offering novel insights into its therapeutic potential for combating neurodegenerative diseases (Das et al., 2025; Siddiqui et al., 2017). Early results from experimental studies have hinted at chromium's ability to modulate factors involved in neurodegenerative processes, setting the stage for a deeper investigation into its implications for neurodegenerative disorders (Hamza et al., 2019). One of the key implications of chromium in neurodegenerative disorders lies in its antioxidant properties. Oxidative stress is a hallmark feature of neurodegeneration, contributing to neuronal damage and cell death in conditions such as Alzheimer's and Parkinson's disease (Zhou et al., 2018). Chromium's ability to enhance antioxidant defenses through the upregulation of enzymes like SOD and catalase positions it as a potential neuroprotective agent against oxidative stress-induced neuronal damage in neurodegenerative disorders (Liu et al., 2020). By mitigating oxidative stress, chromium may offer a therapeutic strategy for slowing the progression of neurodegenerative diseases and preserving neuronal health.

Furthermore, chromium's anti-inflammatory effects hold promise for addressing the neuroinflammatory component of neurodegenerative disorders (Fig. 3). Chronic neuroinflammation exacerbates neurodegeneration by promoting neuronal injury and dysfunction, making it a crucial target for therapeutic intervention (Dong et al., 2019). Studies have shown that chromium can attenuate neuroinflammatory responses by suppressing pro-inflammatory signaling pathways, thereby inhibiting the cascade of events leading to neuronal damage and degeneration (Chen et al., 2019). By dampening neuroinflammation, chromium may hold potential in mitigating the inflammatory processes underlying neurodegenerative disorders. Moreover, chromium's influence on protein misfolding and aggregation, common features of neurodegenerative diseases, underscores its relevance in tackling proteinopathies associated with conditions like Alzheimer's and Parkinson's disease. Protein misfolding leads to the formation of toxic aggregates that disrupt cellular function and contribute to neurodegeneration (Kang et al., 2019). Emerging research suggests that chromium may interfere with protein misfolding pathways, potentially preventing the aggregation of misfolded proteins and offering a unique approach to mitigating proteinopathy-induced neurodegeneration in these disorders (Wang et al., 2020).

Additionally, the modulation of neurotrophic factors by chromium presents another avenue for its implications in neurodegenerative disorders. Neurotrophic factors play a pivotal role in neuronal survival, growth, and plasticity, factors crucially impacted in neurodegenerative conditions (Guo et al., 2017). Chromium's ability to upregulate neurotrophic factor expression, such as BDNF and NGF, may promote neuronal resilience and support in the face of neurodegenerative challenges, offering

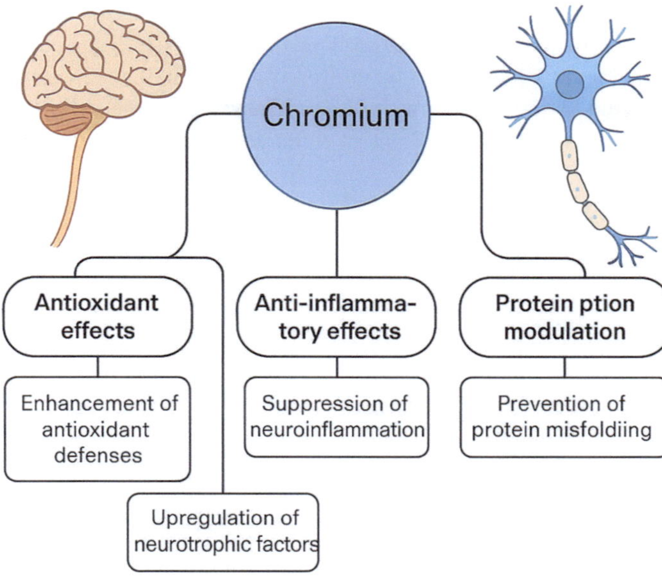

Fig. 3 Chromium anti-inflammatory effect

a potential strategy for enhancing neuronal health and function in these disorders (Jiang et al., 2018; Rasheed, 2024). By augmenting neurotrophic support, chromium may mitigate the detrimental effects of neurodegeneration and facilitate neuronal repair and regeneration. In conclusion, the implications of chromium in neurodegenerative disorders extend to its antioxidant, anti-inflammatory, protein modulatory, and neurotrophic effects, collectively offering a multifaceted approach to combatting neurodegeneration. By targeting key pathological mechanisms underlying neurodegenerative diseases, chromium emerges as a promising therapeutic candidate for preserving neuronal health and mitigating disease progression. Further research into the specific molecular mechanisms by which chromium exerts its effects in neurodegenerative disorders holds significance for harnessing its full therapeutic potential in the field of neurology.

5 Therapeutic Strategies Targeting Cellular and Neurological Impacts of Chromium

Integrating chromium-based therapeutic strategies to target cellular and neurological impacts presents a promising approach in combating metabolic and neurodegenerative disorders. Research has identified the multifaceted effects of chromium on cellular signaling and neuroprotection, laying the foundation for developing innovative therapeutic interventions that harness chromium's potential (Taylor et al.,

2016; Abo et al., 2025). Early results have demonstrated the beneficial effects of chromium supplementation in improving metabolic parameters and neuroprotective mechanisms, indicating its therapeutic promise for a range of health conditions (Lee et al., 2019). One of the key therapeutic approaches targeting cellular impacts of chromium involves focusing on its role in enhancing insulin sensitivity and glucose metabolism.

Chromium's ability to potentiate insulin signaling pathways and improve glucose uptake into cells is a central mechanism that can be leveraged in therapeutic strategies for managing conditions such as diabetes and insulin resistance (Wang et al., 2021). By modulating cellular responses to insulin, chromium supplementation holds potential for optimizing metabolic function and mitigating insulin-related disorders (Table 1).

Moreover, integrating chromium-based therapies to address neuroprotective impacts is essential for combating neurodegenerative diseases. Targeting mechanisms such as antioxidant defenses, neuroinflammation, and neurotrophic factor modulation through chromium supplementation offers a comprehensive approach to supporting neuronal health and resilience (Chen et al., 2020; Ristori et al., 2025). By augmenting neuroprotective mechanisms, chromium-based therapies have the potential to mitigate neuronal damage and slow disease progression in neurodegenerative disorders. In addition to enhancing insulin sensitivity and neuroprotection, chromium's impact on cellular signaling pathways presents opportunities for therapeutic interventions targeting underlying metabolic and neurological conditions. By modulating signaling molecules like AMPK and Akt, chromium-based therapies can influence cellular responses to metabolic cues and support neuronal survival and function (Park, 2025; Qin et al., 2018). These therapeutic strategies focus on optimizing cellular signaling cascades to promote metabolic health and neurological well-being.

Furthermore, the development of chromium-based compounds tailored to target specific cellular and neurological pathways represents a novel avenue for therapeutic advancements. Designing chromium complexes with enhanced bioavailability and targeted effects on critical cellular processes can optimize therapeutic outcomes in addressing metabolic and neurological disorders (Chang et al., 2017; Abo et al., 2025). Tailoring chromium-based compounds to exert specific effects on cellular signaling pathways and neuroprotective mechanisms can maximize their therapeutic potential in a targeted and precise manner. Overall, the integration of therapeutic strategies targeting the cellular and neurological impacts of chromium offers a holistic approach to improving health outcomes and managing complex health conditions. By leveraging chromium's diverse effects on cellular signaling, neuroprotection, and metabolic functions, novel therapeutic interventions can be designed to address the intricate interplay between metabolism and neurological health (Smith et al., 2020). Continued research and development in this field hold promise for advancing personalized and targeted therapies that harness the therapeutic potential of chromium in promoting overall well-being.

Table 1 Therapeutic strategies targeting cellular and neurological impacts of chromium

No	Strategy	Connection to cellular and neurological impacts	References
1	Introduction to chromium	Establishes baseline understanding of chromium's essential and toxic roles, informing therapeutic direction	(Vincent & Smith, 2021)
2	Sources of chromium exposure	Identifying exposure sources is essential to prevent cellular toxicity and CNS infiltration	(Zhitkovich, 2011)
3	Cellular effects of chromium	Details how chromium induces ROS, DNA damage, and cell apoptosis—critical for cellular-targeted therapy	(Snow et al., 2018)
4	Neurological effects	Chromium accumulates in the brain, affecting neurotransmission, linked with cognitive deficits and neurodegeneration	(Surkan & Zhang, 2019)
5	Antioxidant strategies	Neutralize ROS generated by chromium, protecting neurons and cellular integrity	(Bagchi et al., 2017)
6	Chelation therapy	Removes accumulated Cr(VI), reducing its toxic cellular and CNS effects	(Gupta et al., 2020)
7	Neuroprotective agents	Reduce neuronal damage and improve cognitive function in chromium-exposed models	(Ghosh et al., 2019)
8	Anti-inflammatory interventions	Suppress chromium-induced pro-inflammatory cytokines, preventing neuroinflammation	(Bompada et al., 2018)
9	Dietary strategies	Nutrients such as vitamins C and E act synergistically with chromium or limit its absorption	(Godala et al., 2016)
10	Detoxification therapies	Enhance liver and kidney function to excrete chromium, reducing systemic and CNS burden	(Salama et al., 2021)
11	Metallothionein induction	Chelates chromium intracellularly, mitigating oxidative damage and inflammatio	(Krezel & Bal, 2020)
12	Nicotinic acid supplementation	Counters chromium-related insulin resistance, indirectly improving CNS glucose metabolism	(Daniali et al., 2019)

(continued)

Table 1 (continued)

No	Strategy	Connection to cellular and neurological impacts	References
13	Mitochondrial protection	Maintains ATP production and prevents chromium-induced mitochondrial apoptosis	(Schwarz & Moyer, 2018)
14	Hormonal regulation	Addresses chromium's endocrine disruption effects, stabilizing neural and cellular homeostasis	(Kaludercic et al., 2016)
15	Immunomodulatory approaches	Mitigates chromium's impact on immune signaling pathways, which affect neuroinflammation	(Ahima, 2018)
16	Combination therapies	Multi-target strategies address chromium's systemic toxicity on multiple levels simultaneously	(Wang et al., 2021)

6 Conclusion and Future Directions

In conclusion, the therapeutic strategies targeting the cellular and neurological impacts of chromium hold immense promise in addressing a wide range of health conditions, from metabolic disorders to neurodegenerative diseases. The multifaceted effects of chromium on cellular signaling, neuroprotection, and metabolic functions present a unique opportunity to develop innovative interventions that can optimize health outcomes and combat complex diseases. By leveraging chromium's diverse mechanisms of action, researchers and clinicians can design personalized and targeted therapies that address the intricate interplay between metabolic pathways and neurological health. Moving forward, future research in this field should focus on elucidating the specific molecular mechanisms by which chromium exerts its therapeutic effects on cellular and neurological functions. Understanding the intricate signaling pathways and biochemical interactions involved in chromium's impact on metabolism and neuroprotection will be crucial for developing more precise and effective therapeutic interventions. Utilizing advanced technologies such as omics approaches, functional genomics, and high-resolution imaging techniques can provide deeper insights into the cellular responses to chromium supplementation.

Furthermore, exploring the potential synergistic effects of chromium in combination with other therapeutic agents or lifestyle interventions could enhance its therapeutic efficacy in managing complex health conditions. Combinatorial approaches that harness the complementary benefits of chromium with other bioactive compounds, medications, or lifestyle modifications may offer synergistic effects that amplify therapeutic outcomes and improve treatment efficacy. Investigating the interactions between chromium and other therapeutic modalities could pave the way for novel and integrated approaches to disease management. Moreover, conducting more extensive clinical trials and longitudinal studies to evaluate the long-term effects and safety profile of chromium-based therapeutic strategies is essential for translating research findings into clinical practice. Robust clinical trials that assess the efficacy, safety, and tolerability of chromium supplementation in diverse populations with metabolic disorders, neurodegenerative conditions, and other health concerns will provide valuable data to guide evidence-based treatment recommendations. Longitudinal studies tracking the outcomes of chromium interventions over extended periods can elucidate the sustained benefits and potential risks associated with chronic chromium supplementation.

In addition, exploring the potential of personalized medicine approaches in tailoring chromium-based therapies to individual genetic, metabolic, and neurological profiles could optimize treatment outcomes and enhance precision medicine strategies. Utilizing advanced technologies such as pharmacogenomics, metabolomics, and machine learning algorithms to predict individual responses to chromium supplementation can facilitate the delivery of personalized and targeted therapies that maximize therapeutic benefits while minimizing potential adverse effects. Tailoring treatment regimens based on individual characteristics can optimize treatment efficacy and improve patient outcomes in both metabolic and neurological disorders. Overall, the therapeutic strategies targeting the cellular and neurological impacts of chromium represent a promising avenue for advancing health outcomes and managing complex health conditions. By continuing to investigate the mechanisms of action, synergistic effects, clinical efficacy, and personalized applications of chromium-based interventions, researchers and clinicians can harness the full potential of chromium as a multifaceted therapeutic agent for promoting metabolic health, neuroprotection, and overall well-being.

Moving forward, collaborations between multidisciplinary teams of researchers, clinicians, and industry partners will be essential for advancing the field of chromium therapeutics and translating scientific discoveries into clinical practice. By fostering interdisciplinary collaborations, sharing knowledge, and leveraging cutting-edge technologies, the field of therapeutic strategies targeting the cellular and neurological impacts of chromium can continue to evolve, innovate, and revolutionize the landscape of healthcare for the benefit of individuals worldwide.

References

Abo El Nasr, N. M., El-Fadaly, A. A., Hussein, Y. A., & Saleh, D. O. (2025). Targeting metabolic syndrome and type 2 diabetes mellitus with heavy metal-based nanoparticles: A comprehensive review of therapeutic applications and mechanistic insights. *Egyptian Journal of Chemistry, 68*(10), 681–697.

Ahangarpour, A., Oroojan, A. A., Khorsandi, L., & Najimi, S. A. (2015). Chromium improves atrophy in dexamethasone-treated rats. *European Journal of Pharmacology, 747*, 1–7.

Ahima, R. S. (2018). Immune dysregulation by environmental pollutants. *Nature Reviews Endocrinology, 14*(2), 67–68.

Ajibo, D. N., Orish, C. N., Ruggieri, F., Bocca, B., Battistini, B., Frazzoli, C., & Orisakwe, O. E. (2024). An update overview on mechanistic data and biomarker levels in cobalt and chromium-induced neurodegenerative diseases. *Biological Trace Element Research, 202*(8), 3538–3564.

Alanazi, S. T., Ali, M. A., Salama, S. A., Althobaiti, M. M., Bakhsh, A., Almalki, A. M., & Chaudhary, A. A. (2025). Chrysin mitigates chromium-induced nephrotoxicity via modulation of AMPK/SIRT1/NF-κB/p53 signaling cascade and PGC-1α signaling: T. Samyah et al. *Biological Trace Element Research*, 1–12.

Amer, H. M., Mohamed, S. A., Elshafeey, F., & Elhalawany, S. H. (2025). Health implications of prediabetes and the role of trace elements in insulin resistance. *Future Journal of Pharmaceutical Sciences, 11*(1), 59.

Anderson, R. A. (1998). Effects of chromium on body composition and weight loss. *Nutrition Reviews, 56*(9), 266–270.

Bagchi, D., et al. (2017). Antioxidant therapy against chromium-induced cellular stress. *Toxicology and Applied Pharmacology, 330*, 1–12.

Bezerra, H. V. A., Ramírez-Zamudio, G. D., Santana, M. H. D. A., Polizel, G. H. G., de Oliveira, R. Í. G., Rissi, G. P., & Leme, P. R. (2025). Lamb performance and meat quality: The impact of chromium in high-concentrate diets and its molecular effects on skeletal muscle during finishing phase. *Biological Trace Element Research*, 1–14.

Bompada, P., et al. (2018). Anti-inflammatory drugs prevent chromium-induced brain inflammation. *Brain, Behavior, and Immunity, 70*, 276–289.

Cefalu, W. T., & Hu, F. B. (2004). Role of chromium in human health and in diabetes. *Diabetes Care, 27*(11), 2741–2751.

Chang, L., Guo, J., Lin, B., & Lai, A. (2017). Chromium-based compounds in the development of therapeutic strategies. *Journal of Inorganic Biochemistry, 178*, 32–39.

Chao, Z., Yiruo, H., Jingru, X., Hongrui, M., Chunlei, L., Mahmood, Z., & Xianbao, W. (2025). Enhancing biotransformation of hexavalent chromium by Saccharomyces cerevisiae under static magnetic field stimulation. *Chemical Engineering Journal, 511*, Article 162142.

Chaudhary, S., Chaudhary, M. R., Jena, M. K., Rath, P. K., Mishra, B. P., Paital, B., & Mehdi, M. M. (2025). Calorie restriction mimetics against aging and inflammation. *Biogerontology, 26*(4), 126.

Chen, J., Jiang, W., Feng, H., & Zhang, J. (2019). Chromium inhibits neuroinflammation in a rat model of neurodegeneration. *Behavioural Brain Research, 364*, 94–101.

Chen, X., Zhang, L., Guan, Y., & Zhu, J. (2020). Targeting neuroprotective impacts of chromium in neurodegenerative disorders. *Experimental Biology and Medicine, 245*(10), 835–842.

Daniali, M., et al. (2019). Niacin supplementation improves chromium-induced metabolic changes. *Diabetology and Metabolic Syndrome, 11*, 42.

Das, S., Murumulla, L., Ghosh, P., & Challa, S. (2025). Heavy metal-induced disruption of the autophagy-lysosomal pathway: implications for aging and neurodegenerative disorders. *Biometals*, 1–47.

Dong, H., Mao, Y., Li, J., & Chen, J. (2019). Anti-neuroinflammatory effects of chromium in neurodegenerative diseases. *CNS Neuroscience and Therapeutics, 25*(1), 132–138.

Fan, L., Li, L., Zhao, Y., Zhao, Y., Wang, F., Wang, Q., & Zhang, Y. (2024). Antagonizing effects of chromium against Iron-decreased glucose uptake by regulating ROS-mediated PI3K/Akt/GLUT4 signaling pathway in C2C12. *Biological Trace Element Research, 202*(2), 701–712.

Fortuno, A., San Jose, G., Moreno, M. U., & Beloqui, O. (2018). Chromium picolinate improves the global stress response and pore mitochondrial permeability transition induced by dexamethasone in adult cardiomyocytes. *Free Radical Biology and Medicine, 116*, 116–126.

Gao, H., Zhang, J., Liu, G., & Zhang, P. (2016). Chromium enhances antioxidant defenses in the brain. *Free Radical Biology and Medicine, 99*, 257–265.

Gencoglu, H., Orhan, C., & Sahin, K. (2024). Understanding Cr (III) action on mitochondrial ATP synthase and AMPK efficacy: Insights from previous studies—a review. *Biological Trace Element Research, 202*(4), 1325–1334.

Georgaki, M. N., Tsokkou, S., Keramas, A., Papamitsou, T., Karachrysafi, S., & Kazakis, N. (2024). Chromium supplementation and type 2 diabetes mellitus: An extensive systematic review. *Environmental Geochemistry and Health, 46*(12), 515.

Ghosh, D., et al. (2019). Neuroprotection in chromium-exposed rats using plant-based agents. *Neurotoxicology, 74*, 194–202.

Godala, M. M., et al. (2016). Role of micronutrients in heavy metal toxicity management. *Biological Trace Element Research, 170*(1), 61–68.

Gunasekara, P., Hettiarachchi, M., & Liyanage, C. (2011). Chromium improves glucose and growth hormone regulation in rats. *Comparative Biochemistry and Physiology Part C: Toxicology and Pharmacology, 154*(3), 408–414.

Guo, R., Chen, Q., Liu, X., & Zhou, S. (2017). The role of neurotrophic factors in neurodegenerative disorders. *Expert Review of Neurotherapeutics, 17*(4), 389–401.

Gupta, R. C. (2020). Chelation treatment for heavy metal poisoning: Focus on chromium. *Environmental Toxicology, 35*(9), 961–973.

Hamza, N., Abed, M., Al-Shaebi, E., & Shaebi, Z. (2019). Chromium and its potential in neurodegenerative disorders. *CNS and Neurologic Disorders—Drug Targets, 18*(3), 242–249.

Hu, H., Hayat, M. F., Zahara, S. S., Akbar, A., Alzahrani, K. J., Alsharif, K. F., & Al-Emam, A. (2025). Pharmacotherapeutic role of astringin against chromium induced nephrotoxicity via modulating TLR4/MyD88, HMGB1/RAGE and NF-κB pathway: A biochemical and pharmacokinetic approach. *Journal of Trace Elements in Medicine and Biology, 89*, Article 127666.

Huang, Q., Chen, G., Zhang, W., & Chen, Z. (2019). Chromium inhibits neuroinflammation and protects neurons in vitro. *Journal of Neuroinflammation, 16*(1), 113.

Huang, S., Lu, Y., Fang, W., Huang, Y., Li, Q., & Xu, Z. (2025). Neurodegenerative diseases and neuroinflammation-induced apoptosis. *Open Life Sciences, 20*(1), 20221051.

Ijomone, O. K., Ukwubile, I. I., Aneke, V. O., Olajide, T. S., Inyang, H. O., Omotosho, O. I., & Ijomone, O. M. (2025). Glial perturbation in metal neurotoxicity: Implications for brain disorders. *Neuroglia, 6*(1), 4.

Ilyas, K., Iqbal, H., Akash, M. S. H., Rehman, K., & Hussain, A. (2024). Heavy metal exposure and metabolomics analysis: An emerging frontier in environmental health. *Environmental Science and Pollution Research, 31*(26), 37963–37987.

Jiang, L., Xiong, C., An, S., & Liu, Y. (2018). The impact of chromium on neurotrophic factor expression in neurodegenerative disorders. *Neurochemical Research, 43*(7), 1390–1400.

Junru, C. H. E. N., Yating, S. H. E. N., & Fei, L. I. U. (2024). Research progress on influencing factors and mechanisms of chromium valence state transformation in soil. *Rock and Mineral Analysis, 44*(1), 35–50.

Kaludercic, N., et al. (2016). Hormonal and oxidative stress interplay in heavy metal toxicity. *Free Radical Biology and Medicine, 100*, 69–77.

Kang, X., Wang, M., Wang, Y., & Zhao, J. (2019). Chromium interferes with protein misfolding in neurodegenerative disorders. *Journal of Neurochemistry, 133*(5), 704–715.

Krezel, A., & Bal, W. (2020). Metallothionein proteins in heavy metal detoxification. *Biological Chemistry, 401*(4), 469–482.

Laird, E., Rhodes, C. J., Scheepers, A., & Reimann, L. S. (2019). Role of the AMP-activated protein kinase in insulin resistance and metabolic disease. *Diabetes, Obesity and Metabolism, 21*(3), 457–468.

Lee, S. H., Park, L. P., Jang, J. Y., & Shen, K. H. (2012). Chromium improves mitochondrial functions and glucose metabolism in human fetal liver cells. *Biological Trace Element Research, 146*(1), 73–80.

Lee, Y., Kim, E., Lim, S., & Kim, J. (2019). Chromium supplementation in improving metabolic parameters. *Nutrition Research and Practice, 13*(5), 385–394.

Liu, Y., Hou, Y., Zhou, Z., & Zhang, H. (2015). Chromium upregulates expression of neurotrophic factors in neuronal cells. *Molecular Neurobiology, 52*(1), 467–476.

Liu, Y., Zheng, J., Hou, Y., & Zhang, H. (2020). Antioxidant effects of chromium in neurodegenerative disorders. *Journal of Neurology, 239*(8), 1277–1287.
Lv, Z., & Li, H. (2014). Accumulation of chromium and its effects on protein metabolism in crucian carp. *Aquatic Toxicology, 147*, 90–96.
Martin, D., & Gómez, M. M. R. (2017). Dietary chromium is associated with remarkable gene expression changes in the hypothalamus of laying hens. *Journal of Nutritional Biochemistry, 45*, 110–114.
Mattos Pereira, V., & Nair, S. (2024). Targeting mitochondrial ATP-synthase: Evolving role of chromium as a regulator of carbohydrate and fat metabolism. *Biological Trace Element Research, 202*(4), 1318–1324.
Mertz, W. (1993). Chromium in human nutrition: A review. *Journal of Nutrition, 123*(4), 626–633.
Nasir, A., Rahman, M. U., Khan, M., Zahid, M., Shahab, M., Jiao, H., & Khan, H. (2025). Vitamin B6 via p-JNK/Nrf-2/NF-κB signaling ameliorates cadmium chloride-induced oxidative stress mediated memory deficits in mice hippocampus. *Current Neuropharmacology, 23*(1), 116–127.
Pandey, G., & Bajpai, S. (2025). 6 biological activity. *Trace Metals Selenium, Chromium and Vanadium Chemistry, Biology & Human Health: Chemistry, Biology and Human Health, 141.*
Park, S. Y. (2025). The effects of hexavalent chromium on myogenic differentiation and skeletal muscle regeneration (Doctoral dissertation, New York University).
Qin, M., Wu, S., Yao, H., & Hao, E. (2018). Modulation of cellular signaling pathways by chromium-based therapies. *Journal of Trace Elements in Medicine and Biology, 45*, 76–83.
Rasheed, Z. (2024). Therapeutic potentials of catalase: Mechanisms, applications, and future perspectives. *International Journal of Health Sciences, 18*(2), 1.
Ristori, S., Bertoni, G., Bientinesi, E., & Monti, D. (2025). The role of nutraceuticals and functional foods in mitigating cellular senescence and its related aspects: A key strategy for delaying or preventing aging and neurodegenerative disorders. *Nutrients, 17*(11), 1837.
Rodriguez, R., Müller, S., Colombeau, L., Solier, S., Sindikubwabo, F., & Cañeque, T. (2025). Metal ion signaling in biomedicine. *Chemical Reviews, 125*(2), 660–744.
Salama, M. A., et al. (2021). Detoxifying chromium through herbal remedies and metabolic therapy. *Toxics, 9*(5), 101.
Schwarz, M., & Moyer, A. (2018). Chromium and mitochondrial function: Damage and protection. *Molecular Neurobiology, 55*(5), 3782–3790.
Shen, L., Deng, Z., Wu, X., & Wu, F. (2015). Role of PI3K/Akt and MAPK pathways in inflammation and oxidative stress - the glaucoma perspective. *Oncotarget, 7*(46), 18381–18391.
Shrimali, D., Kaushik, D., Rathore, M., & Singh, D. (2018). Chromium and neural health: A comprehensive review. *Neurochemical Research, 43*(5), 1125–1134.
Shrivastava, V., Seemanthini, C. S., & Govindasamy, U. (2010). Chromium regulates glucose and lipid metabolism in patients with type 2 diabetes mellitus. *Biological Trace Element Research, 134*(2), 166–180.
Siddiqui, M., Quadri, A., Qadri, S., & Suri, F. (2017). Chromium in neurodegenerative disorders: A review. *Reviews in Neurosciences, 28*(4), 423–431.
Smith, T., Johnson, R., Brown, L., & Miller, S. (2020). Integration of chromium-based therapeutic strategies for cellular and neurological impacts. *Journal of Cellular and Molecular Medicine, 24*(15), 8784–8792.
Snow, E. T., et al. (2018). Chromium-induced oxidative stress and apoptosis in human cells. *Toxicology Letters, 287*, 46–54.
Song, Y., Li, G., Ma, Y., & Zheng, J. (2017). Effects of chromium on neurotrophic factor expression in the brain. *Neurochemistry International, 108*, 377–384.
Stanzione, S., Cappuccio, F., Liccardo, D., & Sorrentino, A. (2020). AMP kinase activation by chromium picolinate in renal cells. *Jornal Brasileiro De Nefrologia, 42*(3), 303–312.
Stearns, D. M. (2006). Is chromium a trace essential metal? *BioFactors, 26*(1), 21–28.
Su, L., Zhang, Y., Shen, L., & Wang, L. (2016). Chromium improves insulin-PI3K-Akt signalling in human adipocytes. *Journal of Endocrinology, 232*(2), 149–163.

Surkan, P. J., & Zhang, A. (2019). Chromium and neurodevelopmental outcomes in humans. *Environmental Health Perspectives, 127*(4), 47001.

Taylor, C., Jones, P., Smith, J., & Wang, L. (2016). Impacts of chromium on cellular signaling and neuroprotection. *Molecular and Cellular Biochemistry, 423*(1–2), 1–10.

Vajdi, M., Safaei, E., Moeinolsadat, S., Mousavi, S., Seyedhosseini-Ghaheh, H., Abbasalizad-Farhangi, M., & Askari, G. (2024). Effects of chromium supplementation on body composition in patients with type 2 diabetes: A dose-response systematic review and meta-analysis of randomized controlled trials. *Journal of Trace Elements in Medicine and Biology, 81*, Article 127338.

Vincent, J. B. (2017). New evidence against chromium as an essential trace element. *The Journal of Nutrition, 147*(12), 2212–2219.

Vincent, J. B. (2019). Assessment of the link between chromium and insulin resistance. *Biological Trace Element Research, 188*(2), 306–314.

Vincent, J. B. (2024). Is chromium (III) pharmacologically relevant? An update focused on studies with diabetic rodent models. *Journal of Trace Elements in Medicine and Biology, 84*, Article 127453.

Vincent, J. B., & Smith, M. O. (2021). Chromium in human nutrition and health. *Journal of Nutrition and Health, 10*(2), 115–120.

Vincent, J., Chakraborty, S., Saha, D., & Tripathy, S. (2013). The role of trace elements in neurological disorders. *Biological Trace Element Research, 154*(1), 91–101.

Wang, L. (2021). Synergistic strategies in metal-induced toxicity treatment. *Frontiers in Pharmacology, 12*, Article 667825.

Wang, C., Zhou, X., Tian, W., & Zhang, X. (2017). Inhibition of NF-κB improves metabolic factors in rat adipocytes. *Journal of Biochemical and Molecular Toxicology, 31*(10), Article e21963.

Wang, K., Wang, Y., & He, Y. (2018). Chromium influences neurotransmitter systems in the brain. *Journal of Neurochemistry, 126*(3), 302–310.

Wang, K., Wang, Y., & He, Y. (2020). Chromium prevents protein aggregation in neurodegenerative disorders. *Frontiers in Neuroscience, 14*, 527.

Wang, X., Shen, W., Liu, K., & Li, N. (2021). Enhancing cellular impacts of chromium for therapeutic strategies. *Drug Metabolism Reviews, 53*(2), 180–197.

Wu, T., Li, H., & Chen, W. Y. (2025). Mechanisms and targeted treatments of cobalt-encephalopathy from cobalt-chromium orthopedic implants. *Neuroscience.*

Zhao, L., Wang, Y., Zuo, Z., & He, M. (2020). Anti-neuroinflammatory effects of chromium in a rodent model of neurodegeneration. *Neuroscience Letters, 730*, Article 134971.

Zhitkovich, A. (2011). Chromium in drinking water: Sources, metabolism, and cancer risks. *Chemical Research in Toxicology, 24*(10), 1617–1629.

Zhou, H., Jiao, J., Huang, L., & Li, Y. (2018). Protective effects of chromium against oxidative stress in neurodegenerative disorders. *Brain Research Bulletin, 141*, 97–104.

Sustainable Mitigation Strategies and Solutions

Bioengineering Strategies for Chromium Remediation: Innovations and Environmental Applications

Abdullah Al Mamun, Md. Amdadul Huq, and Md.Mizanur Rahman

Abstract Chromium (Cr) contamination, especially in its hexavalent form (Cr (VI)), poses a significant threat to environmental and human health due to its high mobility, solubility, and toxicity. Traditional remediation methods are effective in some cases but often fall short in sustainability, cost-effectiveness, and ecological safety. Bioengineering, an integrated biological system with modern genetic and synthetic tools, provides a promising alternative for Cr detoxification. This chapter explores the sources and environmental behavior of Cr, focusing on its industrial origins and ecological impacts. It then delves into biological detoxification mechanisms such as microbial reduction, biosorption, and enzymatic transformation. Central to the discussion are bioengineering strategies involving genetically modified microbes, phytoremediation using transgenic plants, and synthetic biology platforms that enhance bioremediation efficiency. Real-world applications, challenges, and future prospects are discussed, emphasizing the integration of bioengineered systems in large-scale environmental clean-up operations. This chapter provides an accessible yet technical overview for researchers and practitioners working toward sustainable Cr remediation solutions.

Keywords Chromium toxicity · Bioremediation · Phytoremediation · Synthetic biology · Transgenic plant · Environmental biotechnology

---

A. Al Mamun · M. Rahman (✉)
Department of Biotechnology and Genetic Engineering, Faculty of Biological Science, Islamic University, Kushtia, Bangladesh
e-mail: mmrahmanbtg79@hotmail.com

A. Al Mamun
e-mail: aamamunbtge@gmail.com

Md. A. Huq
Department of Life Science, College of BioNano Technology, Gachon University, Seongnam-Si, Gyeonggi-Do, Republic of Korea
e-mail: amdadbge@gmail.com

N. Kumar (ed.), *Chromium Toxicity*, https://doi.org/10.1007/978-3-032-14323-5_7

1 Introduction

Chromium (Cr) is a naturally occurring transition metal that exists in several oxidation states, primarily as trivalent chromium (Cr (III)) and hexavalent chromium (Cr (VI)) (Ayele & Godeto, 2021; Singh et al. 2022a, 2022b, 2022c). While Cr (III) is an essential micro-nutrient involved in glucose metabolism and lipid synthesis, on the other hand, Cr (VI) is highly toxic, carcinogenic, and mutagenic to living organisms. Its high solubility in water, meaning it has the ability to penetrate cell membranes, and strong oxidative properties make it one of the most dangerous environmental pollutants (Saravanan et al., 2022; Xie et al., 2024). Widespread industrialization has significantly increased Cr discharge into the environment, particularly in water, air, and soil (Saxena, 2025). Sources like leather tanning, mining, electroplating, chromate production, and textile dyeing are major contributors to Cr pollution, especially in developing countries with inadequate waste management infrastructure (Islam et al., 2023; Singh et al. 2022a, 2022b, 2022c). The contamination of ecosystems by Cr has become a pressing global issue. Once introduced into the environment, Cr (VI) can persist for long periods due to its stability under oxidizing conditions (Sharma et al., 2022a, 2022b). It poses serious risk to microbial communities, plant systems, aquatic like, and finally human health trough food chain accumulation and water contamination (Li et al., 2021). Exposure to Cr (VI) has been linked to respiratory disorders, kidney and liver damage, genotoxic effects, infertility, and different types of cancer in humans (Shin et al., 2023). Traditional remediation methods such as chemical reduction, ion exchange, precipitation, adsorption, membrane filtration have been employed to treat Cr-contaminated environments (Saleh et al., 2022). However, these approaches often face challenges like high operational costs, low efficiency in low-concentration settings, secondary pollution, and energy intensive processes. These limitations have prompted the search for more eco-friendly, cost-effective, and sustainable alternatives.

Biological remediation, or bioremediation, has emerged as promising strategy that leverages the natural metabolic pathways of microorganisms and plants to detoxify or remove pollutants (Bala et al., 2022). In the context of Cr, various microbial species have demonstrated the ability to reduce toxic Cr (VI) to the less harmful Cr (III) form, immobilize Cr through biosorption and bioaccumulation, or transform it enzymatically (Zha et al., 2024). Plants also play a crucial role in phytoremediation by absorbing and sequestering Cr from soils and water (Gavrilescu et al., 2022). These biological systems often function under ambient conditions and generate minimal secondary waste, making them attractive for long-term environmental management. With the advancement of modern biotechnology, the potential of biological systems for Cr remediation has been significantly enhanced (Dubey et al., 2024). Genetic engineering has enabled the development of microbial strains and plants with superior Cr resistance and detoxification capacities. Genes encoding for chromate reductases, metal transporters, and stress-response regulators have been identified and overexpressed in model organisms to improve bioremediation efficiency (Ranjbar and Malcata 2022). Synthetic biology further expands the toolkit by allowing the construction of custom gene circuits, biosensors, and modular systems that can be tuned for optimal performance under specific environmental conditions (Gallup et al., 2021).

The integration of bioengineering with environmental remediation represents a paradigm shift from passive to active and programmable biological interventions. Engineered microbes can be tailored not only for Cr detoxification but also for sensing and reporting contamination levels (Liu et al., 2022). Transgenic plants can be optimized for deep-rooted absorption and rapid biomass accumulation. Moreover, the coupling of these biological systems with nanotechnology, bioinformatics, and systems biology opens new frontiers in precision bioremediation (Singla, 2021). However, genetically modified organisms (GMOs) used in open environments face regulatory, ecological, and public acceptance barriers. There is also a need to ensure the stability, safety, and functionality of engineered systems in diverse environmental contexts (Lea-Smith et al., 2025; Ngongolo & Mmbando, 2025). Hence, the development of Cr bioremediation strategies must be guided by rigorous risk assessments, field trials, and multidisciplinary collaboration (Finny, 2024).

This chapter provides a comprehensive overview of Cr contamination and its biological remediation, with a special focus on bioengineering approaches. Beginning by examining the sources and environmental behavior of Cr, followed by an in-depth analysis of natural detoxification mechanisms mediated by microbes and plants. The core of the chapter discusses bioengineering interventions, including genetically engineered microbes, transgenic plants, and synthetic biology tools for Cr remediation. Case studies, practical applications, and future directions are also addressed to underscore the potential of bioengineered systems in solving one of the most persistent environmental challenges of our time.

2 Sources and Behavior of Cr in the Environment

2.1 Industrial Source

Cr contamination in the environment largely stems from a variety of industrial processes (Fig. 1). Its widespread use across sectors has made it a significant pollutant, particularly in developing countries where regulatory enforcement may be limited (Saxena, 2025). Among the most prominent contributors is the leather tanning industry, which extensively uses trivalent Cr (III) salts, primarily Cr sulfate (Dubey et al., 2022). Although Cr (III) is relatively less toxic, improper waste disposal practices and oxidation under aerobic conditions can lead to the formation of hexavalent Cr (VI), which is far more toxic and mobile (Liang et al., 2021).

The electroplating industry also significantly contributes to environmental Cr, often utilizing Cr (VI) in plating solutions for its desirable properties such as corrosion resistance and shiny finish. Waste from these processes, if not properly managed, can seep into water bodies and soil systems (Siddiqua et al., 2022). Similarly, the textile

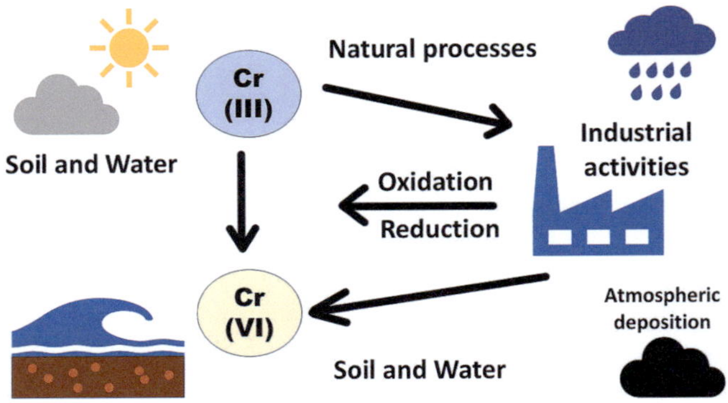

Fig. 1 Different causes of Cr contamination in the environment

and dyeing industries release Cr through the use of Cr-based dyes and mordants, further contaminating industrial effluents. Cr is also released during the mining of chromite ores and their subsequent smelting. The metallurgical processes generate Cr-laden dust, tailings, and slag, all of which can serve as long-term sources of pollution (Piatak and Ettler 2021). Additionally, Cr compounds are employed in wood preservation, notably in the form of chromated copper arsenate (CCA), and in the manufacture of pigments for paints and plastics. The combustion of fossil fuels and cement manufacturing contribute to airborne Cr particles, which eventually settle in terrestrial and aquatic ecosystems (Saxena, 2025).

In industrial zones, the cumulative discharge from these sources often results in elevated Cr levels in surface water, soil, and groundwater (Guo et al., 2024a, 2024b) (Table 1). This not only disrupts ecological balance but also poses severe health risks to local populations who rely on these resources for drinking water and agriculture. Without proper containment and remediation, Cr contamination becomes a persistent and expanding environmental threat.

2.2 Chromium Specification and Mobility

Cr exists in multiple oxidation states, but environmental processes predominantly involve two species: trivalent Cr (III) and hexavalent Cr (VI) (Liang et al., 2021). The behavior, mobility, and toxicity of Cr are strongly influenced by its chemical form and the surrounding physicochemical conditions (Hao et al., 2022). Cr (III) is generally regarded as less toxic and is relatively immobile in soil and water systems. It forms stable complexes with organic matter and often precipitates as Cr hydroxide under neutral to alkaline pH conditions (Liang et al., 2021; Tian et al., 2022).

In contrast, Cr (VI) is highly soluble, mobile, and toxic. It exists primarily as Chromate (CrO_4^{2-}), dichromate ($Cr_2O_7^{2-}$), or hydrogen chromate ($HCrO_4^{-}$), depending

Table 1 Major industrial sources of Cr contamination

Source type	Industrial activities	Cr form released	Environmental entry pathway	Reference
Metallurgical industry	Chrome plating, alloy manufacturing	Cr(VI), Cr(III)	Wastewater discharge, airborne particles	Itankar & Patil, 2022
Leather tanning	Use of basic Cr sulfate (BCS)	Cr(III)	Liquid effluents	Ramalingam et al., 2025
Dye and pigment industry	Manufacturing of Cr-based colors	Cr(VI)	Wastewater, solid waste	Nadarajah et al., 2023
Textile industry	Mordants in dyeing processes	Cr(VI)	Wastewater	Yuan et al., 2024
Wood preservation	Chromium-copper-arsenic (CCA) treatments	Cr(VI)	Leaching into soil and groundwater	Atiang' et al., 2025
Electroplating plants	Cr salts used for coating	Cr(VI)	Rinse water, air emissions	Bratovcic et al., 2022
Mining activities	Cr ore extraction and tailings	Cr(III)	Soil and water runoff	Das et al., 2021
Cement industry	Presence of Cr(VI) in clinker and dust	Cr(VI)	Airborne dust, slag leaching	Guo et al., 2024a, 2024b

on the pH and redox conditions (Carroll, 2022). These oxyanions are capable of penetrating biological membranes through sulfate transport channels, thereby gaining intracellular access where they induce cytotoxic and genotoxic effects (Sokolov et al., 2023). The environmental mobility and persistence of Cr depend on several key factors (Filote et al., 2021). pH plays a crucial role; acidic conditions favor the stability of Cr (VI), while alkaline environments promote the precipitation of Cr (III) (Sun et al., 2023). Redox potential is also significant, with reducing conditions (e.g., waterlogged soils, anaerobic sediments) facilitating the conversion of Cr (VI) to Cr (III). Conversely, oxidative environments can promote the reoxidation of Cr (III) back to its hexavalent form, especially in the presence of manganese oxides (Fu et al., 2021).

Organic matter, clay minerals, and microbial activity further influence Cr speciation. Organic substances can bind with Cr (III), limiting its bioavailability, or act as reducing agents for Cr (VI) (Liang et al., 2021). Microorganisms, both aerobic and anaerobic, play a central role in the redox cycling of Cr through enzymatic and non-enzymatic pathways. This dynamic interplay of chemical and biological processes determines the fate of Cr in natural ecosystems and informs the strategies needed for its remediation (Dubey et al., 2024).

2.3 Toxicokinetic and Eco-Toxicological Effects

The toxicological profile of Cr varies significantly between its two major oxidation states. Cr (VI), in particular, is highly toxic to both humans and the environment due to its strong oxidative potential and ability to generate reactive oxygen species (ROS) (Liang et al., 2021; Singh et al. 2022a, 2022b, 2022c). Upon exposure, Cr (VI) can be absorbed through the respiratory tract, gastrointestinal system, and to a lesser extent, the skin. Once inside the body, it readily enters cells via anion transport mechanisms and undergoes intracellular reduction to Cr (III), a process that generates ROS and induces oxidative damage to DNA, proteins, and cellular membranes (Hossini et al., 2022). Cr toxicity in humans is associated with a range of health issues, including respiratory diseases, skin ulcerations, kidney and liver dysfunction, and an increased risk of cancer, particularly lung cancer (Shin et al., 2023). Chronic occupational exposure, especially among workers in tanneries, electroplating facilities, and mining industries, has been linked to elevated Cr levels in biological tissues and long-term health complications (Dubey et al., 2022). In the environment, Cr exerts detrimental effects across various trophic levels. Soil microbial communities are particularly sensitive, as Cr disrupts microbial diversity, enzymatic activity, and nitrogen cycling processes (Li et al., 2021). In plants, Cr uptake leads to reduced seed germination, stunted root growth, chlorosis, and decreased photosynthetic efficiency. These effects are exacerbated under Cr (VI) exposure, which causes oxidative stress and damages cellular organelles (Saud et al., 2022).

Aquatic ecosystems are also at risk, with Cr accumulation in fish and invertebrates resulting in altered reproductive cycles, enzymatic imbalances, and developmental defects. Furthermore, Cr can bioaccumulate and bio-magnify through food webs, posing indirect risks to higher organisms, including humans. The extent of Cr toxicity is influenced by factors such as exposure concentration, duration, environmental conditions, and the biological sensitivity of the organisms involved (Yoshida, 2025). Understanding these toxicokinetic and ecotoxicological dimensions is essential for evaluating the risks posed by Cr pollution and for guiding the development of bioremediation strategies that are both safe and effective.

3 Biological Mechanisms of Chromium Detoxification

3.1 Microbial Reduction of Cr (VI) to Cr (III)

Microorganisms play a pivotal role in the natural attenuation of Cr through redox transformations, particularly the reduction of Cr (VI) to the less toxic form Cr (III) (Zha et al., 2024). This microbial reduction is an essential mechanism in Cr detoxification, largely mediated by both aerobic and anaerobic bacteria. Aerobic bacteria, such as *Pseudomonas*, *Bacillus*, and *Arthrobacter* species, reduce Cr (VI) using soluble reductases, often as a part of their defense mechanism against oxidative stress (Chen & Tian, 2021). In contrast, anaerobic bacteria, including certain strains of *Desulfovibrio* and *Shewanella*, utilize Cr (VI) as a terminal electron acceptor during respiration. The reduction process can occur both extracellularly and intracellularly. Extracellular reduction is typically facilitated by chromate reductase enzymes secreted by the microbes or bound to the cell surface (Chen & Tian, 2021). Intracellular reduction often leads to the accumulation of Cr (III) within the cell or its immediate vicinity (Table 2), potentially forming stable complexes with cellular ligands (Hao et al., 2022). The efficiency of microbial reduction is influenced by several environmental parameters, including pH, temperature, presence of electron donors (such as glucose or lactate), and the concentration of Cr (Zha et al., 2024).

In many cases, the microbial reduction of Cr (VI) is a detoxification response that protects cellular components from oxidative damage (Singh et al. 2022a, 2022b, 2022c). Recent studies have also explored the genetic basis of Cr resistance and reduction, identifying genes such as *chrA* and *chrR* that encode for chromate transporters and reductases, respectively (Ghosh et al., 2021). These findings have opened avenues for enhancing microbial detoxification capabilities through genetic engineering.

3.2 Biosorption, Bioaccumulation, and Bioprecipitation

In addition to redox transformation, microbes interact with Cr through mechanisms such as biosorption, bioaccumulation, and bioprecipitation (Dubey et al., 2024). Biosorption involves the passive binding of Cr ions onto the cell surface or extracellular polymeric substances (EPS), typically through interactions with functional groups such as hydroxyl, carboxyl, phosphate, and amino groups (Wang et al., 2021). This process does not require metabolic activity and is often rapid and reversible (Table 2).

Bioaccumulation, in contrast, involves the active transport of Cr into the microbial cell. Once internalized, Cr may be sequestered in vacuoles or bound to intracellular ligands such as metallothioneins, thereby reducing its bioavailability and toxicity (Al Mamun et al., 2024; Patil et al., 2024). This mechanism, however, may be limited by the toxicity threshold of Cr (VI) and the microbial species resistance mechanisms.

Bioprecipitation is another detoxification strategy, where Cr is converted into insoluble forms through microbial metabolic processes (Xia et al., 2021). For instance, the production of hydrogen sulfide (H_2S) by sulfate-reducing bacteria can lead to the precipitation of Cr as chromium-sulfide (Wilcox et al., 2024). Similarly, the alkalinization of the local environment by microbial activity can promote the formation of insoluble $Cr(OH)_3$. These processes reduce the mobility and bioavailability of Cr in contaminated environments. Collectively, these mechanisms contribute to the immobilization and detoxification of Cr, making microorganisms valuable agents in bioremediation strategies (Table 2). Their natural ability to interact with heavy metals, coupled with advances in microbiological and genetic tools, provides a robust foundation for the development of environmentally sustainable remediation technologies (Liang et al., 2021).

Table 2 Biological mechanisms for Cr detoxification

Mechanism	Description	Biological agents involved	Chromium form targeted	Notes	References
Microbial reduction	Reduction of toxic Cr(VI) to less toxic Cr(III) via enzymatic pathways	*Pseudomonas spp.*, *Shewanella spp.*	Cr(VI)	Anaerobic/aerobic; key in bioreactors and soil systems	Bakhti & Lashani, 2025
Biosorption	Binding of Cr ions to biomass cell walls via electrostatic interactions	*Bacillus spp.*, *Aspergillus spp.*	Cr(VI), Cr(III)	Fast and passive; suitable for wastewater	Ayele & Godeto, 2021
Bioaccumulation	Uptake and intracellular sequestration of chromium	*E. coli*, *Saccharomyces cerevisiae*	Cr(VI), Cr(III)	Depends on energy and metabolism	Ghosh et al., 2021
Bioprecipitation	Formation of insoluble chromium complexes through microbial activity	Sulfate-reducing bacteria	Cr(III)	Converts dissolved Cr into stable forms	Lin et al., 2023
Enzymatic detoxification	Chromium-specific enzymes catalyze reduction or chelation	Chromate reductase, glutathione reductase	Cr(VI)	Often encoded by plasmids or engineered genes	Adhikary et al., 2024

3.3 Enzymatic Pathways

The enzymatic detoxification of Cr is primarily mediated by chromate reductases, which catalyze the reduction of Cr (VI) to Cr (III) (Fig. 2). These enzymes can be categorized into two main types: soluble reductases, which function within the cytoplasm, and membrane-bound reductases, which are associated with the cell envelope (Chen & Tian, 2021). Soluble chromate reductases, such as those found in *Escherichia coli* and *Pseudomonas putida*, typically use NADH or NADPH as electron donors and exhibit high specificity for Cr (VI) (Tasleem et al., 2023).

The catalytic mechanism generally involves a two-electron transfer, leading to the direct reduction of Cr (VI) to Cr (III) (Cong & Zeng, 2021). However, some pathways may proceed through intermediate valence states such as Cr (V) and Cr (IV), which are highly reactive and can generate ROS. The formation of these intermediates poses a risk of oxidative stress, highlighting the importance of tightly regulated enzymatic systems (Bell et al., 2022). In addition to chromate reductases, other enzymes such as oxidoreductases, peroxidases, and superoxide dismutase also play supportive roles in mitigating the oxidative stress induced by Cr exposure. These enzymes help in scavenging ROS and maintaining cellular redox balance, thus enhancing microbial survival and detoxification efficiency (Shi et al., 2022). Recent advances in omics technologies have facilitated the identification and characterization of novel enzymes involved in Cr metabolism. Proteomic and transcriptomic analyses have revealed differential expression patterns of detoxification-related genes under Cr stress, providing insights into regulatory networks and potential targets for genetic enhancement (Al Mamun et al., 2024).

Understanding these enzymatic pathways is crucial for optimizing bioremediation strategies, particularly in designing genetically engineered microbes with enhanced Cr reduction capabilities. By harnessing the catalytic potential of microbial enzymes, it is possible to develop efficient, eco-friendly solutions to Cr pollution (Table 3).

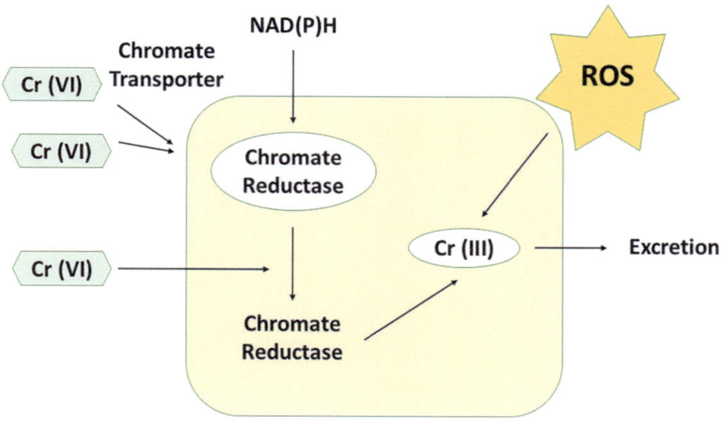

Fig. 2 The enzymatic detoxification of chromium

Table 3 Microorganisms and plants used for Cr remediation

Organism type	Species/strain	Mechanism of action	Chromium form targeted	Notable efficiency/remarks	References
Bacteria	*Pseudomonas putida*	Enzymatic reduction, biosorption	Cr (VI)	Reduces Cr (VI) to Cr (III); adaptable to pH	Tasleem et al., 2023
Bacteria	*Bacillus subtilis*	Bioaccumulation, surface binding	Cr (VI)	Tolerates high metal stress	Li et al., 2023
Fungi	*Aspergillus niger*	Biosorption via cell wall components	Cr (VI)	High binding affinity; low-cost biomass	Sivasamy et al., 2024
Algae	*Chlorella vulgaris*	Biosorption and metabolic transformation	Cr (VI)	Effective in wastewater; high growth rate	Tattibayeva et al., 2022
Plants	*Brassica juncea* (Indian mustard)	Phytoextraction	Cr (VI), Cr (III)	Fast-growing; high biomass; accumulates Cr	Zeremski et al., 2021
Plants	*Helianthus annuus* (sunflower)	Rhizofiltration and phytoaccumulation	Cr (VI)	Root exudates support microbial action	Ullah et al., 2023
Genetically modified microbes	*E. coli* expressing *chrR* gene	Enhanced enzymatic reduction	Cr (VI)	Engineered for improved Cr (VI) to Cr (III) conversion	Gupta et al., 2022
Transgenic plants	*Arabidopsis thaliana* with metal transporter genes	Enhanced uptake and detoxification	Cr (VI)	Useful for gene-function studies in phytoremediation	Sharma et al., 2022a, 2022b

4 Bioengineering Approaches

4.1 Genetically Engineered Microbes

Genetic engineering offers an innovative and powerful approach for enhancing the Cr detoxification capacity of microbial systems (Dubey et al., 2024) (Fig. 3). By manipulating specific genes involved in Cr resistance, transport, and reduction, scientists have been able to construct microbial strains with superior remediation efficiency (Xia et al., 2021). One of the most commonly engineered pathways involves the overexpression of chromate reductase genes such as *chrR* and *yieF* (Table 4), which encode enzymes capable of reducing Cr (VI) to Cr (III) with high specificity and turnover rates (Thatoi et al., 2014). Additionally, genes coding for efflux pumps like *chrA* can be introduced or upregulated to increase microbial tolerance by actively exporting intracellular Cr (VI) (Balta et al., 2025). Engineering strains to express metallothioneins or phytochelatins can further enhance intracellular sequestration and detoxification of Cr. Such engineered strains, including modified species of *E. coli*, *P. putida*, and *Bacillus subtilis*, have demonstrated improved survival and chromium removal efficiency in both laboratory and pilot-scale studies (Agrawal et al., 2024).

Moreover, synthetic biology tools allow for the design of regulatory circuits that activate detoxification genes in response to Cr exposure, thereby optimizing energy usage and improving adaptation to fluctuating environmental conditions (Mishra et al., 2024). CRISPR-Cas9 and other genome editing platforms have facilitated

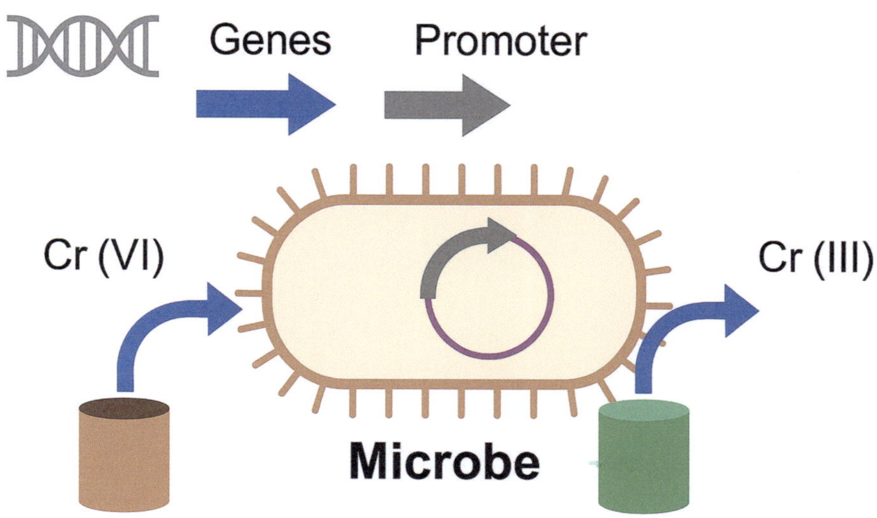

Fig. 3 Genetic engineering approach for enhancing the Cr detoxification capacity of microbial systems

Table 4 Summary of engineered constructs and genes used for Cr detoxification

Organism/host	Gene/construct	Function/role	Source/reference organism	Notes	References
Escherichia coli	*chrR*, *yieF*	Chromate reductase enzymes for Cr(VI) → Cr(III)	*Pseudomonas putida*, *E. coli*	Enhances intracellular reduction of Cr(VI); NAD(P)H-dependent	Anand & Pandey, 2021
Pseudomonas fluorescens	*arsC-chrA* operon	Metal resistance and chromate efflux	*Staphylococcus aureus*	Dual resistance to arsenic and chromium	Kalsoom et al., 2021
Saccharomyces cerevisiae	*GSH1*, *YCF1*	Glutathione synthesis and Cr sequestration in vacuoles	Native	Overexpression improves tolerance and sequestration	Dai et al., 2024
Arabidopsis thaliana	*AtPCS1*, *MT2*, *CAX2*	Phytochelatin synthase, metallothionein, cation transporter	Native or heterologous	Enhances root uptake and vacuolar sequestration	Sethi, 2023
Bacillus subtilis	Recombinant *chr* operon	Improved Cr(VI) resistance and biotransformation	*Ochrobactrum anthropi*	Useful in constructing microbial consortia for wastewater treatment	Sundarraj et al., 2023
Populus spp.	*GST*, *PCS*, *NCR1*	Glutathione metabolism and detoxification pathways	Transgenic constructs	Engineered trees for phytoremediation of Cr-contaminated soils	Vishwakarma et al., 2024

precise and multiplexed editing of microbial genomes, enabling the creation of robust bioremediation agents tailored to specific site conditions (Sharma et al., 2024).

4.2 *Phytoremediation and Transgenic Plants*

Phytoremediation harnesses the natural ability of plants to extract, accumulate, and detoxify heavy metals, including Cr, from contaminated soils and water (Yaashikaa et al., 2022). Several plant species, such as *Brassica juncea*, *Typha latifolia*, and *Pteris vittata*, have demonstrated significant potential in Cr uptake and tolerance (Table 3). These plants utilize mechanisms such as root uptake, chelation, vacuolar sequestration, and translocation to aerial parts for effective detoxification (Bhat et al., 2025). To further improve phytoremediation efficiency, transgenic plants have been developed by introducing genes that enhance metal tolerance and accumulation. For instance, the overexpression of genes encoding for metallothioneins, glutathione-S-transferases, and phytochelatin synthases has led to increased Cr sequestration and reduced oxidative damage (Yang et al., 2023). Genetic manipulation of transporters like ZIP and NRAMP families has also improved Cr uptake and compartmentalization (Reddy et al., 2022).

Transgenic approaches have been complemented by the use of plant growth-promoting rhizobacteria (PGPR) (Table 3), which facilitate Cr mobilization and uptake through the production of siderophores, organic acids, and enzymes (Mukherjee et al., 2024). These synergistic plant–microbe interactions form the basis of rhizoremediation, a promising strategy for in situ bioremediation of Cr-contaminated environments (Mishra et al., 2023).

4.3 *Synthetic Biology Tools*

Synthetic biology has revolutionized the field of environmental biotechnology by enabling the design and construction of standardized genetic modules, pathways, and entire microbial systems tailored for Cr remediation (Ezeako et al., 2024). By integrating tools such as genetic circuits, biosensors, and metabolic engineering, synthetic biology allows for the precise control of detoxification processes.

One of the key innovations in this field is the development of whole-cell biosensors that can detect and quantify Cr (VI) in environmental samples. These biosensors typically employ reporter genes, such as those encoding for green fluorescent protein (GFP) or luciferase, linked to Cr responsive promoters (Del Valle et al., 2021; Liu et al., 2022). Such systems enable real-time monitoring of contamination and assessment of bioremediation efficiency. Furthermore, metabolic engineering strategies have been employed to optimize co-factor availability, electron donor supply, and cellular redox balance, thereby enhancing the catalytic efficiency of chromate reductases (Srivastava, 2021). Modular genetic systems, such as the BioBrick standard, have facilitated the rapid prototyping and deployment of engineered strains (Radde et al., 2024).

Synthetic biology also supports the development of microbial consortia with complementary functions, such as Cr reduction, biofilm formation, and pollutant degradation (Chaudhary et al., 2023). These engineered consortia mimic natural ecological interactions and offer improved resilience and adaptability under field conditions. The integration of bioengineering approaches holds significant promise for addressing the complex challenge of Cr contamination (Xing et al., 2025). By combining the strengths of genetic engineering, plant biotechnology, and synthetic biology, it is possible to design robust, efficient, and scalable solutions for environmental remediation.

5 Conclusion

The environmental burden of chromium (Cr), particularly in its hexavalent form Cr (VI), poses a serious threat to both ecological systems and human health. Traditional remediation techniques, while effective to some extent, are often associated with high operational costs, energy requirements, and secondary pollution. In contrast, bioengineering approaches rooted in microbial and plant-based systems offer sustainable, cost-effective, and eco-friendly alternatives for Cr detoxification and remediation. The biological mechanisms of Cr transformation, ranging from microbial reduction, biosorption, bioaccumulation, to enzymatic detoxification, form the foundation of these bioremediation strategies. Recent advances in genetic engineering, synthetic biology, and transgenic technology have significantly enhanced the detoxification capacity of microbes and plants, enabling the development of tailored organisms for site-specific Cr remediation. Engineered biosensors and microbial consortia further add to the arsenal of tools available for real-time monitoring and adaptive remediation strategies Despite these advances, challenges remain in scaling up bioengineered systems for field applications, ensuring biosafety, and gaining public acceptance. Continuous research, supported by multidisciplinary collaboration, is essential to overcome these barriers. Future directions should emphasize the integration of omics technologies, environmental genomics, and data-driven modeling to optimize bioremediation processes. In summary, bioengineering represents a promising frontier in the quest for sustainable solutions to Cr toxicity. By leveraging the inherent capabilities of biological systems and enhancing them through modern biotechnological tools, we can pave the way for a cleaner, safer, and more resilient environment.

References

Adhikary, S., Saha, J., Dutta, P., & Pal, A. (2024). Bacterial homeostasis and tolerance to potentially toxic metals and metalloids through diverse transporters: Metal-specific insights. *Geomicrobiology Journal, 41*(5), 496–518.

Agrawal, K., Ruhil, T., Gupta, V. K., & Verma, P. (2024). Microbial assisted multifaceted amelioration processes of heavy-metal remediation: A clean perspective toward sustainable and greener future. *Critical Reviews in Biotechnology, 44*(3), 429–447.

Al Mamun, A., Rahman, M. M., Huq, M. A., Rahman, M. M., Rana, M. R., Rahman, S. T., & Alam, M. K. (2024). Phytoremediation: A transgenic perspective in omics era. *Transgenic Research, 33*(4), 175–194.

Anand, V., & Pandey, A. (2021). Bioremediation of Cr (VI)-Contaminated soil using bacteria. In *Removal of Refractory Pollutants from Wastewater Treatment Plants* (pp. 205–232). CRC Press.

Atiang', S., Ndunda, E. N., & Okello, V. A. (2025). Advances in removal of chromated copper arsenate elements in wood waste, contaminated water and soils. *Frontiers in Environmental Chemistry, 6*, 1452837.

Ayele, A., & Godeto, Y. G. (2021). Bioremediation of chromium by microorganisms and its mechanisms related to functional groups. *Journal of Chemistry, 2021*(1), 7694157.

Bakhti, A., & Lashani, E. (2025). Microbial Bioremediation for Industrial Waste Valorization. In *Biotechnological Applications in Industrial Waste Valorization* (pp. 437–479). Springer Nature Singapore, Singapore.

Bala, S., Garg, D., Thirumalesh, B. V., Sharma, M., Sridhar, K., Inbaraj, B. S., & Tripathi, M. (2022). Recent strategies for bioremediation of emerging pollutants: A review for a green and sustainable environment. *Toxics, 10*(8), 484.

Balta, I., Lemon, J., Gadaj, A., Cretescu, I., Stef, D., Pet, I., & Corcionivoschi, N. (2025). The interplay between antimicrobial resistance, heavy metal pollution, and the role of microplastics. *Frontiers in Microbiology, 16*, 1550587.

Bell, J., Ma, X., McDonald, T. J., Huang, C. H., & Sharma, V. K. (2022). Overlooked role of chromium (V) and chromium (IV) in chromium redox reactions of environmental importance. *Acs Es&t Water, 2*(6), 932–942.

Bhat, B. A., Rather, M. A. R. A., Pirzadah, T. B., Nazir, R., Mir, R. A., & Qadir, R. U. (2025). Plant hyperaccumulators: A state-of-the-art review on mechanism of heavy metal transport and sequestration. *Frontiers in Plant Science, 16,* 1631378.

Bratovcic, A., Buksek, H., Helix-Nielsen, C., & Petrinic, I. (2022). Concentrating hexavalent chromium electroplating wastewater for recovery and reuse by forward osmosis using underground brine as draw solution. *Chemical Engineering Journal, 431*, Article 133918.

Carroll, J. (2022). *Formation of Chromium (VI) in Surface Soils During Simulated Bushfire Temperatures* (Doctoral dissertation, Southern Cross University).

Chaudhary, P., Xu, M., Ahamad, L., Chaudhary, A., Kumar, G., Adeleke, B. S., & Abou Fayssal, S. (2023). Application of synthetic consortia for improvement of soil fertility, pollution remediation, and agricultural productivity: A review. *Agronomy, 13*(3), 643.

Chen, J., & Tian, Y. (2021). Hexavalent chromium reducing bacteria: Mechanism of reduction and characteristics. *Environmental Science and Pollution Research, 28*, 20981–20997.

Cong, X., & Zeng, X. (2021). Mechanistic diversity of low-valent chromium catalysis: Cross-coupling and hydrofunctionalization. *Accounts of Chemical Research, 54*(8), 2014–2026.

Dai, H., Lv, C., Huang, Z., Shen, Y., & Shi, P. (2024). The involvement of glutathione in cadmium detoxification of Saccharomyces cerevisiae. *Toxicological & Environmental Chemistry, 106*(1–10), 117–131.

Das, P. K., Das, B. P., & Dash, P. (2021). Chromite mining pollution, environmental impact, toxicity and phytoremediation: A review. *Environmental Chemistry Letters, 19*(2), 1369–1381.

Del Valle, I., Fulk, E. M., Kalvapalle, P., Silberg, J. J., Masiello, C. A., & Stadler, L. B. (2021). Translating new synthetic biology advances for biosensing into the earth and environmental sciences. *Frontiers in Microbiology, 11*, Article 618373.

Dubey, R., Verma, P., & Kumar, S. (2022). Cr (III) genotoxicity and oxidative stress: An occupational health risk for leather tannery workers of South Asian developing countries. *Toxicology and Industrial Health, 38*(2), 112–126.

Dubey, P., Farooqui, A., Patel, A., & Srivastava, P. K. (2024). Microbial innovations in chromium remediation: Mechanistic insights and diverse applications. *World Journal of Microbiology and Biotechnology, 40*(5), 151.

Ezeako, E. C., Itam, Y. B., Osuagwu, G. O., Ogbodo, C. R., Oyibo, O. N., Olokor, A. N., & Nwokafor, C. V. (2024). Prospects of synthetic biology in the actualization of green chemistry and environmental solutions.

Filote, C., Roşca, M., Hlihor, R. M., Cozma, P., Simion, I. M., Apostol, M., & Gavrilescu, M. (2021). Sustainable application of biosorption and bioaccumulation of persistent pollutants in wastewater treatment: Current practice. *Processes, 9*(10), 1696.

Finny, A. S. (2024). 3D bioprinting in bioremediation: A comprehensive review of principles, applications, and future directions. *PeerJ, 12*, Article e16897.

Fu, Y., Wang, L., Peng, W., Fan, Q., Li, Q., Dong, Y., & Wang, Z. (2021). Enabling simultaneous redox transformation of toxic chromium (VI) and arsenic (III) in aqueous media—a review. *Journal of Hazardous Materials, 417*, Article 126041.

Gallup, O., Ming, H., & Ellis, T. (2021). Ten future challenges for synthetic biology. *Engineering Biology, 5*(3), 51–59.

Gavrilescu, M. (2022). Enhancing phytoremediation of soils polluted with heavy metals. *Current Opinion in Biotechnology, 74*, 21–31.

Ghosh, S., Jasu, A., & Ray, R. R. (2021). Hexavalent chromium bioremediation with insight into molecular aspect: An overview. *Bioremediation Journal, 25*(3), 225–251.

Guo, H., Liu, C., Yan, S., Yin, J., & Shan, J. (2024). Source, distribution, and geochemical processes of geogenic high chromium groundwater around the world: A critical review. *Journal of Hydrology*, 131480.

Guo, X., Zeng, M., Yu, H., Lin, F., Li, J., Wang, W., & Chen, G. (2024). Critical review for the potential analysis of material utilization from inorganic industrial solid waste. *Journal of Cleaner Production*, 142457.

Gupta, N., Sundar, R. D. V., & Arunachalam, S. (2022). A review on the genetically engineered microbes for bioremediation of THMs namely Hg and Cr. *ECS Transactions, 107*(1), 11509.

Hao, Y., Ma, H., Wang, Q., Zhu, C., & He, A. (2022). Complexation behaviour and removal of organic-Cr (III) complexes from the environment: A review. *Ecotoxicology and Environmental Safety, 240*, Article 113676.

Hossini, H., Shafie, B., Niri, A. D., Nazari, M., Esfahlan, A. J., Ahmadpour, M., & Hoseinzadeh, E. (2022). A comprehensive review on human health effects of chromium: Insights on induced toxicity. *Environmental Science and Pollution Research, 29*(47), 70686–70705.

Islam, M. M., Mohana, A. A., Rahman, M. A., Rahman, M., Naidu, R., & Rahman, M. M. (2023). A comprehensive review of the current progress of chromium removal methods from aqueous solution. *Toxics, 11*(3), 252.

Itankar, N., & Patil, Y. (2022). Assessing physicochemical technologies for removing hexavalent chromium from contaminated waters—an overview and future research directions. *Water, Air, & Soil Pollution, 233*(9), 355.

Kalsoom, A., Batool, R., & Jamil, N. (2021). Highly Cr (vi)-tolerant Staphylococcus simulans assisting chromate evacuation from tannery effluent. *Green Processing and Synthesis, 10*(1), 295–308.

Lea-Smith, D. J., Hassard, F., Coulon, F., Partridge, N., Horsfall, L., Parker, K. D., & Krasnogor, N. (2025). Engineering biology applications for environmental solutions: Potential and challenges. *Nature Communications, 16*(1), 3538.

Li, D., Li, G., & Zhang, D. (2021). Field-scale studies on the change of soil microbial community structure and functions after stabilization at a chromium-contaminated site. *Journal of Hazardous Materials, 415*, Article 125727.

Li, T., He, Y., Wang, J., Xiang, H., Xu, X., Li, C., & Wu, Z. (2023). Bioreduction of hexavalent chromium via Bacillus subtilis SL-44 enhanced by humic acid: An effective strategy for detoxification and immobilization of chromium. *Science of the Total Environment, 888*, Article 164246.

Liang, J., Huang, X., Yan, J., Li, Y., Zhao, Z., Liu, Y., & Wei, Y. (2021). A review of the formation of Cr (VI) via Cr (III) oxidation in soils and groundwater. *Science of the Total Environment, 774*, Article 145762.

Lin, H., Zhou, M., Li, B., & Dong, Y. (2023). Mechanisms, application advances and future perspectives of microbial-induced heavy metal precipitation: A review. *International Biodeterioration & Biodegradation, 178*, Article 105544.

Liu, C., Yu, H., Zhang, B., Liu, S., Liu, C. G., Li, F., & Song, H. (2022). Engineering whole-cell microbial biosensors: Design principles and applications in monitoring and treatment of heavy metals and organic pollutants. *Biotechnology Advances, 60*, Article 108019.

Mishra, P., Mishra, J., & Arora, N. K. (2023). Biofortification revisited: Addressing the role of beneficial soil microbes for enhancing trace elements concentration in staple crops. *Microbiological Research, 275*, Article 127442.

Mishra, S., Patel, A., Bhatt, P., Chen, S., & Srivastava, P. K. (2024). Perspective evaluation of synthetic biology approaches for effective mitigation of heavy metal pollution. *Reviews of Environmental Contamination and Toxicology, 262*(1), 21.

Mukherjee, P., Dutta, J., Roy, M., Thakur, T. K., & Mitra, A. (2024). Plant growth-promoting rhizobacterial secondary metabolites in augmenting heavy metal (loid) phytoremediation: An integrated green in situ ecorestorative technology. *Environmental Science and Pollution Research, 31*(44), 55851–55894.

Nadarajah, R., Hossain, M. S., Siddique, M. B. M., Arafath, M. A., Naushad, M., Lim, J. W., & Ahmad, H. (2023). A review on environmental chemodynamics, isothermal, kinetics, and thermodynamics modeling for the adsorptive removal of Cr (VI) from the industrial effluent using magnetic nanoparticles as a bio-sorbent. *Environmental Science: Water Research & Technology, 9*(7), 1764–1782.

Ngongolo, K., & Mmbando, G. S. (2025). Necessities, environmental impact, and ecological sustainability of genetically modified (GM) crops. *Discover Agriculture, 3*(1), 29.

Patil, A., Chakraborty, S., Yadav, Y., Sharma, B., Singh, S., & Arya, M. (2024). Bioremediation strategies and mechanisms of bacteria for resistance against heavy metals: a review. *Bioremediation Journal*, 1–33.

Piatak, N. M., & Ettler, V. (Eds.). (2021). *Metallurgical slags: environmental geochemistry and resource potential.* Royal Society of Chemistry.

Radde, N., Mortensen, G. A., Bhat, D., Shah, S., Clements, J. J., Leonard, S. P., & Barrick, J. E. (2024). Measuring the burden of hundreds of BioBricks defines an evolutionary limit on constructability in synthetic biology. *Nature Communications, 15*(1), 6242.

Ramalingam, S., Das, I., Ganesan, S., Rajendiran, N., Rathinam, A., & Srinivasan, S. V. (2025). Chromium chemistry and chromium management in tanneries. In *Advanced Technologies in Wastewater Treatment* (pp. 93–123). Elsevier.

Ranjbar, S., & Malcata, F. X. (2022). Is genetic engineering a route to enhance microalgae-mediated bioremediation of heavy metal-containing effluents?. *Molecules, 27*(5), 1473.

Reddy, S. S., Kumar, P., & Dwivedi, P. (2022). Heavy metal transporters, phytoremediation potential, and biofortification. In *Plant metal and metalloid transporters* (pp. 387–405). Springer Nature Singapore, Singapore.

Saleh, T. A., Mustaqeem, M., & Khaled, M. (2022). Water treatment technologies in removing heavy metal ions from wastewater: A review. *Environmental Nanotechnology, Monitoring & Management, 17*, Article 100617.

Saravanan, A., Kumar, P. S., Hemavathy, R. V., Jeevanantham, S., Harikumar, P., Priyanka, G., & Devakirubai, D. R. A. (2022). A comprehensive review on sources, analysis and toxicity of environmental pollutants and its removal methods from water environment. *Science of the Total Environment, 812*, Article 152456.

Saud, S., Wang, D., Fahad, S., Javed, T., Jaremko, M., Abdelsalam, N. R., & Ghareeb, R. Y. (2022). The impact of chromium ion stress on plant growth, developmental physiology, and molecular regulation. *Frontiers in Plant Science, 13*, Article 994785.

Saxena, V. (2025). Water quality, air pollution, and climate change: investigating the environmental impacts of industrialization and urbanization. *Water, Air, & Soil Pollution, 236*(2), 1–40.

Sethi, S. (2023). Phytochelatins: heavy metal detoxifiers in plants. In *Advanced and innovative approaches of environmental biotechnology in industrial wastewater treatment* (pp. 361–379). Springer Nature Singapore, Singapore.

Sharma, P., Singh, S. P., Parakh, S. K., & Tong, Y. W. (2022b). Health hazards of hexavalent chromium (Cr (VI)) and its microbial reduction. *Bioengineered, 13*(3), 4923–4938.

Sharma, B., Chaudhary, T., & Shukla, P. (2022). Combinatorial genetic engineering approaches in phytoremediation of pollutants. In *Current developments in biotechnology and bioengineering* (pp. 55–71). Elsevier.

Sharma, V., Sehgal, R., Angra, V., & Gupta, R. (2024). Applications of genome editing in bioremediation. In *Genome Editing in Bacteria (Part 2)* (pp. 150–183). Bentham Science Publishers.

Shi, L., Zhao, X., Zhong, K., Jia, Q., Shen, Z., Zou, J., & Chen, Y. (2022). Physiological mechanism of the response to Cr (VI) in the aerobic denitrifying ectomycorrhizal fungus Pisolithus sp. 1. *Journal of Hazardous Materials, 429*, 128318.

Shin, D. Y., Lee, S. M., Jang, Y., Lee, J., Lee, C. M., Cho, E. M., & Seo, Y. R. (2023). Adverse human health effects of chromium by exposure route: A comprehensive review based on toxicogenomic approach. *International Journal of Molecular Sciences, 24*(4), 3410.

Siddiqua, A., Hahladakis, J. N., & Al-Attiya, W. A. K. (2022). An overview of the environmental pollution and health effects associated with waste landfilling and open dumping. *Environmental Science and Pollution Research, 29*(39), 58514–58536.

Singh, S., Naik, T. S. K., Chauhan, V., Shehata, N., Kaur, H., Dhanjal, D. S., & Ramamurthy, P. C. (2022a). Ecological effects, remediation, distribution, and sensing techniques of chromium. *Chemosphere, 307*, Article 135804.

Singh, V., Singh, N., Verma, M., Kamal, R., Tiwari, R., Sanjay Chivate, M., & Mishra, V. (2022b). Hexavalent-chromium-induced oxidative stress and the protective role of antioxidants against cellular toxicity. *Antioxidants, 11*(12), 2375.

Singla, P. (2021). Plant synthetic biology: a paradigm shift targeting stress mitigation, reduction. *Plant Stress Biology: Strategies and Trends*, 435.

Sivasamy, S., Karuppiah, P. S., Kamaraj, K., Mahendran, K., Chandran, A. D., & Prabhakaran, R. (2024). Bioremediation of hexavalent chromium using Beauveria bassiana SSR5: adsorption mechanisms, kinetics, and characterization. *Bioremediation Journal*, 1–21.

Sokolov, S., Zyrina, A., Akimov, S., Knorre, D., & Severin, F. (2023). Toxic effects of penetrating cations. *Membranes, 13*(10), 841.

Srivastava, N. (2021). Remediation of heavy metals through genetically engineered microorganism. *Environmental Pollution and Remediation*, 315–366.

Sun, S., Deng, T., Ao, M., Mo, Y., Li, J., Liu, T., & Tang, Y. (2023). Release of chromium from Cr (III)-and Ni (II)-substituted goethite in presence of organic acids: Role of pH in the formation of colloids and complexes. *Science of the Total Environment, 904*, Article 166979.

Sundarraj, S., Sudarmani, D. N. P., Samuel, P., & Sevarkodiyone, S. P. (2023). Bioremediation of hexavalent chromium by transformation of Escherichia coli DH5α with chromate reductase (ChrR) genes of Pseudomonas putida isolated from tannery effluent. *Journal of applied microbiology, 134*(1), lxac019.

Tasleem, M., El-Sayed, A. A. A., Hussein, W. M., & Alrehaily, A. (2023). Bioremediation of chromium-contaminated groundwater using chromate reductase from Pseudomonas putida: An in silico approach. *Water, 15*(1), 150.

Tattibayeva, Z., Tazhibayeva, S., Kujawski, W., Zayadan, B., & Musabekov, K. (2022). Peculiarities of adsorption of Cr (VI) ions on the surface of Chlorella vulgaris ZBS1 algae cells. *Heliyon, 8*(9).

Thatoi, H., Das, S., Mishra, J., Rath, B. P., & Das, N. (2014). Bacterial chromate reductase, a potential enzyme for bioremediation of hexavalent chromium: A review. *Journal of Environmental Management, 146*, 383–399.

Tian, Y., Yu, T., Shen, J., Zheng, G., Li, H., & Zhao, W. (2022). Cr release after Cr (III) and Cr (VI) enrichment from different layers of cast iron corrosion scales in drinking water distribution systems: the impact of pH, temperature, sulfate, and chloride. *Environmental Science and Pollution Research*, 1–15.

Ullah, S., Liu, Q., Wang, S., Jan, A. U., Sharif, H. M. A., Ditta, A., & Cheng, H. (2023). Sources, impacts, factors affecting Cr uptake in plants, and mechanisms behind phytoremediation of Cr-contaminated soils. *Science of the Total Environment, 899*, Article 165726.

Vishwakarma, D., Kannaujiya, D., Awasthi, S., & Shikha. (2024). Phytoremediation of Contaminated water, its mechanisms, and advancements. In *Phytoremediation: Biological Treatment of Environmental Pollution* (pp. 37–56). Springer Nature Switzerland, Cham.

Wang, Y., Liu, Y., Zheng, K., Xie, L., Ren, Q., Yang, Z., Liao, Q., Wang, Y., & Chen, R. (2021). The role of extracellular polymeric substances (EPS) in the reduction of Cr (VI) by Pannonibacter phragmitetus BB. *Journal of Environmental Chemical Engineering, 9*(5), 106163.

Wilcox, S. M., Mulligan, C. N., & Neculita, C. M. (2024). Bioprecipitation as a Bioremediation Strategy for Environmental Cleanup. In *Bioremediation for Sustainable Environmental Cleanup* (pp. 19–39). CRC Press.

Xia, X., Wu, S., Zhou, Z., & Wang, G. (2021). Microbial Cd (II) and Cr (VI) resistance mechanisms and application in bioremediation. *Journal of Hazardous Materials, 401*, Article 123685.

Xie, S. (2024). Water contamination due to hexavalent chromium and its health impacts: Exploring green technology for Cr (VI) remediation. *Green Chemistry Letters and Reviews, 17*(1), 2356614.

Xing, Y., Zheng, Y., & Wang, X. (2025). Integrated strategies for effective remediation of chromium-contaminated soils: Advancements, challenges, and sustainability implications. *Environmental Advances*, 100614.

Yaashikaa, P. R., Kumar, P. S., Jeevanantham, S., & Saravanan, R. (2022). A review on bioremediation approach for heavy metal detoxification and accumulation in plants. *Environmental Pollution, 301*, Article 119035.

Yang, X., Ren, J., Lin, X., Yang, Z., Deng, X., & Ke, Q. (2023). Melatonin alleviates chromium toxicity in maize by modulation of cell wall polysaccharides biosynthesis, glutathione metabolism, and antioxidant capacity. *International Journal of Molecular Sciences, 24*(4), 3816.

Yoshida, M. (2025). Microbial methylation of mercury in wetlands and its environmental risks. In *Heavy Metals in the Environment-Contamination, Risk, and Remediation*. IntechOpen.

Yuan, Y., Tian, Q., Hou, L., Rao, R., Yao, C., & Zhu, H. (2024). The self-boosting ultrafast removal of Cr (VI) and organic dye in textile wastewater through sulfite-induced redox processes. *Environmental Pollution, 355*, Article 124182.

Zeremski, T., Ranđelović, D., Jakovljević, K., Marjanović Jeromela, A., & Milić, S. (2021). Brassica species in phytoextractions: Real potentials and challenges. *Plants, 10*(11), 2340.

Zha, S., Yu, A., Wang, Z., Shi, Q., Cheng, X., Liu, C., & Zhou, L. (2024). Microbial strategies for effective hexavalent chromium removal: a comprehensive review. *Chemical Engineering Journal, 489*, Article 151457.

Role of Nanotechnology in Remediation of Chromium from the Environment

Nitish Dhingra and Khushi

Abstract Environmental contamination driven by accelerated industrialization, population expansion, and intensive agricultural activity poses serious ecological and human health challenges. Among these, hexavalent chromium (Cr(VI)) stands out due to its extreme toxicity, mobility, and persistence in natural systems. Conventional remediation approaches such as adsorption, chemical treatment, reverse osmosis, coagulation, and ion exchange are often laborious, resource-intensive, and sometimes ineffective across diverse media including soil, water, and air. In response, nanotechnology presents novel and promising alternatives in this direction. Engineered nanomaterials, with their high surface area, enhanced reactivity, and tunable properties, offer innovative and efficient strategies for Cr(VI) mitigation. This chapter delves into the principles, nanomaterial classifications, and recent technological breakthroughs using nano-scale interventions for chromium remediation from the environment.

Keywords Chromium · Environment · Nanotechnology · Remediation · Water

1 Introduction

The accelerated pace of industrialization and urban expansion over the past century has led to increased human exposure to heavy metals. These heavy metals are naturally occurring elements with atomic densities exceeding 4–5 g/cm^3. Metals such as chromium (Cr), lead (Pb), mercury (Hg), cadmium (Cd), and arsenic (As) are among the most prevalent and significant environmental contaminants (Balali-Mood et al., 2021). Despite their toxicity, these metals are widely utilized in various

N. Dhingra (✉)
Electron Microscopy and Nanoscience Laboratory, Directorate of Research, Punjab Agricultural University, Ludhiana, Punjab 141004, India
e-mail: nitishdhingra@pau.edu

Khushi
Department of Mathematics, Statistics and Physics, College of Basic Sciences and Humanities, Punjab Agricultural University, Ludhiana, Punjab 141004, India

N. Kumar (ed.), *Chromium Toxicity*, https://doi.org/10.1007/978-3-032-14323-5_8

sectors including battery manufacturing, semiconductor production, nuclear technology, and metallurgical and electroplating industries, owing to their mechanical strength, electrical conductivity, durability, and corrosion resistance (Dagdag et al., 2023). However, improper handling, disposal, and lack of stringent regulatory oversight have resulted in widespread environmental contamination. This poses serious risks to ecosystems and human health, necessitating the development of effective and sustainable remediation strategies. Among these metals, chromium, particularly in its hexavalent form [Cr(VI)], is of special concern due to its high toxicity, mobility, and persistence in the environment (Mohanty et al., 2023; Zhigalenok et al., 2025). Rapid and effective removal of chromium using novel techniques is extremely important for restoring water resources and safeguarding the environment (Cao et al., 2024).

Chromium is a distinctive heavy metal known for its toxicity to all forms of life. Due to its carcinogenic properties and environmental persistence, chromium removal is essential to safeguard plant ecosystems and aquatic biodiversity. Chromium is one of the most widely used metals, and for over 15 years, it has consistently ranked among the top 20 hazardous substances on the Superfund Priority List (Hassan et al., 2016). It enters the environment primarily through mining activities, industrial discharge, and certain geochemical processes (Zulfiqar et al., 2023). Despite its hazardous nature, chromium remains industrially indispensable due to the lack of viable substitutes. Chromium exists in multiple oxidation states, with hexavalent chromium [Cr(VI)] being the most toxic, mobile, and environmentally persistent form. Although complete elimination of Cr(VI) from the environment is challenging, its harmful effects can be mitigated through reduction to the less toxic trivalent form [Cr(III)] or immobilization via adsorption and containment strategies (Acharyya et al., 2023). Chromium(VI) compounds, like hydrogen chromate ($HCrO^{4-}$), chromate (CrO_4^{2-}), and dichromate ($Cr_2O_7^{2-}$) are highly soluble and diffusible which permit these molecules to cross the biological membrane tissues easily (Pakade et al., 2019).

A variety of physical, chemical, and biological techniques have been developed for the removal of chromium from the environment, each presenting distinct advantages and limitations. Commonly employed methods include coagulation–precipitation, adsorption using materials such as iron oxides and activated carbon, membrane filtration, and bioremediation strategies. Coagulation–precipitation remains a widely used and cost-effective approach for trivalent chromium [Cr(III)] removal. However, its application to hexavalent chromium [Cr(VI)] requires a two-step process involving reduction followed by precipitation, which results in the generation of substantial sludge volumes (Njoya et al., 2021). Traditional approaches used for the reduction of hexavalent chromium [Cr(VI)] to less toxic trivalent state [Cr(III)] employ sulfur-based compounds, such as sulfur dioxide gas or sodium bisulfite and iron salts like ferrous sulfate or sodium sulfite. These processes are highly effective under acidic conditions and often followed by the precipitation of trivalent chromium [Cr(III)] as chromium hydroxide. However, both methods have their own limitations. Sulfur-based reduction techniques generate large volumes of sludge which is difficult to manage, handle, and transport. The final disposal is also a costly affair. On the other hand, iron-based methods produce solid waste that require further treatment or disposal. When ferrous sulfate acts as the reducing agent, it results in the formation of ferric hydroxide (solid waste). The process requires extremely low pH levels to achieve efficient Cr(VI) reduction. When sodium sulfate acts as a reducing agent under acidic conditions, it generates sulfur dioxide. Sulfur dioxide gas contributes to air pollution due to its highly toxic, strong odor and pungent nature. Moreover, both approaches are less effective in treating low concentrations of Cr(VI), often requiring excessive chemical input to compensate. These traditional methods also pose risks of secondary pollution (Barrera-Díaz et al., 2012).

Adsorption techniques are effective for both Cr(III) and Cr(VI) removal, offering lower sludge production and operational simplicity. Adsorption also does not produce any additional harmful substances. Nevertheless, the disposal of spent adsorbents remains a critical concern (Fouda-Mbanga et al., 2024). Many adsorbents including synthetic polymers, activated carbons, biomass, and graphene oxides have been used for Cr(VI) remediation (Pakade et al., 2019). Adsorbents in their natural form as well as generated composite materials are used for the removal process. Recent studies have explored advanced adsorbents such as oxygen-doped porous carbon and biogenic iron nanoparticles, which demonstrate enhanced removal efficiencies and reusability (Beig et al., 2023; Prema et al., 2021). Membrane technologies, including reverse osmosis and nanofiltration, are considered among the most effective for chromium removal due to their high selectivity and broad contaminant removal spectrum. However, these systems are capital-intensive and prone to fouling, which diminishes long-term performance and increases maintenance costs (Irshad et al., 2023). Biological methods, including microbial reduction and biosorption, offer eco-friendly alternatives, particularly for Cr(VI) detoxification. Strains such as *Bacillus cereus* and sulfate-reducing bacteria have shown promising results in recent studies (Sun et al., 2021). In soil environments, chromium typically exists as insoluble chromium hydroxide or as chromate salts of calcium, iron, and aluminum, depending on pH and redox conditions (Bao et al., 2022). These forms influence its mobility and bioavailability, necessitating tailored remediation strategies for terrestrial systems.

To perform a systematic investigation of the research studies performed on the remediation of chromium from the environment, this chapter delves into the evolution, implementation, and future trajectory of nanotechnology-based approaches for mitigating chromium contamination in environmental media. In addition, it highlights technical and operational challenges that may hinder large-scale deployment, providing a critical overview of the practical considerations surrounding their real-world application.

2 Environmental Chemistry of Chromium

Discovered in the year 1797 by French chemist Nicolas-Louis Vauquelin, chromium is a lustrous, steel-gray metal known for its hardness and brittleness. Classified under Group VI-B of the transition metals, it has an atomic number of 24 and an estimated atomic weight of 51.9961 g/mol. Though found naturally in the Earth's crust, it exists only in trace concentrations, typically between 0.1 and 0.3 mg kg^{-1}. In the environment, chromium can exist in several oxidation states ranging from –II to + VI (Kholisa et al., 2021). Figure 1 summarizes general properties of chromium. Elemental Cr [Cr(0)] is chemically inert and does not occur naturally. The zero-valent state exists as a metallic solid, typically produced industrially. Due to its high melting point, it is commonly utilized in the fabrication of steel and various metal alloys. The monovalent [Cr(I)] and divalent [Cr(II)] forms are exceptionally rare, with Cr(II) being particularly unstable and readily oxidizing into Cr(III) compounds. Tetravalent and pentavalent oxidation states [Cr(IV) and Cr(V)] do not occur naturally. The trivalent [Cr(III)] and hexavalent [Cr(VI)] species are the most prevalent and environmentally stable forms of chromium in nature.

Trivalent chromium [Cr(III)], typically found in forms such as oxides, hydroxides, and sulfates, exhibits relatively low toxicity due to its limited solubility in water and reduced mobility. It predominantly associates with organic matter in soil and aquatic systems. Additionally, under common groundwater pH conditions, Cr(III) compounds tend to co-precipitate with iron as hydroxides. Hexavalent chromium [Cr(VI)], commonly present as chromate (CrO_4^{2-}), dichromate ($Cr_2O_7^{2-}$), and chromium trioxide (CrO_3), is regarded as the most toxic form of chromium due to its strong oxidizing nature, high solubility in water, and significant mobility across biological membranes and within environmental systems. Table 1 provides an overview of different chromium species and their natural occurrences.

Chromium is extensively found throughout geological formations, freshwater bodies, and marine environments (Sharma et al., 2023). A visual summary of its concentration and distribution across diverse environmental samples is presented in

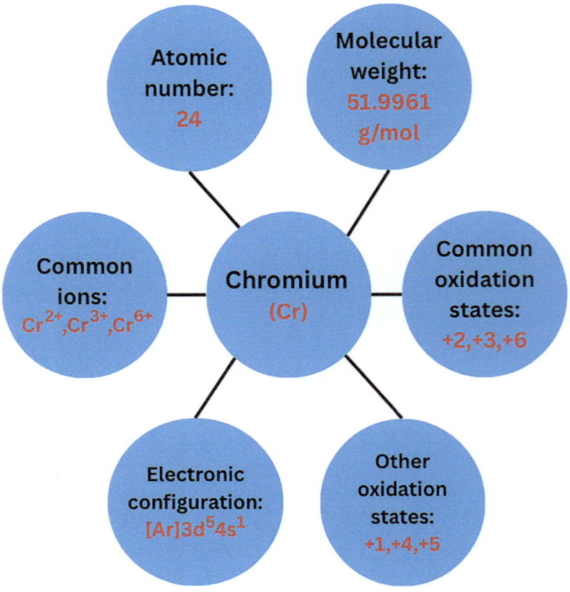

Fig. 1 General properties of chromium

Table 1 Different chemical species of chromium

Chemical species	Oxidation states	Features
Elemental Cr	Cr(0)	• Not found in nature, produced industrially
Monovalent Cr	Cr(I)	• Unstable
Divalent Cr	Cr(II)	• Relatively unstable • Easily oxidized to trivalent state
Trivalent Cr	Cr(III)	• Forms stable compounds • Commonly found in form of ores like ferrochromite ($FeCr_2O_4$)
Tetravalent Cr	Cr(IV)	• Not naturally occurring • Relatively rare • Forms compounds that are not very stable
Pentavalent Cr	Cr(V)	• Does not exist in nature • Forms only limited stable compounds containing Cr(V)
Hexavalent Cr	Cr(VI)	• Second most stable form of Cr • Exists as chromate, dichromate, and chromium trioxide

Source Khatua & Dey, 2023

Fig. 2. As outlined by Krishnamurthy and Wilkens (1994) and Pawlisz (1997), recommended concentration limits for chromium vary depending on the environmental context as:

- For the freshwater ecosystems, Cr(VI) should not exceed 1 $\mu g\ L^{-1}$, while Cr(III) should be below 8 $\mu g\ L^{-1}$.
- For marine environments, the permissible limits are 1 $\mu g\ L^{-1}$ for Cr(VI) and up to 50 $\mu g\ L^{-1}$ for Cr(III).
- For agricultural irrigation, Cr(VI) concentrations should be limited to 8 $\mu g\ L^{-1}$, with Cr(III) not exceeding 5 $\mu g\ L^{-1}$.
- For drinking water, the acceptable concentration for Cr(VI) is up to 50 $\mu g\ L^{-1}$.

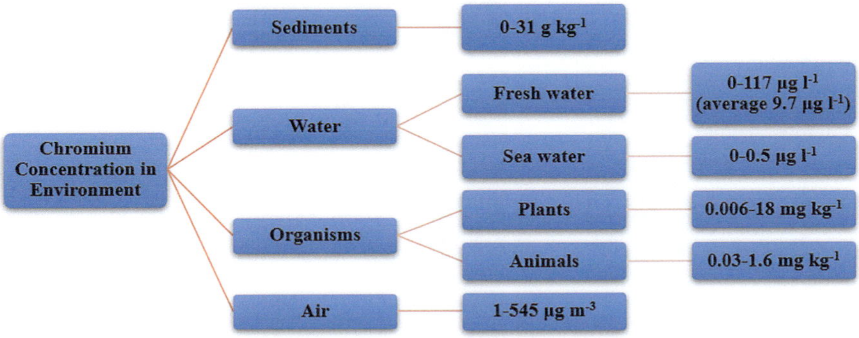

Fig. 2 Chromium concentrations in the environment

3 Sources of Chromium Contamination and Toxicity in Environment

Chromium contamination in the environment, largely driven by anthropogenic activities, poses serious threats to both public health and ecological integrity. Among various sources, chromite mining is particularly concerning due to its role toward environmental pollution and ecological degradation. Hexavalent chromium is far more hazardous than trivalent chromium, especially under prolonged exposure. To address this issue, mitigation efforts have focused on the reduction and removal of Cr(VI) from affected environments, with one critical approach involving its conversion to the less harmful Cr(III) form. Ferrous sulfate has demonstrated efficacy in this process, serving as a reducing agent that transforms hexavalent chromium into its non-carcinogenic trivalent state (Das et al. 2010).

Industrial activities such as dyeing, electroplating, paint manufacturing, and leather tanning are known to cause traces of Cr(VI) into wastewater streams (Singh & Kalamdhad, 2011). Effluents from textile industries, particularly those involving synthetic dyes, are also significant contributors to Cr(VI) contamination (Mitra et al., 2017). Figure 3 provides a schematic overview illustrating the role of various industrial processes in the release and dissemination of Cr(VI) pollutants into the environment (Kerur et al., 2021).

Cr(III) is a naturally occurring element. It is considered to be less harmful and is regarded as essential for humans and animals up to a certain amount. It helps in glucose metabolism and insulin functioning. But it is toxic at higher concentrations and has a greater tendency to form stable complexes. However, understanding the exact biological behavior of Cr(III) is challenging for scientists. In contrast to Cr(III), hexavalent chromium [Cr(VI)] exhibits substantially higher toxicity and poses a greater threat to both environmental and human health. Its high solubility and mobility enable it to contaminate water sources and infiltrate the food chain. Importantly, Cr(VI) and Cr(III) can interconvert under varying environmental conditions. Cr(VI) is typically formed through the oxidation of Cr(III), particularly during the extraction and refining of chromium-containing ores (chromite), and is also introduced into natural ecosystems via industrial activities. Toxic amounts of Cr^{6+} ions in humans can damage the skin and cause irritations, rashes, nasal and hearing impairment, lung cancer, and respiratory diseases.

3.1 Toxicity of Hexavalent Chromium in Human Beings and Animals

Cr(VI) is recognized as a potent carcinogen affecting a wide range of living organisms. The ill-effects of Cr(VI) exposure on human health are summarized in Table 2. Its toxicity stems from its ability to infiltrate biological systems via ingestion, inhalation, or dermal absorption. Once inside the body, Cr(VI) can trigger a range of

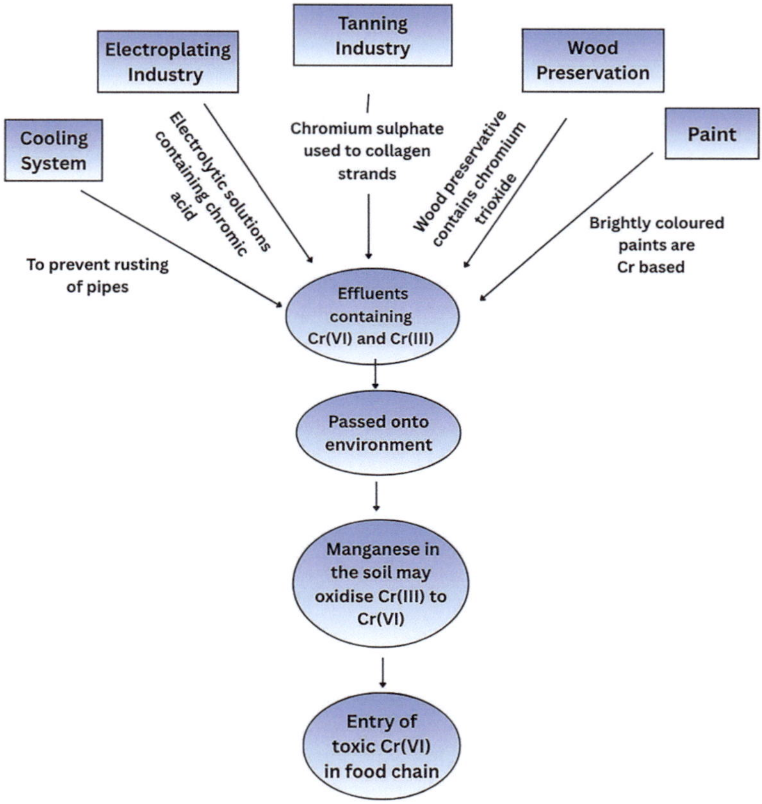

Fig. 3 Sources of Cr(VI) pollution in the environment

adverse health effects, including allergic reactions, skin rashes, nasal irritation and bleeding, immunosuppression, as well as renal and hepatic dysfunction. In severe cases, exposure can lead to fatal outcomes (Mishra & Bharagava, 2016).

3.1.1 Toxicity of Hexavalent Chromium in Water

Chromium and its chemical compounds can originate from both natural and human-made activities (Ramli et al., 2023). Most of the Cr(VI) comes from human activities like mining, agricultural practices, and industrial processes. The mining operation often discharges large amounts of untreated waste directly into the water bodies. The runoff from these activities carries a high level of toxic contaminants. It further contaminates rivers and streams by meeting nearby water bodies and causing a threat to aquatic life. Water contamination by Cr arising from natural processes include rock weathering and volcanic activity (Yunus et al., 2020). In nature, chromium usually exists in a harmless state (trivalent chromium), but certain environmental chemical

Table 2 Impact of hexavalent chromium on human health

Cr(VI) exposure medium for human body	Impacts of Cr(VI) on human health
Through air	• Lungs damage • Paranasal sinuses • Respiratory issues such as asthma, nasal irritation, and various breathing disorders • Skin-related problems including ulcers, allergic reactions and dermatitis marked by itching, burning, and redness • Cancer-related symptoms linked with mouth, breast, lung, and uterus
Through water	• May lead to cancer of stomach and bladder • Trigger gastrointestinal issues resulting in ulcers within digestive tract • Blood-related disorders • Neurological complications • DNA damage

reactions convert it into the toxic hexavalent state. The interplay between natural processes and anthropogenic activities has significantly contributed to the proliferation of harmful substances in the ecosystems. Industrial effluents often contain Cr(VI) along with additional co-pollutants such as polychlorinated biphenyls (PCBs), which commonly originate from manufacturing facilities and municipal sewage discharge. While PCBs exert minimal impact on the chemical speciation of Cr(VI), certain bacterial strains possess the remarkable ability to concurrently eliminate both pollutants. This dual-remediation potential renders these microbial species valuable candidates for the treatment of wastewater contaminated with Cr(VI) (Yasir et al., 2021).

3.1.2 Toxicity of Hexavalent Chromium in Soil

Elevated levels of Cr(VI) in soil, primarily due to chromite mining activities, pose a serious threat to plant productivity (Majhi & Samantaray, 2020). Plants adsorb Cr(VI) through their root systems, and those that thrive in chromium-rich soils (serpentine), often accumulate substantial concentrations of chromium in their foliage. These plants are referred to as Cr-accumulators. A notable example is *Leptospermum scoparium*, which can sequester up to 20,000 mg kg^{-1} of chromium in its leaves. Once absorbed, chromium is translocated throughout various plant tissues. The presence of Cr(VI) in edible plants facilitates its entry into the human body through dietary consumption. Exposure to heavy metals like chromium disrupts cellular integrity in plants, impairs physiological functions, and ultimately diminishes their overall growth and yield.

3.1.3 Toxicity of Hexavalent Chromium in Air

Chromium contamination in the atmosphere is primarily attributed to the emissions from industrial chimneys, chromite mining operations, coal-fired power plants, and various other anthropogenic sources. These activities frequently elevate airborne chromium levels beyond the permissible threshold of 0.1 ppm. Among these, mining-related dust emerges as a dominant contributor to the air pollution, typically comprising fine chromite particles and residual traces of toxic hexavalent chromium. Large volumes of particulate matter are released during the extraction, processing, and transportation stages of mining. Poorly maintained adjacent roadways, marked by deep indentations, exacerbate the issue by facilitating additional dust dispersal into nearby surroundings (Mohanty et al., 2023).

The cumulative effect of these airborne particulates significantly deteriorates air quality, posing serious health hazards through inhalation exposure. Moreover, airborne dust settles on surrounding vegetation, impeding light penetration and thereby disrupting photosynthesis. This not only diminishes plant vitality but also adversely affects the broader ecological balance of local flora and fauna.

4 Convention Cr Remediation Techniques and Their Limitations

A range of remediation approaches such as chemical precipitation, reverse osmosis, coagulation, ion exchange, and adsorption have been effectively employed to eliminate chromium ions from diverse environmental matrices, including water, soil, and atmospheric particulates (Fig. 4).

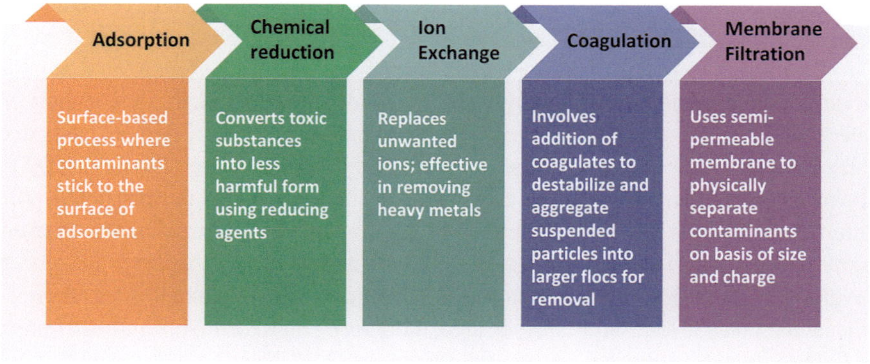

Fig. 4 Conventional Cr remediation techniques

4.1 Adsorption

Adsorption relies on a mass transfer process between solid and liquid phases, utilizing materials known as sorbents (Rafique et al., 2022). Solutes adhere to the surface of these sorbents (which may be solid or liquid) through a sequential mechanism involving initial diffusion of contaminants to the sorbent's surface, attachment of solutes onto the external surface, and eventually the further diffusion into the internal porous structure (Rafique et al., 2022). Among remediation techniques, adsorption stands out due to its operational simplicity, low cost, scalability, and the potential for regeneration and reuse (Pakade et al., 2019). Sorbents are often employed in their native state, but recent advancements include modifications such as chemical grafting, cross-linking, and the development of composite materials to improve performance.

Carbon-based adsorbents including activated carbon (AC), carbon microspheres (CMS), carbon nanotubes (CNTs), graphene, and graphene oxide have gained prominence for their efficiency in chromium remediation (Liu et al., 2023). AC and CMS are particularly well-studied, owing to their large surface areas and favorable physicochemical properties, which enhance their affinity for Cr(VI) ions (Assefa et al., 2024). The microporous architecture of activated carbon facilitates physical adsorption via weak Van der Waals forces. However, this alone often falls short in achieving high removal efficiency. To address this, AC is frequently functionalized with specific chemical groups that increase the number and variety of active sites for adsorption. These functional groups interact effectively with both positively charged metal ions and negatively charged Cr(VI) species (Boustila et al., 2022). In general, sorbents exhibiting a positive surface charge tend to demonstrate superior capacity for Cr(VI) removal due to enhanced electrostatic interactions (Liu et al., 2023).

In recent years, biosorption technique has gained significant momentum as a sustainable and environmentally friendly approach for eliminating toxic metals. This process involves interactions between heavy metal ions and the ligands or functional groups present in biomaterials, leading to the formation of stable complexes (Peng & Guo, 2020). Compared to conventional techniques, biosorption offers several notable advantages, including cost-effectiveness, high adsorption capacity, and enhanced selectivity. Among the widely adopted biosorbents are biomass-derived activated carbons, which have shown promise in metal removal applications (Liu et al., 2023). However, traditional ACs often suffer from limited pore size and restricted pore volume, which can slow down the transport and adsorption of contaminants. To overcome this limitation, mesoporous materials such as carbon microspheres have been developed. These advanced sorbents feature larger pores, expanded internal channels, and increased accessible surface area, facilitating rapid diffusion and adsorption of Cr(VI) ions. Mesoporous carbon microspheres (MPCMS), in particular, demonstrate impressive performance, with adsorption capacities reaching up to 165.3 mg/g of Cr(VI) (Liu et al., 2023; Zhang et al., 2019), highlighting their potential for efficient environmental remediation.

4.2 Chemical Reduction

The chemical reduction of chromium involves two primary stages. Initially, highly toxic hexavalent chromium [Cr(VI)] is converted into its less toxic trivalent form [Cr(III)] using suitable reducing agents under controlled pH conditions. Subsequently, Cr(III) precipitates as an insoluble hydroxide when the medium becomes alkaline.

One of the key strengths of this method lies in its versatility due to which it can be effectively integrated with other remediation strategies such as chemical leaching, adsorption, or biological treatment. The simplicity and high efficiency of chemical reduction make it a widely adopted technique. Nonetheless, a notable drawback is the generation of hazardous solid sludge during the precipitation of chromium compounds (Assefa et al., 2024), which poses challenges for waste management and environmental safety.

4.3 Ion Exchange Method

Ion exchange is a reversible process involving the transfer of ions between a liquid medium and a solid-phase sorbent. The solid materials, referred to as ion exchangers, maintain their structural integrity throughout multiple cycles of operation. These exchangers are typically categorized as either cationic or anionic, based on the specific type of ions they are designed to exchange (Homem & Santos, 2011). This technique is widely recognized for its efficiency in replacing liquid-phase cations and anions with corresponding ions on the solid sorbent. An effective ion exchanger must exhibit several desirable characteristics, including high hydrophilicity, robust chemical and physical stability, rapid exchange kinetics, sufficient ion exchange capacity, and optimal particle size distribution (Hariganesh et al., 2020).

Both adsorption and ion exchange are considered simple, cost-effective, and site-specific approaches for contaminant removal. Notably, ion exchange demonstrates great efficacy in eliminating charged natural organic matter (NOM) from aqueous solutions (Homem & Santos, 2011). However, its limitations include the generation of secondary waste (sludge) and the eventual need to replace saturated sorbents. Additionally, factors such as the solution's pH and the presence of competing ions can significantly influence the exchange efficiency.

Despite these drawbacks, ion exchange offers several advantages over traditional chemical precipitation methods. These include the recovery of valuable metals, improved selectivity, reduced sludge generation, and compliance with stringent discharge standards (Assefa et al., 2024; Esfandiar et al., 2022).

4.4 Coagulation

Coagulation and flocculation are widely employed techniques for the removal of suspended solids from wastewater effluents. The efficiency of these processes largely depends on the physical characteristics of the particles, such as their size, shape, and density. Suspended particles in water often carry similar surface charges, which cause mutual repulsion and hinder aggregation. As a result, the particles stay dispersed in water. Coagulation and flocculation counteract this by neutralizing surface charges, promoting inter-particle collisions, and facilitating the formation of larger aggregates capable of settling out of the water. Coagulants play a pivotal role by neutralizing the negative charges on fine, non-settleable particles, initiating the formation of tiny clusters called microflocs (Kerur et al., 2021).

4.5 Membrane Filtration

Advanced membrane filtration techniques, such as nanofiltration (NF) and reverse osmosis (RO), are highly effective in isolating and removing hexavalent chromium [Cr(VI)] from water. These methods are frequently employed as pretreatment steps for eliminating suspended solids within the size range of 100 to 1000 nm (Kerur et al., 2021). During the operation, water is pressurized and directed through a semi-permeable membrane that selectively permits the passage of water molecules while retaining ions, particulates, and larger contaminants including Cr(VI).

Reverse osmosis excels in removing a broad spectrum of pollutants, whereas nanofiltration is particularly adopted for targeting divalent and polyvalent ions. For optimal performance, the selected membrane must possess high wet strength and the capacity to handle large volumes of liquid. Despite their effectiveness, both NF and RO systems present operational challenges, notably high energy consumption and the risk of membrane fouling, which must be carefully addressed to maintain efficient operation and effectiveness.

5 Nanotechnology-Based Cr Remediation

Nanotechnology is a science that operates at the nano-scale (1 to 100 nm), enabling the manipulation and restructuring of matter at atomic and molecular dimensions. The rising interest in nanoremediation stems from the exceptional characteristics of nanoparticles (NPs), such as their high surface-area-to-mass ratio, enhanced chemical reactivity, and catalytic efficiency. Due to their minuscule size, NPs can penetrate confined spaces, making them highly effective for cleaning polluted soils and water bodies. Nanomaterial-based membranes have also been integrated into nanofiltration processes for the targeted removal of chromium particles and other contaminants from environmental systems. Frequently employed nanomaterials include metal oxides, carbon-derived structures such as graphene and carbon nanotubes, and polymeric nanoparticles, all known for their capability to adsorb, degrade, or neutralize heavy metals and microbial pollutants, facilitating the provision of clean and safe water.

Nanoparticles can be synthesized via a variety of techniques, broadly categorized into physical, chemical, and biological methods. Among chemical approaches, the sol–gel method is prominent, encompassing five distinct stages: hydrolysis, polycondensation, aging, drying, and thermal decomposition (Danks et al., 2016). This method offers precise control over nanoparticle size, porosity, and composition, though it demands elevated temperatures, extended processing time, and the use of organic solvents (Niu et al., 2018). Alternatively, physical synthesis methods avoid solvent contamination and promote uniform nanoparticle dispersion, offering advantages in terms of environmental safety and processing efficiency (Dhanapal et al., 2024).

While conventional remediation techniques remain prevalent, nanoremediation presents a more cost-effective and efficient strategy for environmental detoxification, positioning nanotechnology as a transformative tool for future sustainability (Dhingra, 2024). Over the past few years, research on application of nanomaterials for environmental remediation has witnessed substantial growth. As clean water is vital for sustaining life, the contamination of groundwater, particularly by pollutants like chromium, has become a global concern. In response, nanomaterial-based solutions have gained traction due to the distinctive properties of nanoparticles, including their high surface area, enhanced reactivity, and tunable physical characteristics.

Historically, the transition of human beings toward foraging and agriculture marked the beginning of widespread environmental disruption. This transformation led to the extinction of certain plant and animal species, along with escalating issues such as soil erosion and pollution. The subsequent rise of urbanization and industrialization further intensified soil degradation (Kumar et al., 2021). In light of these challenges, scientists have turned to nanomaterials as a promising tool for soil remediation. Owing to their unique physicochemical properties such as customizable size, shape, and porosity, NPs offer improved efficiency and specificity in pollutant removal. Notably, in-situ remediation techniques, where nanoparticles are introduced directly into the soil, enable large-scale decontamination in a cost-effective and time-efficient manner.

Nanotechnology offers an effective solution for addressing environmental pollution, owing to its distinctive advantages in remediation applications (Prado-Audelo et al., 2021). The unique physicochemical attributes of NPs, including their ability to target pollutants even at trace concentrations, enable precise and efficient cleanup across expansive areas. This capability often results in faster and more effective remediation compared to traditional techniques (Alsammarraie et al., 2018). A key benefit of nanomaterials lies in their ability to be directly deployed to contaminated environments such as soil and groundwater. This direct application minimizes ecological disruption and avoids the generation of harmful byproducts, while significantly reducing the use of large volumes of chemicals, thereby mitigating the risk of secondary pollution. Moreover, nanotechnology harmonizes seamlessly with green innovations, including solar-powered photocatalysis and plant-based synthesis approaches. These integrations contribute to sustainable, eco-conscious practices, positioning nanotechnology as a contemporary and environmentally responsible solution for restoring degraded ecosystems.

The subsequent section highlights a selection of widely studied and commonly used nanomaterials that are employed in the removal of hazardous environmental pollutants, with a particular focus on chromium.

5.1 Metal-Based Nanomaterials

Metal oxide nanoparticles such as iron oxide (Fe_2O_3/Fe_3O_4) and zinc oxide (ZnO) are widely utilized in water purification technologies. Among them, nano zero-valent iron (nZVI) has emerged as a particularly effective material for mitigating Cr(VI) contamination from the environment (Zhou et al., 2018). Due to its strong electron-donating capacity and negative redox potential, nZVI has become a staple in small-scale environmental remediation efforts (Cheng et al., 2021). To enhance their adsorption capabilities, these nanoparticles are frequently incorporated with substrates like activated carbon, silica, and alumina, forming composite or core–shell nanostructures. The nZVI core, composed of elemental iron in its zero oxidation state, is encased within a shell of mixed oxides of iron. This core–shell configuration facilitates both the adsorption and reduction of chromium through surface complexation and electrostatic interactions. In this process, Cr(VI) ions are effectively captured and converted to the less toxic trivalent form [Cr(III)].

Despite its high efficacy, nZVI requires cautious handling due to its intrinsic reactivity and potential toxicity (Baby et al., 2022; Inobeme et al., 2022). Nonetheless, studies have consistently reported that chromium removal efficiencies can exceed 90% under optimized conditions, particularly when slightly acidic pH levels, adequate contact time, and appropriate nanoparticle dosages are maintained (Arora et al., 2024).

5.2 Carbon-Based Nanomaterials

Porous carbon-based materials such as activated carbon and carbon nanotubes (CNTs), including both multi-walled (MWCNTs) and single-walled (SWCNTs) forms, are considered highly effective for the removal of heavy metals like chromium from aqueous environments. Their high specific surface area, excellent adsorption capacity, and chemical modifiability make them particularly suitable for water treatment applications. Studies have shown that MWCNTs outperform SWCNTs in removing Cr(VI), owing to their structural advantages and the presence of multiple active adsorption sites (Mpouras et al., 2021).

Typically, MWCNTs exhibit lengths ranging from 100 to 1000 nm and diameters between 1 and 3 nm (Inobeme et al., 2022). However, while an increased surface area enhances reactivity, it can also promote flocculation, thereby diminishing their overall nanoremediation efficiency. To counteract this, surfactants such as poloxamer 407 have proven effective in maintaining the stability of MWCNTs in water-based systems (Matos et al., 2017). Moreover, functionalizing CNT surfaces with chemical groups like hydroxyl, amino, and carboxylic acids significantly boosts their chromium adsorption performance. Beyond removing Cr(VI) from contaminated water and soils, CNTs have also been shown to enhance microbial activity and promote plant growth, thereby contributing to the broader success of environmental remediation strategies (Shan et al., 2015).

5.3 Polymer-Based Nanomaterials

Polymer-based nanotechnologies encompass nanoparticles, nanocomposites, and nanofiltration (NF) membranes, all of which have demonstrated efficacy in environmental remediation. Polymeric nanomembranes are particularly effective at removing contaminants such as chromium by leveraging their fine pore structures to trap and immobilize chromium particles (Abdelbasir & Shalan, 2019; Bassyouni et al., 2019). These polymeric materials offer key advantages including excellent mechanical strength, biodegradability, reusability, and the presence of functional groups that facilitate pollutant binding (Inobeme et al., 2022). Among various polymer-based systems, both organic and biopolymer-derived nanocomposites have shown considerable promise (Renugopal et al., 2019).

Common biopolymers employed in chromium remediation include cellulose, chitosan, and alginate:

- **Cellulose** contains hydroxyl groups and glucose units that provide abundant binding sites for chromium ions, making cellulose-based nanocomposites efficient adsorbents.
- **Chitosan**, a biodegradable and eco-friendly natural polymer, is frequently used in membrane fabrication. Its structure, enriched with hydroxyl and amino groups, offers strong affinity for Cr(VI) ions.
- **Alginate** is known for its biocompatibility, non-toxicity, and biodegradability, further reinforcing its role in sustainable remediation (da Silva Alves et al., 2021).

A notable study by Alzahrani et al. (2021) involved the development of a magnetic polymer nanocomposite by modifying magnetic material with polypyrrole. This engineered composite was tested for its ability to remove hexavalent chromium from contaminated water and was thoroughly characterized using techniques like scanning electron microscopy (SEM), X-ray diffraction (XRD), Fourier transform infrared spectroscopy (FTIR), and thermogravimetric analysis (TGA) to assess its structural and functional attributes. The investigation revealed that chromium adsorption occurred in a monolayer configuration, conforming to the Langmuir isotherm model, with a maximum Cr(VI) adsorption capacity of 298 mg/g. These findings underscored the high efficiency and potential of polymer-based nanocomposites in environmental detoxification.

5.4 Graphene-Based Nanomaterials

Graphene is a two-dimensional nanomaterial composed of carbon atoms. It has garnered significant attention in environmental applications due to its remarkable mechanical strength, flexibility, electrical conductivity, and thermal stability. Among its derivatives, graphene oxide (GO) stands out for its enhanced ability to adsorb heavy metal ions, attributed to its abundant oxygen-containing functional groups namely hydroxyl, carboxyl, and carbonyl, which serve as active binding sites for metal ion adsorption (Khan et al., 2017).

Experimental findings indicate that graphene oxide demonstrates considerable adsorption capacity across various heavy metals. However, its efficiency in removing chromium was comparatively lower, with the following observed adsorption ranking: Ni > Zn > Pb > Cd > Cr (Inobeme et al., 2022). Despite this, graphene oxide remains a viable and effective material for chromium removal, showcasing potential for selective adsorption and environmental cleanup.

Din et al. (2023) synthesized GO nanoparticles using a modified Hummer's method and subsequently evaluated their potential for Cr(VI) removal from aqueous solutions. The results demonstrated that GO nanoparticles are highly effective adsorbents, achieving a remarkable maximum removal efficiency of 92% under acidic conditions within a short contact time. The study revealed that these nanoparticles can be applied as an efficient adsorbent for heavy metals adsorption from water, especially for Cr(VI) removal. These findings underscore the suitability of GO nanoparticles as a promising and efficient material for heavy metal adsorption in water treatment applications, particularly targeting, Cr(VI).

5.5 Silica-Based Nanomaterials

Silica-based nanomaterials represent another important class of adsorbents recognized for their effectiveness in chromium remediation, attributed to their high surface area, tunable pore structure, and biocompatibility (Shukla et al., 2020). The surfaces of these nanoparticles can be chemically functionalized with groups such as –SH and $-NH_2$ to form nanocomposites that improve their metal ion adsorption capabilities. Despite their promising attributes, silica nanoparticles present certain challenges for chromium removal. In aqueous environments, they are prone to agglomeration due to interparticle surface interactions, which compromises their adsorption efficiency. Furthermore, some reaction byproducts generated during their application may lack biodegradability, introducing the potential for secondary contamination that could adversely affect human health and ecological systems. Consequently, thorough cytotoxicity evaluations are essential before implementing silica-based nanomaterials in large-scale chromium remediation initiatives (Nayl et al., 2022).

5.6 *Bionanomaterials*

The environmentally sustainable approaches are being increasingly integrated into the nanotechnology sector with the aim of minimizing resource and energy consumption while fostering economic advancement (Rajakumar et al., 2023). Among these innovations, green nanotechnology has emerged as a versatile and transformative field, seamlessly integrating into daily life through a wide array of practical applications that are inspiring new developments and reshaping modern human activities (Sarkar et al., 2023). Green synthesis techniques, particularly plant-extract-based nanoparticle synthesis, have gained much attention due to their advantages over microorganism-based methods which involve macromolecules or other components that synthesize the nanoparticles (Samrot et al., 2020). Plant-derived nanoparticles have demonstrated high efficiency in the sustainable removal of pollutants such as chromium, presenting a safer and cleaner alternative to conventional remediation methods.

A major contributor to the long-term viability of nanotechnology is the regenerative and reusable nature of nanoparticles, which enhances both cost-effectiveness and environmental compatibility (Irshad et al., 2023). Bio-nanoadsorbents sourced from agricultural waste, fungi, algae, and microorganisms are particularly valuable due to their biocompatibility, biodegradability, and low environmental persistence. These features make them highly suitable for biomedical applications (Jeevanandam et al., 2022). Furthermore, microbial systems such as bacteria, fungi, and algae continue to play an integral role in eco-friendly remediation strategies (Znad et al., 2022).

6 Factors Affecting the Efficiency of Nanomaterials for Cr(VI) Removal

The efficacy of nanomaterials in removing chromium is significantly affected by a range of environmental parameters that govern their sorption behavior. Sorption, which includes both absorption and adsorption mechanisms, is fundamental to how

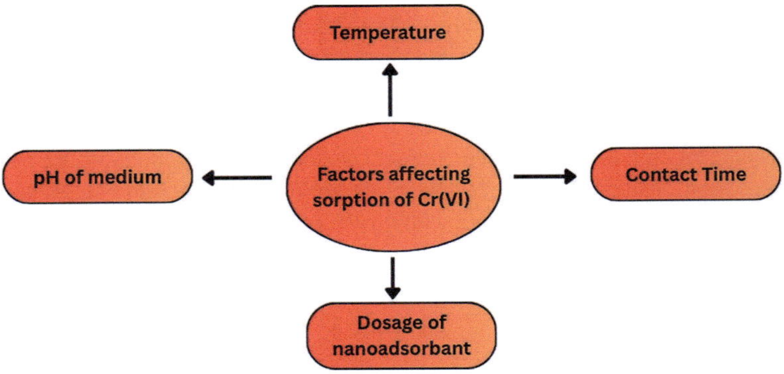

Fig. 5 Key factors influencing Cr(VI) sorption process

chromium species interact with the surface and internal architecture of the nanomaterial. Figure 5 illustrates key factors that modulate the overall sorption dynamics of the nanomaterials.

6.1 pH of the Medium

The pH of the surrounding medium plays a pivotal role in modulating the effectiveness of nanomaterials in chromium remediation. This is primarily because pH not only affects the speciation of chromium ions but also alters the surface properties of the nanomaterial itself. The adsorption capability of nanomaterials is highly sensitive to pH fluctuations, as different pH conditions can shift the surface charge of the material either toward a positive or negative state, thus influencing its affinity for metal ions. A study conducted by Wu et al. (2017) demonstrated that increasing the pH from 4 to 6 significantly enhanced the removal efficiency of chromium ions by iron sulfide nanoparticles, improving performance from 92.51 to 99.65%.

6.2 Temperature

Temperature serves as another key parameter influencing the adsorption behavior of chromium ions on nanoadsorbent surfaces. It directly impacts the kinetic energy of chromium ions and alters the activation energy required for effective adsorption. Elevated temperatures typically reduce the spacing between nanoparticles and supply additional energy, thereby enhancing the mobility of Cr(VI) ions and facilitating their interaction with reactive sites on the nanomaterial surface (Maksoud et al., 2020). Moreover, temperature variations can shift the adsorption equilibrium, further affecting the overall efficiency of the nanomaterial-based removal process.

6.3 Contact Time

The duration of contact between chromium metal ion and nanoadsorbents significantly influences the overall efficiency of chromium remediation during the adsorption process. Studies consistently indicate that the adsorption occurs quickly during the initial phase and progressively decelerates until a state of equilibrium is achieved. In their investigation, Cao et al. (2018) reported that chromium removal efficiency surged quickly within the first two hours, followed by a gradual decline in adsorption rate until equilibrium was reached.

6.4 Dose of Nanomaterial

The amount of nanoadsorbent utilized also plays a crucial role in determining the effectiveness of chromium removal from contaminated environments. Increasing the dosage of nanomaterials typically results in a greater number of active adsorption sites and an expanded surface area, thereby enhancing overall removal efficiency. A study conducted by Fu et al. (2014) reported that increasing the dosage of a sepiolite-based nanomaterials from 0.05 to 3.2 g/L led to a substantial improvement in chromium removal efficiency from 56.2 to 99.9%.

7 Conclusions and Outlook

Chromium is a naturally occurring element that can exist in multiple oxidation states, among which trivalent chromium [Cr(III)] and hexavalent chromium [Cr(VI)] are the most environmentally relevant. Cr(III) is relatively stable, less toxic, and serves as a trace essential nutrient. In contrast, Cr(VI) is highly toxic, soluble, and mobile, posing a significant environmental and public health risk due to its carcinogenic properties. Its presence in the environment arises both from natural processes and a range of human activities, including electroplating, leather tanning, textile production, and mining operations.

Given the harmful and accumulative nature of Cr(VI), its removal or reduction in contaminated sites is imperative. A widely adopted approach involves the transformation of Cr(VI) into its less toxic and more stable form, Cr(III), which is considered environmentally safer. Various conventional treatment techniques have been employed to achieve this conversion, such as coagulation, precipitation, adsorption, membrane filtration, and ion exchange. Among these, adsorption, particularly using advanced carbon-based materials, has gained prominence due to its high efficiency, cost-effectiveness, reusability, and operational simplicity. However, the selection of a suitable treatment method depends on parameters such as initial chromium concentration, economic feasibility, and ecological impact.

Despite their effectiveness, conventional methods often generate secondary waste, raising sustainability concerns in long-term applications. Consequently, the focus has shifted toward eco-friendly and cost-effective alternatives. In this context, nanotechnology has emerged as a promising avenue. Nanomaterials possess high surface area and elevated reactivity, enabling enhanced adsorption and reduction of Cr(VI) with minimal secondary pollution. The performance of nanomaterials in chromium remediation is influenced by several factors, including solution pH, contact time, temperature, and the dosage of the nanoadsorbent.

Building on their promising attributes, nanomaterials present a transformative potential in the field of chromium remediation. However, for large-scale environmental deployment, further investigation into their long-term stability, environmental fate, and potential ecotoxicological risks is essential. Addressing these challenges requires a multidisciplinary approach that integrates nanotechnology, material science, environmental engineering, and policy frameworks. Future research should focus on the development of biocompatible and biodegradable nanomaterials to minimize unintended consequences while optimizing removal efficiency. Additionally, advances in nanocomposite design and functionalization may unlock targeted removal mechanisms that can operate effectively across diverse environmental conditions.

References

Abdelbasir, S. M., & Shalan, A. E. (2019). An overview of nanomaterials for industrial wastewater treatment. *Korean Journal of Chemical Engineering, 36*(8), 1209–1225. https://doi.org/10.1007/s11814-019-0306-y

Acharyya, S., Das, A., & Thaker, T. P. (2023). Remediation processes of hexavalent chromium from groundwater: A short review. *AQUA-Water-Infrastructure-Ecosystems-and-Society, 72*(5), 648–662. https://doi.org/10.2166/aqua.2023.123

Alsammarraie, F. K., Wang, W., Zhou, P., Mustapha, A., & Lin, M. (2018). Green synthesis of silver nanoparticles using turmeric extracts and investigation of their antibacterial activities. *Colloids and Surfaces. B, Biointerfaces, 171*, 398–405. https://doi.org/10.1016/j.colsurfb.2018.07.059

Alzahrani, F. M., Alsaiari, N. S., Katubi, K. M., Amari, A., Rebah, F. B., & Tahoon, M. A. (2021). Synthesis of polymer-based magnetic nanocomposite for multi-pollutants removal from water. *Polymers (Basel), 13*(11), 1742. https://doi.org/10.3390/polym13111742

Arora, D., Arora, A., Panghal, V., & Singh, A. (2024). Unleashing the feasibility of nanotechnology in phytoremediation of heavy metal-contaminated soil: a critical review towards sustainable approach. *Water, Air, and Soil Pollution, 235*, 57. https://doi.org/10.1007/s11270-023-06874-9

Assefa, H., Singh, S., Olu, F. E., Dhanjal, D. S., et al. (2024). Advances in adsorption technologies for hexavalent chromium removal: Mechanisms, materials, and optimization strategies. *Desalination and Water Treatment, 319*, Article 100576. https://doi.org/10.1016/j.dwt.2024.100576

Baby, R., Hussein, M., Abdullah, A., & Zaina, Z. (2022). Nanomaterials for the treatment of heavy metal contaminated water. *Polymers, 14*(3), 583. https://doi.org/10.3390/polym14030583

Balali-Mood, M., Naseri, K., Tahergorabi, Z., Khazdair, M. R., & Sadeghi, M. (2021). Toxic mechanisms of five heavy metals: Mercury, lead, chromium, cadmium, and arsenic. *Frontiers in Pharmacology, 12*, Article 643972. https://doi.org/10.3389/fphar.2021.643972

Bao, Z., Feng, H., Tu, W., Li, L., & Li, Q. (2022). Method and mechanism of chromium removal from soil: A systematic review. *Environmental Science and Pollution Research International, 29*(24), 35501–35517. https://doi.org/10.1007/s11356-022-19452-z

Barrera-Díaz, C. E., Lugo-Lugo, V., & Bilyeu, B. (2012). A review of chemical, electrochemical and biological methods for aqueous Cr(VI) reduction. *Journal of Hazardous Materials, 223–224*, 1–12. https://doi.org/10.1016/j.jhazmat.2012.04.054

Bassyouni, M., Abdel-Aziz, M. H., Zoromba, M. S., Abdel-Hamid, S. M. S., & Drioli, E. (2019). A review of polymeric nanocomposite membranes for water purification. *Journal of Industrial and Engineering Chemistry, 73*, 19–46. https://doi.org/10.1016/j.jiec.2019.01.045

Beig, S. U. R., Dar, U. A., Sheergugri, S. A., & Shah, S. A. (2023). Remediation of hexavalent chromium and pharmaceuticals from aquatic environments by employing an oxygen-doped porous carbon adsorbent. *New Journal of Chemistry, 47*, 8693–8713. https://doi.org/10.1039/D2NJ05841C

Boustila, H., Boutillara, Y., Velasco, L. F., Djellali, A., & Tazibet, S. (2022). Tailoring activated carbon properties for Pb(II) and Cr(VI) removal from water in continuous mode. *Chemical Engineering and Technology, 45*, 258–265. https://doi.org/10.1002/ceat.202100331

Cao, Y., Zhang, S., Zhong, Q., Wang, G., Xu, X., Li, T., Wang, L., Jia, Y., & Li, Y. (2018). Feasibility of nanoscale zero-valent iron to enhance the removal efficiencies of heavy metals from polluted soils by organic acids. *Ecotoxicology and Environmental Safety, 162*, 464–473. https://doi.org/10.1016/j.ecoenv.2018.07.036

Cao, D. X., Zhou, Y. J., Jiang, H. X., Feng, X. N., Liu, X. Y., Li, W., Liu, J. Q., Tang, A. N., & Kong, D. M. (2024). Always positive covalent organic nanosheet enabling pH-independent adsorption and removal of Cr(VI). *Journal of Hazardous Materials, 465*, Article 133420. https://doi.org/10.1016/j.jhazmat.2023.133420

Cheng, P., Zhang, S., Wang, Q., Feng, X., Zhang, S., Sun, Y., & Wang, F. (2021). Contribution of nano-zero-valent iron and arbuscular mycorrhizal fungi to phytoremediation of heavy metal-contaminated soil. *Nanomaterials (Basel), 11*(5), 1264. https://doi.org/10.3390/nano11051264

Dagdag, O., Quadri, T. W., Haldhar, R., Kim, S. C., Daoudi, W., Berdimurodov, E., Akpan, E. D., & Ebenso, E. E. (2023). An overview of heavy metal pollution and control. In: D. K. Verma, C. Verma, P. Kumar Mahish (eds.), *Heavy metals in the environment: management strategies for global pollution.* ACS Symposium Series, vol. 1456 (pp. 3–24). https://doi.org/10.1021/bk-2023-1456.ch001

Danks, A. E., Hall, S. R., & Schnepp, Z. (2016). The evolution of 'sol-gel' chemistry as a technique for materials synthesis. *Materials Horizons, 3*, 91–112. https://doi.org/10.1039/C5MH00260E

Das, A. P., & Mishra, S. (2010). Biodegradation of the metallic carcinogen hexavalent chromium Cr(VI) by an indigenously isolated bacterial strain. *Journal of Carcinognesis, 9*, 6. https://doi.org/10.4103/1477-3163.63584

Dhanapal, A. R., Thiruvengadam, M., Vairavanathan, J., Venkidasamy, B., Easwaran, M., & Ghorbanpour, M. (2024). Nanotechnology approaches for the remediation of agricultural polluted soils. *ACS Omega, 9*(12), 13522–13533. https://doi.org/10.1021/acsomega.3c09776

Dhingra, N. (2024). Applications of nanomaterials for the effective elimination of fluoride from water. In: N. Kumar (ed.), *Fluoride and fluorocarbon toxicity.* Environmental Science and Engineering (pp. 275–291). Springer, Singapore. https://doi.org/10.1007/978-981-97-7733-4_9

Din, I., Khan, I., Aziz, F., Sher, A., & Hakim, S. (2023). Graphene oxide nanoparticles a remedy for chromium pollution. *Current Research in Green and Sustainable Chemistry, 7*, Article 100372. https://doi.org/10.1016/j.crgsc.2023.100372

Esfandiar, N., Suri, R., & McKenzie, E. R. (2022). Competitive sorption of Cd, Cr, Cu, Ni, Pb and Zn from stormwater runoff by five low-cost sorbents; effects of co-contaminants, humic acid, salinity and pH. *Journal of Hazardous Materials, 423*, Article 126938. https://doi.org/10.1016/j.jhazmat.2021.126938

Fouda-Mbanga, B. G., Onotu, O. P., & Tywabi-Ngeva, Z. (2024). Advantages of the reuse of spent adsorbents and potential applications in environmental remediation: A review. *Green Analytical Chemistry, 11*, Article 100156. https://doi.org/10.1016/j.greeac.2024.100156

Fu, F., Dionysiou, D. D., & Liu, H. (2014). The use of zero-valent iron for groundwater remediation and wastewater treatment: A review. *Journal of Hazardous Materials, 267*, 194–205. https://doi.org/10.1016/j.jhazmat.2013.12.062

Hariganesh, S., Vadivel, S., Maruthamani, D., & Rangabhashiyam, S. (2020). Disinfection by-products in drinking water: detection and treatment methods. In: M.N.V. Prasad (ed.), Disinfection By-products in Drinking Water (pp. 279–304). Elsevier, Netherlands. https://doi.org/10.1016/B978-0-08-102977-0.00013-5

Hassan, Z., Ali, S., Farid, M., Rizwan, M., & Shahid, M. (2016). Effect of Chromium (Cr) on the microbial and enzymatic activities in the soil: A review. In: M. M. Abid Ali Khan, M. Abid, A. M. Omar, S. Nazeer Haider Zaidi, & R. K. Maheshwari (eds.), Biodiversity Conservation in Changing Climate (pp. 305–323). Lenin Media Private Limited, Delhi, India.

Homem, V., & Santos, L. (2011). Degradation and removal methods of antibiotics from aqueous matrices—a review. *Journal of Environmental Management, 92*(10), 2304–2347. https://doi.org/10.1016/j.jenvman.2011.05.023

Inobeme, A., Mathew, J. T., Adetunji, C. O., Ajia, A. I., Inobeme, J., Maliki, M., Okonkwo, S., Adekoya, M. A., et al. (2022). Recent advances in nanotechnology for remediation of heavy metals. *Environmental Monitoring and Assessment, 195*, 111. https://doi.org/10.1007/s10661-022-10614-7

Irshad, M. A., Sattar, S., Nawaz, R., Al-Hussain, S. A., Rizwan, M., Bukhari, A., et al. (2023). Enhancing chromium removal and recovery from industrial wastewater using sustainable and efficient nanomaterial: A review. *Ecotoxicology and Environmental Safety, 263*, Article 115231. https://doi.org/10.1016/j.ecoenv.2023.115231

Jeevanandam, J., Ling, J. K. U., Barhourm, A., Chan, Y. S., & Danquah, M. K. (2022). Bionanomaterials: Definitions, sources, types, properties, toxicity, and regulations. In: A. Barhoum., J. Jeevanandam, & M. K. Danquah (eds.), *Fundamentals of bionanomaterials* (pp. 1–29). Elsevier, Netherlands. https://doi.org/10.1016/B978-0-12-824147-9.00001-7

Kerur, S. S., Bandekar, S., Hanagadakar, M. S., Nandi, S. S., Ratnamala, G. M., & Hegde, P. G. (2021). Removal of hexavalent Chromium-Industry treated water and wastewater: A review. *Mater Today: Proceedings, 42*(2), 1112–1121. https://doi.org/10.1016/j.matpr.2020.12.492

Khan, I., Saeed, K., & Khan, I. (2017). Nanoparticles: Properties, applications and toxicities. *Arabian Journal of Chemistry, 12*(7), 908–931. https://doi.org/10.1016/j.arabjc.2017.05.011

Khatua, S., & Dey, S. K. (2023). The chromium and toxicity of chromium pollution: An overview. *Asian Journal of Agricultural and Horticultural Research, 10*(2), 1-14. https://doi.org/10.9734/ajahr/2023/v10i2221

Kholisa, B., Matsena, M., & Chirwa, E. M. N. (2021). Evaluation of Cr(VI) reduction using indigenous bacterial consortium isolated from a municipal wastewater sludge: Batch and kinetic studies. *Catalysts, 11*(9), 1100. https://doi.org/10.3390/CATAL11091100/S1

Krishnamurthy, S., & Wilkens, M. M. (1994). Environmental chemistry of Cr. *Northeastern Geology, 16*(1), 14–17.

Kumar, L., Ragunathan, V., Chugh, M., & Bharadvaja, N. (2021). Nanomaterials for remediation of contaminants: a review. *Environmental Chemistry Letters, 19*(4), 3139–3163. https://doi.org/10.1007/s10311-021-01212-z

Liu, B., Xin, Y. N., Zuo, J., Khoso, F. M., Liu, Y. P., et al. (2023). Removal of chromium species by adsorption: fundamental principles, newly developed adsorbents and future perspectives. *Molecules, 28*(2), 639. https://doi.org/10.3390/molecules28020639

Majhi, P., & Samantaray, S. M. (2020). Effect of hexavalent chromium on paddy crops (Oryza sativa). *Journal of Pharmacognosy and Phytochemistry, 9*(2), 1301–1305.

Maksoud, M. I. A. A., Elgarahy, A. M., Farrell, C., Al-Muhtaseb, A. H. A., Rooney, D. W., & Osman, A. I. (2020). Insight on water remediation application using magnetic nanomaterials

and biosorbents. *Coordination Chemistry Reviews, 403*, Article 213096. https://doi.org/10.1016/j.ccr.2019.213096
Matos, M., Correia, A. A. S., & Rasteiro, M. G. (2017). Application of carbon nanotubes to immobilize heavy metals in contaminated soils. *Journal of Nanoparticle Research, 19*, 126. https://doi.org/10.1007/s11051-017-3830-x
Mishra, S., Bharagava, R. N. (2016). Toxic and genotoxic effects of hexavalent chromium in environment and its bioremediation strategies. *Journal of Environmental Science and Health. Part C, Environmental Carcinogenesis and Ecotoxicology Reviews, 34*(1), 1–32. https://doi.org/10.1080/10590501.2015.1096883
Mitra, S., Sarkar, A., Sen, S. (2017). Removal of chromium from industrial effluents using nanotechnology: a review. *Nanotechnology for Environment Engineering, 2*(11). https://doi.org/10.1007/s41204-017-0022-y
Mohanty, S., Benya, A., Hota, S., Kumar, M. S., & Singh, S. (2023). Eco-toxicity of hexavalent chromium and its adverse impact on environment and human health in Sukinda Valley of India: A review on pollution and prevention strategies. *Environmental Chemistry Ecotoxicology, 5*, 46–54. https://doi.org/10.1016/j.enceco.2023.01.002
Mpouras, T., Polydera, A., Dermatas, D., Verdone, N., & Vilardi, G. (2021). Multi wall carbon nanotubes application for treatment of Cr(VI)-contaminated groundwater; modeling of batch & column experiments. *Chemosphere, 269*, Article 128749. https://doi.org/10.1016/j.chemosphere.2020.128749
Nayl, A. A., Abd-Elhamid, A. I., Aly, A. A., & Bräse, S. (2022). Recent progress in the applications of silica-based nanoparticles. *RSC Advances, 12*(22), 13706–13726. https://doi.org/10.1039/D2RA01587K
Niu, B., Wang, X., Wu, K., He, X., & Zhang, R. (2018). Mesoporous titanium dioxide: Synthesis and applications in photocatalysis, energy and biology. *Materials, 11*(10), 1910. https://doi.org/10.3390/ma11101910
Njoya, O., Zhao, S., Kong, X., Shen, J., Kang, J., Wang, B., & Chen, Z. (2021). Efficiency and potential mechanism of complete Cr(VI) removal in the presence of oxalate by catalytic reduction coupled with membrane filtration. *Separation and Purification Technology, 275*, Article 118915. https://doi.org/10.1016/j.seppur.2021.118915
Pakade, V. E., Tavengwa, N. T., & Madikizela, L. M. (2019). Recent advances in hexavalent chromium removal from aqueous solutions by adsorptive methods. *RSC Advances, 9*, 26142–26164. https://doi.org/10.1039/C9RA05188K
Pawlisz, A.V. (1997). Canadian water quality guidelines for Cr. *Environmental Toxicology and Water Quality, 12*, 123–161. https://doi.org/10.1002/(SICI)1098-2256(1997)12:2%3C123::AID-TOX4%3E3.0.CO;2-A
Peng, H., & Guo, J. (2020). Removal of chromium from wastewater by membrane filtration, chemical precipitation, ion exchange, adsorption electrocoagulation, electrochemical reduction, electrodialysis, electrodeionization, photocatalysis and nanotechnology: A review. *Environmental Chemistry Letters, 18*, 2055–2068. https://doi.org/10.1007/s10311-020-01058-x
Prado-Audelo, M. L. D., Kerdan, I. G., Escutia-Guadarrama, L., Reyna-González, J. M., Magaña, J. J., & Leyva-Gómez, G. (2021). Nanoremediation: nanomaterials and nanotechnologies for environmental cleanup. *Frontiers in Environmental Science, 9*, Article 793765. https://doi.org/10.3389/fenvs.2021.793765
Prema, P., Nguyen, V. H., Venkatachalam, K., Murugan, J. M., Ali, H. M., Salem, M. Z. M., Ravindran, B., & Balaji, P. (2021). Hexavalent chromium removal from aqueous solutions using biogenic iron nanoparticles: Kinetics and equilibrium study. *Environmental Research, 205*, Article 112477. https://doi.org/10.1016/j.envres.2021.112477
Rafique, M., Hajra, S., Tahir, M. B., Gillani, S. S. A., & Irshad, M. (2022). A review on sources of heavy metals, their toxicity and removal technique using physico-chemical processes from wastewater. *Environmental Science and Pollution Research, 29*, 16772–16781. https://doi.org/10.1007/s11356-022-18638-9

Rajakumar, G., Sudha, P. N., & Thiruvengadam, M. (2023). Synthesis, bioactivity evaluation and application of plant-based nanoparticles. *Molecules, 28*(12), 4783. https://doi.org/10.3390/molecules28124783

Ramli, N. N., Kurniawan, S. B., Ighalo, J. O., Said, N. S. M., Marsidi, N., Buhari, J., Shah, R. A. R., Zulkifi, M., Alias, J., Daud, N. M., Ahmad, J., Othman, A. R., Abdullah, S. R. S., & Hasan, H. A. (2023). A review of the treatment technologies for hexavalent chromium contaminated water. *BioMetals, 36*, 1189–1219. https://doi.org/10.1007/s10534-023-00512-x

Renugopal, L., Kow, K., Kiew, P., Yeap, S., Chua, H., Chan, C., & Yusoff, R. (2019). Selective adsorption of copper and cadmium ions using nano-particles aligned in silica gel matrix. *AIP Conference Proceedings, 2124*(1), Article 020001. https://doi.org/10.1063/1.5117061

Samrot, A., Sahithva, S., Selvarani, J., Purayi, S., Ponnish, P. (2020). A review on synthesis, characterization and potential biological applications of superparamagnetic iron oxide nanoparticles. *Current Research in Green and Sustainable Chemistry, 4,* Article 10042. https://doi.org/10.1016/j.crgsc.2020.100042

Sarkar, A., Masud, M. A. A., Deepo, D. M., Das, K., Nandi, R., Ansary, M. W. R., Islam, A. R. M. T., & Islam, T. (2023). Biological and green remediation of heavy metal contaminated water and soils: A state-of-the-art review. *Chemosphere, 332*, Article 138861. https://doi.org/10.1016/j.chemosphere.2023.138861

Shan, J., Ji, R., Yu, Y., Xie, Z., & Yan, X. (2015). Biochar, activated carbon, and carbon nanotubes have different effects on fate of (14)C-catechol and microbial community in soil. *Science and Reports, 5*, 16000–16011. https://doi.org/10.1038/srep16000

Sharma, D., Ahmed, A., & Lodhi, S. (2023). Chromium toxicity in aquatic ecosystem: A review. *International Journal of Creative Research Thoughts, 11*(8), 540–547.

Shukla, S., Khan, R., & Hussain, C. M. (2020). Nanoremediation. In: C. M. Hussain (ed.), *The handbook of environmental remediation* (pp. 443–467). RSC, UK. https://doi.org/10.1039/9781788016261-00443

da Silva Alves, D. C., Healy, B., Pinto, L. A. d. A., Cadava, T. R. S., Jr., & Breslin, C. B. (2021). Recent developments in chitosan-based adsorbents for the removal of pollutants from aqueous environments. *Molecules, 26*(3), 594. https://doi.org/10.3390/molecules26030594

Singh, J., & Kalamdhad, A. S. (2011). Effects of heavy metals on soil, plants, human health and aquatic life. *International Journal of Research in Chemistry and Environment, 1(2)*, 15–21.

Sun, Y., Jin, J., Li, W., Zhang, S., & Wang, F. (2021). Hexavalent chromium removal by a resistant strain *Bacillus cereus* ZY-2009. *Environmental Technology, 44*(9), 1–28. https://doi.org/10.1080/09593330.2021.2016994

Wu, J., Wang, X. B., & Zeng, R. J. (2017). Reactivity enhancement of iron sulfide nanoparticles stabilized by sodium alginate: Taking Cr (VI) removal as an example. *Journal of Hazardous Materials, 333*, 275–328. https://doi.org/10.1016/j.jhazmat.2017.03.023

Yasir, M. W., Siddique, M. B. A., Shabbir, Z., Ullah, H., Riaz, L., Nisa, W. U., Rahman, S. U., & Shah, A. A. (2021). Biotreatment potential of co-contaminants hexavalent chromium and polychlorinated biphenyls in industrial wastewater: Individual and simultaneous prospects. *Science of the Total Environment, 779*, Article 146345. https://doi.org/10.1016/j.scitotenv.2021.146345

Yunus, K., Zuraidah, M. A., & John, A. (2020). A review on the accumulation of heavy metals in coastal sediment of Peninsular Malaysia. *Ecofeminism Climate Change, 1*, 21–35. https://doi.org/10.1108/EFCC-03-2020-0003

Zhang, L., Fu, F., & Tang, B. (2019). Adsorption and redox conversion behaviors of Cr(VI) on goethite/carbon microspheres and akaganeite/carbon microspheres composites. *Chemical Engineering Journal, 356*, 151–160. https://doi.org/10.1016/j.cej.2018.08.224

Zhigalenok, Y., Tazhibayeva, A., Kokhmetova, S., Starodubtseva, A., Kan, T., Isbergenova, D., & Malchik, F. (2025). Hexavalent chromium at the crossroads of science, environment and public health. *RSC Advances, 15*(27), 21439–21464. https://doi.org/10.1039/D5RA03104D

Zhou, L., Li, R., Zhang, G., Wang, D., Cai, D., & Wu, Z. (2018). Zero-valent iron nanoparticles supported by functionalized waste rock wool for efficient removal of hexavalent chromium. *Chemical Engineering Journal, 339*, 85–96. https://doi.org/10.1016/j.cej.2018.01.132

Znad, H., Awual, M. R., & Martini, S. (2022). The utilization of algae and seaweed biomass for bioremediation of heavy metal-contaminated wastewater. *Molecules, 27*(4), 1275. https://doi.org/10.3390/molecules27041275

Zulfiqar, U., Haider, F. U., Ahmad, M., Hussain, S., Maqsood, M. F., Ishfaq, M., Shahzad, B., Waqas, M. M., Ali, B., Tayyab, M. N., Ahmad, S. A., Khan, I., & Eldin, S. M. (2023). Chromium toxicity, speciation, and remediation strategies in soil-plant interface: A critical review. *Frontiers in Plant Science, 13*, 1081624. https://doi.org/10.3389/fpls.2022.1081624

Phytoremediation of Chromium: A Review

Simmi Goel

Abstract The usage of chemical fertilizers, fungicides, pesticides, hormones, soil regulators, sludge in the agriculture sector resulted in toxicity, depletion of nutrients, and pollution of soil. The basic reason of using these is rapid and uncontrolled rise in population, urbanization, deforestation and industrialization. The result of all these activities finally led to metal contamination of the ecosystem especially chromium, nickel, lead, mercury, zinc, and cadmium. Various plants can be successfully used for removal of metals from the ecosystem through the process of phytoremediation. In this review paper, sources of chromium contamination, health effects, mechanism of phytoremediation including phytoextraction, phytostabilization, rhizofiltration, phytovolatilization, rhizodegradation, properties of plants used for phytoremediation, factors controlling rate of phytoremediation, advantages, and limitations of the phytoremediation technique have been summarized.

Keywords Phytoremediation · Chromium toxicity · Soil contamination · Plants · Phytoextraction · Metal-binding proteins · Phytostabilization · Rhizofiltration

1 Introduction

Chromium exists in the form of oxides having two major oxidation states out of which hexavalent chromium is more toxic than trivalent chromium. Some plant species have demonstrated exceptional ability to remove heavy metals from soil and show strong resistance to heavy metal toxicity. Chromium (Cr), a naturally occurring element found widely in the Earth's crust, mainly exists in two oxidation states: trivalent chromium (Cr [III]) and the more hazardous hexavalent chromium (Cr [VI]). In general, chromium levels in soil range from 0.1 to 250 ppm, but in severely polluted sites, they can reach as high as 400 ppm. Major contributors to chromium contamination include industrial processes such as steel production, cement manufacturing, and the burning of fossil fuels (Langard & Norseth, 1979). Industrial processes,

S. Goel (✉)
Department of Biotechnology, Mata Gujri College, Sri Fatehgarh Sahib, Punjab, India
e-mail: simmibiotech@matagujricollege.org

N. Kumar (ed.), *Chromium Toxicity*, https://doi.org/10.1007/978-3-032-14323-5_9

including the production of steel, cement, and fossil fuel burning, are the main causes of chromium pollution. Chromium has the ability to slow down plant growth. It can limit photosynthesis by causing ultrastructural alterations in the chloroplasts of Lemna minur, Pistia sp., and Taxithelium nepalense (Bassi et al., 1990). According to Vajpayee et al. (2000), hexavalent chromium can reduce the amount of chlorophyll and replace magnesium ions from the active sites of numerous enzymes.

2 Sources of Chromium

Chromium is commonly released into the environment from several industrial activities, including mining, metalworking, the production of steel and metal alloys, paint manufacturing, and the processing of wood, paper, and textiles. Additionally, the combustion of coal or municipal waste for energy generation releases ash containing chromium, which increases its levels in soil and water (see Table 1). Chrome sulphate is utilized in the leather industry to modify the collagen structure by acting as a crosslinking agent. Anti-corrosion coatings form a protective barrier between metal surfaces and their surroundings, with chromium-based compounds like $ZnCrO_4$ and $SrCrO_4$ offering enhanced protection for vehicles and other metal products. The stainless steel industry employs various chromium compounds such as ammonium dichromate ($(NH_4)_2Cr_2O_7$) and potassium dichromate ($K_2Cr_2O_7$), which contribute to Cr (VI) contamination in wastewater. Chromated copper arsenate (CCA), a pesticide containing Cr (VI), As (V), and Cu (II), is widely used for preserving wood. When treated wood comes into contact with soil or water, these chemicals can gradually leach into the surrounding environment (Singh et al., 2023; Tumolo et al., 2020).

3 Phytoremediation

Phytoremediation is the use of plants to remove a variety of contaminants from soil, sediment, and water. The various advantages of this process include aesthetically pleasing mechanism, cost effectiveness, restore habitat, and clean up contamination. The process of phytoremediation consists of various methods including phytovolatilization, microorganism stimulation, phytostabilization, phytoextraction, and phytodegradation by plants (Cunningham et al., 1996; Flathman & Lanza, 1998; Khan et al., 2023).

Table 1 Sources including activities responsible for chromium addition in environment

S. No.	Sources	Activities	References
1	Geological sources	Ultramafic rocks, chromite ore processing residues, weathering of ultramafic igneous and metamorphic rocks	Awa & Hadibarata, 2020
2	Volcanic activity	Volcanic dust and gases	Sharma et al., 2023
3	Leather tanning	Chrome sulphate in structural modification in collagen protein present in leather acting as a crosslinking agent	Adiloğlu et al., 2021
4	Metal processing	Steel, metal alloy. electroplating, chrome plating	Han et al., 2023
5	Petroleum refining	Refining processes utilize chromium containing catalysts in catalytic cracking or hydrocracking, crude oil contain trace amounts of chromium	Mandal et al., 2014
6	Pulp and paper processing, dyeing, paints and pigments	Bleaching agents for pulp, Chromium-based pigments and dyes to colour paper products, used as mordants which help dyes bind to fabrics	Fu & Wang, 2011; Mazumdar & Das, 2015
7	Landfilling/ improper disposal	Rainwater percolating through landfills can dissolve chromium, Decomposition of organic matter promote the mobilization of chromium leading to leaching	Testa, 2004
8	Burning of fossil fuels/coal	Extraction, processing and combustion of fuels, with chromium found in coal, fly ash and bottom ash	Babu et al., 2021; Bhatt et al. 2022
9	Wood preservatives	chromated copper arsenate used in wood treatment	Singh et al., 2023; Tumolo et al., 2020
10	Fertilizers/soil additives	Phosphate fertilizers contain chromium as a contaminant, soil conditioners or amendments may contain chromium	Chaudhary et al., 2018
11	Cooling towers	Chromium-based corrosion inhibitors	Fuller & Norman, 1984

3.1 Characteristics of Plants Making Them Suitable for Phytoremediation

Various characteristics of plants which enable them to be highly metal tolerant are their short life cycle, broad distribution, and large biomass (Mazumdar & Das, 2015). Apart from these certain other characteristics of plants making them suitable for phytoremediation are as follows:

- Rapidly growing woody species like *Populus* and *Salix* can be easily regenerated due to their extensive root systems.

- These plants should exhibit high evapotranspiration rates, which can help stabilize pollutants and support growth in contaminated soils.
- They must also produce high biomass, allowing them to tolerate a wide range of harmful elements in their root zones.
- Key plant species are often those that naturally colonize polluted environments, such as vegetation found at mining sites.
- Selecting appropriate plants for phytoremediation largely depends on factors like the type and concentration of contaminants, their depth and volatility, the site's age, and how quickly the contaminants degrade.

3.2 Mechanism of Phytoremediation

Plants have evolved intricate strategies to regulate the uptake and accumulation of heavy metals. These strategies include the chelation and sequestration of metal ions using ligands such as phytochelatins (PCs) and metallothioneins (MTs) (Cobbett & Goldsbrough, 2002). A variety of metal-binding proteins play key roles in processes such as absorption, transport, accumulation, and detoxification of heavy metals, thereby enhancing the plant's tolerance to metal stress (Feki et al., 2021; Sharma et al., 2023). These proteins include phytochelatins, metallothioneins, and metal transporter proteins like heavy metal ATPases. (Chaudhary et al., 2018; Mathur & Chauhan, 2020). Various processes of phytoremediation along with list of plants involved is represented in Table 2.

3.2.1 Phytoextraction

Plants absorb chromium through their roots and accumulate it in their above-ground parts (shoots and leaves). Phytoextraction is a method where plants absorb pollutants through their roots and transport them to their aboveground tissues, where the contaminants accumulate. These parts are then typically harvested and disposed of. This approach is effective for removing metals, radionuclides, metalloids, non-metals, and certain hydrocarbons. During the process, plants form complexes with these elements—often through chelation using their own metabolites—which helps reduce the elements' toxicity. For a plant to be effective in phytoextraction, it should possess three key traits: (i) the ability to tolerate toxic metals, (ii) the capacity to produce substantial biomass, and (iii) the efficiency to accumulate heavy metals in parts that are easy to harvest (Sharma et al., 2023).

3.2.2 Phytostabilization

Plants reduce the bioavailability and mobility of chromium in the soil, preventing its spread to other areas or into the food chain. This process involves immobilizing

Table 2 Various processes of phytoremediation along with list of plants involved

S. No.	Process	Name of plants	References
1	Phytoextraction	*Brachiaria mutica, Leptochloa fusca, Chrysopogon zizanioides Typha angustifolia L., Canna indica L., Hydrocotyle umbellate L Bixa orellana, Gynura pseudochina and Leersia hexandra*	Han et al., 2023
2	Phytostabilization	*Tetraena qataranse, Calotropis procera*	Han et al., 2023
3	Rhizofiltration	*Zea mays, Sunflower*	Benavides et al. 2018; de la Rosa et al. 2014
4	Metal binding Proteins	*Nicotiana langsdorffii, Barley*	DelBubba et al. 2013; Brunetti et al. 2012
5	Plant–microbe interaction	*Vetiveria zizanioides/ Bacillus cereus, Sesbania sesban/Bacillus xiamenensisPM14*	Nayak et al. 2018

contaminants in the soil through either absorption or adsorption by plant roots or through precipitation within the rhizosphere of the growing medium. It works by stabilizing contaminants and limiting their movement, achieved by establishing a vegetative cover around the plant root zone, which helps prevent contaminants from reaching the groundwater. Several factors influence the effectiveness of this stabilization, including the type of plant species, soil characteristics, the nature of the contaminant, the use of bioinoculants like plant growth-promoting bacteria, the addition of organic and inorganic amendments (such as lime, clay, and phosphate compounds), as well as environmental conditions like rainfall and temperature.

3.2.3 Rhizofiltration

Plants filter chromium metal from contaminated water as it passes through their root zone. Rhizofiltration is a method used to absorb pollutants from a solution near the root zone by concentrating and precipitating them onto or into plant roots through biological or non-biological processes. This technique is primarily applied to clean contaminated groundwater. Plant roots release secondary metabolites called root exudates, which trigger various biogeochemical reactions, leading to the deposition of pollutants on roots or into surrounding water. Subsequently, contaminants are adsorbed by the root system and may be transported to the plant's aboveground parts, depending on the plant species as well as the type and concentration of pollutants. The efficiency of rhizofiltration in removing metals from polluted sites is largely influenced by the plant species, nutrient levels, chemical nature of the contaminants, and groundwater conditions.

3.2.4 Phytovolatilization

Through transpiration, plants release chromium into the atmosphere after converting it into a less hazardous form. A modified form of the pollutant is released into the nearby atmosphere after being absorbed, translocated, and evaporated by a plant species. Growing trees and other plant species that concurrently consume water and pollutants cause phytovolatilization, which is the volatilization of their roots, stems, or foliage.

3.2.5 Rhizodegradation

Microorganisms located in the plant's rhizosphere, or root zone, are responsible for breaking down chromium. Rhizoremediation refers to the microbial-driven breakdown of various organic pollutants in soil that is actively growing, and this process can be improved by the presence of the root zone. Among methods for treating polycyclic aromatic hydrocarbons (PAHs) in soil, rhizodegradation is considered the most effective. The degradation of contaminants is facilitated by the combined effects of the microbial community and the plant root system within the rhizosphere. This includes the release of root exudates such as amino acids, sugars, and organic acids, as well as substances like hydrogen cyanide (HCN), siderophores, phosphatases, and phytohormones produced by plant growth-promoting rhizobacteria (PGPR). Various factors including direct phytohormonal activity, plant–microbe interactions, soil parameters, increased nutrient availability, and regulating microbial load increase the rate of degradation.

3.2.6 Metal Binding Proteins in Plants

Plants contain various metal-binding proteins such as metallothioneins, phytochelatins, metalloenzymes, along with several metal storage, carrier, and channel proteins (Memon & Schröder, 2009). These proteins have the ability to bind metals including Fe, Cr, Zn, As, Cd, Ni, and Pb (Mejáre & Bülow, 2001). Different cadmium-binding proteins in plants, characterized by fewer cysteine residues, are introduced and overexpressed to enhance metal-binding capacity, tolerance, or accumulation in both bacteria and plants.

3.2.7 Phytochelatins

Phytochelatins are low-molecular weight polypeptides rich in cysteine, synthesized enzymatically, with their production triggered by heavy metals (Chia, 2021). They bind heavy metals through their thiol groups, forming metal-phytochelatin complexes that are stored within vacuoles. The ability of phytochelatins to aid in phytoremediation is closely linked to the degree of their polymerization.

3.2.8 Metallothionein

In order to deal with the high concentration of metal ions in the soil, plants have evolved certain adaptations. There are four distinct types of MT-encoding genes that can be discovered in rice, sugarcane, and Arabidopsis. MTs are characterized by their high cysteine content, which allows them to bind to various metal ions, including zinc, copper, cadmium, and mercury. They play a vital role in maintaining the balance of essential metals like zinc and copper within cells, preventing their accumulation or depletion. MT expression is regulated by various factors, including metal availability, cellular stress, and hormonal signals.

3.2.9 Transporter Proteins

Plant transporter proteins have the ability to absorb, move, and store heavy metals, which helps the plants to tolerate these metals and ultimately aids in cleaning up contaminated soil. The transporter proteins responsible for taking up heavy metals like cadmium (Cd) are categorized into various families according to their sequence similarities. These Cd transporters include the ZIP family (zinc- and iron-regulated transporter proteins), metal tolerance proteins, and natural resistance-associated macrophage proteins.

4 Factors Responsible for Maximum Phytoremediation Efficiency

Different phytoremediation techniques are used based on the specific plant species and the kind of pollutant involved. Selecting the appropriate method depends on factors such as the plant's ability to absorb and eliminate the pollutant, the chemical and physical characteristics of the contaminant, how well the phytoremediation approach matches the pollutant, as well as the pollutant's concentration, its depth in the soil, and the climate conditions. The effectiveness of phytoremediation is influenced by a number of soil and plant parameters, such as the physical and chemical characteristics of the soil, plant and microbial exudates, metal bioavailability, and the capacity of plants to absorb, accumulate, move, sequester, and detoxify metals (Wang et al. 2019). The following are some variables that may affect how well phytoremediation works:

4.1 Type of Plant Species

Different plant species show distinct capabilities in absorbing and tolerating chromium. Based on these abilities, they can be classified into three categories: metal excluders, metal indicators, and hyperaccumulators.

4.1.1 Metal Excluders

Metal excluders are plants that prevent metals from reaching their aerial parts or maintain consistently low metal levels in their tissues, regardless of varying soil concentrations. Notable examples include *Oenothera biennis*, *Commelina communis*, *Salix* species, *Silene maritime* and *Populus* species.

4.1.2 Metal Indicators

These plants reflect the metal concentration in the surrounding soil within their tissues. Metals are absorbed either by forming complexes (chelates) or being stored within plant biomass. Examples include *Pluchea dioscoridis*, *Agave sisalana*, and *Cyperus articulatus*.

4.1.3 Hyperaccumulators

Hyperaccumulators absorb and concentrate high levels of metals such as Co, Cd, Zn, Ni, Pb, and Mn often 100 to 1000 times more than excluder species. These metals can accumulate in the roots, stems, or leaves. Woody plants like poplars and willows are commonly used in phytoextraction due to their large biomass, which enables greater metal uptake. Their deep root systems also help minimize soil erosion, preventing the spread of contaminants.

4.2 Chromium Levels in Soil and Water

The initial concentration of chromium in soil or water directly influences plant growth and chromium uptake. Phytoextraction is dependent on the use of specific hyperaccumulators and how they metabolize contaminants.

4.3 *Soil and Water Conditions*

Environmental factors such as pH, nutrient levels, and the presence of other contaminants can impact phytoremediation effectiveness. Temperature affects plant metabolism and root development. Agronomic practices like the application of fertilizers, chelating agents, pH adjustment, and vermicompost can improve metal uptake by enhancing soil conditions.

4.4 *Use of Chelating Agents*

Chelating agents can significantly enhance chromium uptake in plants, though their effectiveness may vary. Biodegradable compounds such as micronutrients and chelators like EDTA (Ethylene diamine tetraacetic acid) and NTA (nitrilotriacetic acid) are commonly used to improve metal phytoextraction from soils.

4.5 *Role of Microorganisms*

Microbial activity in the rhizosphere, i.e. the area around plant roots can influence the transformation and uptake of chromium by plants.

4.6 *Root Zone Function*

Roots play a critical role in pollutant accumulation and transformation. They can store or metabolize contaminants and release exudates that help break down pollutants in the surrounding soil.

4.7 *Metal Bioavailability*

The bioavailability of metals refers to the fraction of total metals in soil or water that is accessible to organisms. Metal uptake by plants depends on this availability, which is influenced by the retention time of metals and their interaction with other substances in the soil.

5 Benefits of Phytoremediation

- Phytoremediation is effective both on organic and inorganic pollutants. The process can be easily handled and economically applied to accumulate chromium from the contaminated soil.
- Cost effective: The process is both economical and cost effective. It is generally more affordable than traditional remediation methods. The process can be applied successfully both on laboratory and pilot scale treatments.
- Environmentally friendly: Phytoremediation is a natural process that doesn't require harsh chemicals or intensive energy input.
- Sustainable: Phytoremediation can be a long-term solution for cleaning up contaminated sites.
- Aesthetically pleasing: Plants can improve the appearance of contaminated areas.

6 Conclusion

In this chapter, we have focused on the use of phytoremediation to clean up environmental chromium contamination. Because of its toxicity, chromium must be extracted from the environment using sustainable treatment technologies. A developing technique called phytoremediation aids in the removal of harmful contaminants from soil and water bodies. A sustainable and affordable method of boosting underdeveloped nations' economies is through phytoremediation. In this paper sources of chromium addition in environment, health effects, mechanism and processes of phytoremediation, factors affecting the rate of remediation and characteristics of plants for assessing their suitability for phytoremediation have been discussed in detail. This type of approach by adopting new technologies can be successfully applied for treatment of chromium-contaminated environment.

References

Adiloğlu, S., Açikgöz, F. E., & Gürgan, M. (2021). Use of phytoremediation for pollution removal of hexavalent chromium-contaminated acid agricultural soils. *Global NEST Journal, 23*(3), 400–406.

Awa, S. H., & Hadibarata, T. (2020). Removal of heavy metals in contaminated soil by phytoremediation mechanism: a review. *Water, Air, and Soil Pollution, 231*, 47.

Babu, S. M. O. F., Hossain, M. B., Rahman, M. S., Rahman, M., Ahmed, A. S. S., Hasan, M. M., Rakib, A., Emran, T. B., Xiao, J., & Simal-Gandara, J. (2021). Phytoremediation of toxic metals: a sustainable green solution for clean environment. *Applied Sciences, 11*, 10348.

Bassi, M., Corradi, M. G., & Realini, M. (1990). Effects of Cr(VI) on two freshwater plants, Lemna minor and Pistia straiotes: Morphological observations. *Cytobios, 62*, 27–38.

Benavides, L. C. L., Pinilla, L. A. C., Serrezuela, R. R. amp; Serrezuela, W. F. R. (2018). Extraction in laboratory of heavy metals through rhizofiltration using the Plant Zea Mays (maize). *International Journal of Applied Environmental Science, 13*(1), 9–26.

Bhat, S. A., Bashir, O., Haq, S. A. U., Amin, T., Rafiq, A., Ali, M., Americo-Pinheiro, J. H. P., & Sher, F. (2022). Phytoremediation of heavy metals in soil and water: An eco-friendly, sustainable and multidisciplinary approach. *Chemosphere, 303*, Article 100774.

Brunetti, G., Farrag, K., Rovira, P. S., Ferrara, M., Nigro, F., & Senesi, N., (2012). The effect of compost and Bacillus Licheniformis on the phytoextraction of Cr, Cu, Pb and Zn by three brassicaceae species from contaminated soils in the Apulia region, Southern Italy. *Geoderma, 170,* 322–330.

Chaudhary, K., Agarwal, S., & Khan, S. (2018). Role of Phytochelatins (PCs), Metallothioneins (MTs), and Heavy Metal ATPase (HMA) genes in heavy metal tolerance, Chapter 2. In R. Prasad (ed.), Springer International Publishing AG, part of Springer Nature, Mycoremediation and Environmental Sustainability, Fungal Biology (pp. 39–60).

Chia, J. C. (2021). Phytochelatin synthase in heavy metal detoxification and xenobiotic metabolism. In K. F. Mendes, R. N. de Sousa, & K. C. Mielke (eds.), *Book Biodegradation Technology of Organic and Inorganic Pollutants.* https://doi.org/10.5772/intechopen.99077

Cobbett, C., & Goldsbrough, P. (2002). Phytochelatins and metallothioneins: Roles in heavy metal detoxification and homeostasis. *Annual Review of Plant Biology, 53*, 159–182.

Cunningham, S.D., Anderson, T.A., Schwab, A.P., Hsu, F.C. (1996). Phytoremediation of soils contaminated with organic pollutants. In *Advances in Agronomy*, Vol. 56 (pp. 55–114). Academic Press Inc.: Cambridge, MA, USA.

Del Bubba, M., Ancillotti, C., Checchini, L., Ciofi, L., Fibbi, D., Gonnelli, C., & Mosti, S. (2013). Chromium accumulation and changes in plant growth, selected phenolics and sugars of wild type and genetically modified Nicotiana langsdorffii. *Journal of Hazardous Materials, 262,* 394–403.

De la Rosa, G., Castillo-Michel, H., Cruz-Jiménez, G., Bernal-Alvarado, J., Córdova-Fraga, T., Laura López-Moreno, L., Cotte, M. (2014). Cr localization and speciation in roots of Chromate Fed Helianthus annuus L. Seedlings using synchrotron techniques. *International Journal of Phytoremediation, 16*(11), 1073-1086.

Feki, K., Tounsi, S., Mrabet, M., Mhadhbi, H., & Brini, F. (2021). Recent advances in physiological and molecular mechanisms of heavy metal accumulation in plants. *Environmental Science and Pollution Research International, 28*(46), 64967–64986.

Flathman, P. E., & Lanza, G. R. (1998). Phytoremediation: Current views on an emerging green technology. *Journal of Soil Contamination, 7*(4), 415–432.

Fu, F., & Wang, Q. (2011). Removal of heavy metal ions from wastewaters: A review. *J. Environ. Manag., 92*, 407–418.

Fuller, E. J., & Norman, N. L. (1984). Extraction of chromium and zinc from cooling tower blowdown by liquid membranes. *Journal of Membrane Science, 18*, 251–271.

Han, L., Gu, H., Lu, W., Li, H., Peng, W., Ma, N. L., Lam, S. S., & Sonne, C. (2023). Progress in phytoremediation of chromium from the environment. *Chemosphere, 344*, 40307.

Khan, S., Masoodi, T.H., Pala, N.A., Murtaza, S., Mugloo, J.A., Sofi, P.A., Zaman, M.U., Kumar, R., & Kumar, A. (2023). Phytoremediation prospects for restoration of contamination in the natural ecosystems. *Water*, *15*(8), 1498.

Langard, S., & Norseth, T. (1979). Chromium. In L. Friberg, G.F. Nordberg, & V.B. Vouk (eds.), *Handbook on the Toxicology of Metals* (383–398). Elsevier/North-Holland Biomedical, Amsterdam.

Mandal, A., Purakayastha, T. J., Ramana, S., Neenu, S., Bhaduri, D., Chakraborty, K., Manna, M. C., & Rao, A. S. (2014). Status on phytoremediation of heavy metals in India—a review. *International Journal of Bio-Resource and Stress Management, 5*, 553–560.

Mathur, J., & Chauhan, P. (2020). Mechanism of toxic metal uptake and transport in plants. In K. Mishra et al. (ed.), *book: Sustainable Solutions for Elemental Deficiency and Excess in Crop Plants* (pp. 335–349).

Mazumdar, K., & Das, S. (2015). Phytoremediation of Pb, Zn, Fe, and Mg with 25 wetland plant species from a paper mill contaminated site in North East India. *Environmental Science and Pollution Research International, 22*(1), 701–710.

Mejáre, M., & Bülow, L. (2001). Metal-binding proteins and peptides in bioremediation and phytoremediation of heavy metals. *Trends in Biotechnology, 9*(2), 67–73.

Memon, A. R., & Schroder, P. (2009). Implications of metal accumulation mechanisms to phytoremediation. *Environmental Science Pollution Research Int. 16*(2), 162–175.

Nayak, A. K., Panda, S. S., Basu, A., & Dhal, N. K. (2018). Enhancement of toxic Cr (VI), Fe, and other heavy metals phytoremediation by the synergistic combination of native Bacillus cereus strain and Vetiveria zizanioides L. *International Journal of Phytoremediation, 20*(7), 682–691.

Sharma, J. K., Kumar, N., Singh, N. P., & Santal, A. R. (2023). Phytoremediation technologies and their mechanism for removal of heavy metal from contaminated soil: An approach for a sustainable environment. *Frontiers in Plant Science, 14*, 1–13.

Singh, V., Abhishek, K., Nand Rai, S., Singh, S. K., Vamanu, E., & Kumar, A. (2023). Source of Cr (VI) in the aquatic ecosystem, its genotoxic effects and microbial removal from contaminated water. *Green Chemistry Letters and Reviews, 16*(1).

Testa, S. M. (2004). Sources of chromium contamination in soil and groundwater. In *book: Cr (VI) Handbook* (pp. 143–163), CRC Press LLC.

Tumolo, M., Ancona, V., De Paola, D., Losacco, D., Campanale, C., Massarelli, C., & Uricchio, V.F. (2020) Chromium pollution in European water, sources, health risk, and remediation strategies: an overview. *International journal of environmental research and public health, 17*(15), 5438.

Vajpayee, P., Tripathi, D., Rai, U. N., & Ali, M. B. (2000). Chromium (VI) accumulation reduces chlorophyll biosynthesis, nitrate reductase activity and protein content in Nymphaea alba L. *Chemosphere, 41*(7), 1075–1082.

Wang, L., Hou, D., Shen, Z., Zhu, J., Jia, X., Sik, Y., Tack, F. M. G., & Rinklebe, J. (2019). Field trials of Phyto mining and phytoremediation: A critical review of influencing factors and effects of additives. *Critical Reviews in Environmental Science and Technology, Taylor and Francis,*. https://doi.org/10.1080/10643389.2019.1705724

Microbiome Mediated Remediation of Chromium

Simmi Goel and Aarti

Abstract Significant amounts of heavy metals are introduced into the environment due to human activities and industrial processes such as the combustion of fossil fuels, mining, electroplating, and waste discharge from textile, automotive, chemical, steel, iron, and petrochemical industries. Additionally, natural phenomena like volcanic eruptions, sea spray, organic matter decomposition, and wind erosion also contribute to heavy metal release. A sustainable, eco-friendly approach using biological agents like bacteria, algae, yeasts, and plants has emerged as an effective method for treating heavy metal contamination. These biological techniques are not only cost-effective and widely available but also offer high metal removal efficiency. The success of these methods depends on several factors, including the type of biomass used (living or dead), the nature of the metal being targeted, the specific binding sites involved in metal capture, and environmental conditions such as pH, temperature, contact time, metal ion concentration, the type of biological agent used, the composition of the metal solution, and the presence of other competing ions. In view of this, the present paper focuses on the sources of chromium, health effects, mechanism of microbial remediation, biosorbents, various bioreactors, and the factors affecting the process along with the future prospects in field of treatment technology.

Keywords Chromium toxicity · Heavy metals · Bioremediation · Carcinogenic · Biological sorbents · Biosorption · Active and passive uptake

1 Introduction

Chromium is a naturally occurring trace element found in a variety of foods, including whole grains, meats, poultry, fish, and certain fruits and vegetables. In nature, the trivalent form of chromium (Cr (III)) is commonly present in complex forms with chromite ore ($FeCr_2O_4$), a mineral typically located in mafic and ultramafic rocks. These rocks contain varying levels of magnesium, iron, aluminum, and chromium.

S. Goel (✉) · Aarti
Department of Biotechnology, Mata Gujri College, Sri Fatehgarh Sahib, Punjab, India
e-mail: simmibiotech@matagujricollege.org

N. Kumar (ed.), *Chromium Toxicity*, https://doi.org/10.1007/978-3-032-14323-5_10

Through microbial activity and geochemical processes, Cr (III) can be transformed into its more toxic hexavalent form, Cr (VI). One of the primary natural sources of chromium pollution in the environment is geologic in origin, such as from exposed rock formations and volcanic activity. Generally, chromium is a widely dispersed geochemical element found in rocks, mineral-rich soils, and freshwater systems. Natural sources also include soil, rocks, and volcanic dust. While Cr (III) is often present in ores like ferrochromite and considered less harmful, Cr (VI) is highly toxic and mainly results from human industrial activities. However, natural processes involving mafic and ultramafic rocks can also lead to Cr (VI) contamination in groundwater, in addition to mining and other industrial operations (Nath & Kumar, 1988). Anthropogenic sources of chromium come from various industrial sectors, where it is used in manufacturing processes such as metallurgy, refractory materials, oil well drilling, metal plating, alloy production, textiles, pulp and paper, petroleum refining, pigment and dye production, paints, inks, polymers, chemicals, and leather goods. Effluents containing chromium are commonly discharged from activities like textile dyeing, steel and fertilizer production, film and photography, electroplating, chromite ore processing, packaging, leather tanning, and wood treatment, all of which contribute to the degradation of aquatic ecosystems (Tumolo et al., 2020; Zayed et al., 2003). Cr (III) is considered essential for mammals, playing a key role in the proper metabolism of glucose, lipids, and proteins. In contrast, Cr (VI) is known to be toxic and potentially carcinogenic. Therefore, distinguishing between Cr (III) and Cr (VI) is crucial for understanding their respective toxicological impacts (Mamyrbaev et al., 2015; Murthy et al., 2022).

Cr contamination has emerged as a serious issue in the environment because of its greater concentration in different agricultural and industrial processes. The tolerable limit of Cr in soil from the viewpoint of human and environmental health protection has been estimated at about 64 mg/kg. For instance, considerable quantities of Cr are emitted in various environmental segments through various industrial processes like cement and steel manufacturing units, leather industries, electroplating, paints and dyes, metal plating, wood processing, paper manufacture, and leaching processes. Furthermore, the ash fallout of burning fossil fuels or sewage waste for bioenergy production and fertilizer industries also has greatly elevated the chromium contamination (Nigam et al., 2015).

2 Human Health Risks

Individuals residing in chromium-contaminated regions had a high likelihood of having chronic disease conditions including nasal and lung cancer, lung abnormality induced by toxic substances (Beaver et al., 2009). The genotoxic and mutagenic properties of Cr (VI) induce cell transformation, sister chromatid exchange, DNA damage, and gene mutation, thus causing the development of cancer. All these abnormalities in the body can lead to serious medical conditions in humans (Remy et al., 2017). The occupational exposure showed a very high level of Cr (VI) in red blood cells, resulting in DNA damage and induced oxidative stress, targeting blood cells and ultimately disrupting the hematological system (Mamyrbaev et al., 2015; Ray et al., 2016).

Humans exposed to chromium contamination had skin allergic reactions. Toxic Cr (VI) brought about injurious effects on the kidney, liver, and blood (Mishra & Bharagava, 2016). Human exposure to chromium causes allergies, irritations, eczema, ulceration, skin and nasal irritations, eardrum perforation, respiratory tract disorders, and lung cancer (Duarte et al., 2018). Exposure to Cr (VI) resulted in oxidative protein alterations, DNA mutations, chromosomal aberrations and the cytotoxicity, genotoxicity, and carcinogenicity. Furthermore, breathing in excessive amounts of hexavalent chromium resulted in anemia, nose ulcers, irritation of the nose lining, stomach and small intestine irritation and ulcers, and other respiratory issues like coughing, wheezing, nasal blockage, and facial erythema (Table 1). Chromium exposure was linked to a number of detrimental health effects, including protein denaturation, DNA damage, aberrant enzymatic activity, congenital malformations, metabolic syndrome, chromosomal aberrations, low birth weight, cancer, back pain, asthma, dermatitis, chronic bronchitis, hemoglobin changes, and hypertension (Sharma et al., 2022). Occupational exposure to chromium causes adverse health effects such as bronchitis, asthma, respiratory tract irritation, and ulceration and perforation of the nasal septum. Cr can affect the development of the embryo by causing weight loss, delayed fetal development, skeletal defects, dead fetus, malformations by crossing the placenta, and reaching the fetus (Alimba et al., 2016; Banu et al., 2017; Pan et al., 2017).

3 Microbes Used in Removal of Chromium

Several microorganisms have evolved different strategies to neutralize the toxic action of chromium. Common physicochemical methods for removing Cr (VI) are very costly and ineffective at low concentrations (Anah & Astrini, 2017). Factors such as cell density, pH, temperature, salt concentration, redox potential, electron donors, oxyanions, metabolic inhibitors, and various heavy metals affect the efficiency of hexavalent chromium removal (Al Battashi et al., 2016).

3.1 Bacterial Remediation of Chromium

Bacteria are ubiquitous in nature and are often utilized for the removal of heavy metal pollutants (Carolin et al., 2017). Under anaerobic conditions, species like *Pseudomonas* can reduce hexavalent chromium [Cr (VI)], a toxic form, through various bioremediation processes such as adsorption, accumulation, and intracellular transformation. Certain *Lactobacillus* species which are common in the human microbiome and fermented foods also exhibit metal-binding properties. *Arthrobacter* and

Table 1 Health effects of chromium exposure to humans

S. No	Type of system	Health implications of chromium exposure	References
1	Respiratory system	Asthma, dyspnea, coughing, wheezing, nasal irritation, chronic rhinorrhea, soreness, epistaxis, nasal mucosal atrophy, bronchitis, nasal septal perforation and ulceration, pneumonia, reduced pulmonary function, laryngeal, tracheal, and bronchus hyperplasia and metaplasia, emphysema	(Huvinen et al., 2002; Nigam et al., 2015)
2	Reproductive system	Spermatogenesis, decreased fertility, increased testes and ovarian weight and decreased uterine weight, altered skeletal development, low birth weight, miscarriages ductal obstruction, depletion of germ cells and decreased sperm count and motility, Sertoli cell fibrosis and leydig cell hyperplasia	(Banu et al., 2017; Pagano et al., 1983; Pan et al., 2017)
3	Gastrointestinal tract	Intestinal irritation, ulcers, blood anemia, bloody diarrhea, ulceration, hemorrhage, necrosis, vomiting, abdominal pain, glandular stomach epithelial hyperplasia and metaplasia, Stomach and intestinal cancers, liver and kidney damage	(Mamyrbaev et al., 2015; Sharma et al., 2022)
4	Skin and eyes	Edema, mouth ulcers, pharyngitis, scaly, dry, and fissured skin, corneal vesication, conjunctival congestion, discharge, corneal scarring, allergic contact dermatitis, erythema, edema, papules, vesicles, acute and chronic dermatitis, and palate gingivitis and oral inflammation	(Duarte et al., 2018; Shelnutt et al., 2007)

(continued)

Table 1 (continued)

S. No	Type of system	Health implications of chromium exposure	References
5	Lung infections	Lung cancer, chronic pulmonary inflammation, bronchiolar cell apoptosis, pneumonitis, chronic peribronchiolar inflammation, neutrophil infiltration, fibrosis, bronchiolar epithelial hyperplasia, cellular atypia, cytotoxic damage to alveolar capillary barrier	(Beaver et al., 2009)
6	Hematologic system	Impaired hemoglobin synthesis, reduced hematologic indices, less oxygen-carrying capacity, hematocrit–reduced proportion of red blood cells in blood, smaller red blood cells, less hemoglobin per cell, anemia, blood cell abnormalities and systemic toxicity, potentially leading to hemorrhage and death	(Nickens et al., 2010; Ray et al., 2016)
7	Genetic/cellular system	Genotoxicity: double-strand DNA breaks, chromosomal abnormalities, histone ubiquitination, epigenetic changes, mutagenesis, carcinogenesis, inhibition of apoptosis, loss of cell cycle control	(Alimba et al., 2016; Chen et al., 2019; Trzeciak et al., 2000)

Bacillus species, both isolated from tannery wastewater, demonstrated the ability to convert Cr (VI) into the less toxic trivalent chromium [Cr (III)] using both whole-cell suspensions and cell extracts. This reduction is facilitated by the bacterial enzyme chromate reductase, which transforms soluble and harmful Cr (VI) into insoluble and less toxic Cr (III). Another strain, *Bacillus cereus* S6, also from tannery effluent, has been applied for this detoxification process. Reduction experiments have utilized whole cells, along with both free and immobilized enzymes, in batch and continuous modes within specialized bioreactors. The Gram-negative bacterium *Stenotrophomonas maltophilia* achieved complete (100%) Cr (VI) removal via bioreduction. In a separate study, 97% of Cr (VI) from industrial sources was eliminated through bioaugmentation using *Aeromonas hydrophila* and *Alcaligenes faecalis* within 12 to 24 h. *Morganella morganii* was able to remove 92% of 4600 mg/L Cr (VI) at 37 °C.

Several Gram-positive bacteria, including *Bacillus*, *Staphylococcus* and *Pseudomonas*, have shown over 90% Cr (VI) removal efficiency. Metal-resistant strains such as *Pseudomonas putida*, *Bacillus subtilis* and *Azotobacter* S8 can tolerate high chromium concentrations, with *P. putida* enduring up to 400 μg/mL of Cr (III) and *Azotobacter* S8 up to 300 mg/L of Cr (VI). *Bacillus subtilis* has also been reported to survive Cr (III) concentrations of up to 400 μg/mL (Aryal, 2020; Malaviya & Singh, 2016; Wu et al., 2017).

3.2 *Fungal Remediation of Chromium*

Fungi are highly adaptable organisms capable of thriving in extreme environments, including those with elevated metal concentrations. Over time, they have developed mechanisms to survive in such harsh conditions. Mycoremediation is the use of fungi to detoxify environmental pollutants involving either macrofungi (mushrooms) or filamentous fungi (molds). Their high biomass yield and robust tolerance to stress make fungi ideal biosorbents for heavy metal removal (Akhtar & Mannan, 2020). Fungi are particularly effective in biosorption of Cr (VI) due to their ability to withstand concentrations exceeding 10,000 mg/L. Cr (VI) binds to the fungal cell surface through chemical interactions with functional groups on proteins, lipids, and polysaccharides like galactosamine, chitin, and glycan. Key functional groups involved include carboxyl (−COOH), phosphate (PO_4^{3-}), amine ($-NH_2$), thiol (−SH), and hydroxyl (-OH) (Benila Smily & Sumithra, 2017).

Filamentous fungi not only tolerate Cr (VI) but also effectively colonize and remediate contaminated sites. For instance, *Aspergillus niger* secretes enzymes that facilitate Cr (VI) reduction, while its cell wall shows strong metal-binding capabilities. *Aspergillus flavus* has also demonstrated successful detoxification of Cr (VI) from tannery wastewater (Srivastava & Thakur, 2006). Similarly, *Fusarium* species have been observed to biotransform Cr (VI) into Cr (III). Compared to bacteria, fungal biomass can be significantly larger, enabling the removal of substantial quantities of heavy metals like copper, cadmium, and chromium, up to 300 mg/L. Fungi employ multiple mechanisms like biosorption, bioaccumulation, biotransformation, biomineralization, and extracellular precipitation for remediation. Through biosorption, *Aspergillus flavus* and *Cladosporium* spp. have achieved nearly complete (∼100%) removal of Cr (VI) (Dinakarkumar et al., 2024).

3.3 *Algal Remediation of Chromium*

Algae and cyanobacteria are ubiquitous microbes that are present in nearly all environmental habitats including those with extreme aquatic or terrestrial conditions (El-Sikaily et al., 2007). Phycoremediation is termed as the application of algae and cyanobacteria in decontaminating heavy metals either through removal or breakdown of toxins. Algae have various chemical groups on their surfaces such as amide, carboxyl, hydroxyl, and phosphate that act as binding sites for metals. Among them, *Spirogyra* species demonstrated the highest efficiency in removing and adsorbing Cr (VI) ions. Additionally, the blue-green alga *Phormidium* sp. and the biomass of *Desmonostoc muscorum* have shown potential as biosorbents for Cr (VI) removal from wastewater. Algae are highly recommended for wastewater treatment due to their affordability, low energy consumption, and strong bioaccumulation capabilities (Al Homaidan et al., 2018; Yen et al., 2017). A recent study found that after more than two days of treatment via combustion, the green alga *Chlorella vulgaris* achieved an 88% reduction in Cr (VI). Moreover, brown and red algae such as *Sargassum* sp. and *Pterocladia capillacea* removed over 80% of Cr (VI) through adsorption (Pakshirajan et al., 2013; Tripathi & Sumathy, 2018).

3.4 Mycorrhizal Associations

By facilitating the uptake and translocation of pollutants within the plant system, fungi that develop symbiotic relationships with plants such as mycorrhizal fungi improve the bioremediation process. By extending the plant's root system, mycorrhizal fungi increase the surface area available for absorbing nutrients and pollutants. These fungi have the ability to release chelating agents and organic acids that help metals be absorbed and then stabilized in plant tissues. For example, it has been demonstrated that arbuscular mycorrhizal fungi enhance the uptake and immobilization of heavy metals such as lead and arsenic in plant roots, thereby improving phytoremediation of these metals.

3.5 Co-Metabolism and Synergistic Interactions

Additionally, fungi are involved in cometabolism, in which the breakdown of one compound promotes the breakdown of another. This procedure is especially helpful in settings where there are intricate pollutant mixtures. For instance, by changing the pH and redox conditions in the soil, fungi can break down hydrocarbons while also accelerating the breakdown of heavy metals. Additionally, fungi and other microorganisms, like bacteria, can work together to improve the bioremediation process as a whole. Bacterial strains break down complex molecules into simpler intermediates, which are then further degraded by fungi causing the sequential degradation of pollutants.

4 Mechanism of Bioremediation of Chromium

Chemical precipitation, oxidation–reduction, ion exchange, filtration, membrane separation, evaporation, and adsorption utilizing materials like activated carbon, alum, kaolinite, and ash are examples of conventional techniques for extracting metals from contaminated environments. These methods, however, frequently call for large amounts of energy or chemicals, which could result in secondary contamination. On the other hand, bioremediation provides a more environmentally friendly option by using microorganisms' metabolic processes to remove harmful metals from contaminated areas. Metals like chromium have been effectively removed from industrial wastewater using microbial remediation strategies such as biosorption, bioaccumulation, biotransformation, and bioleaching (Garg et al., 2012). Bioaccumulation, also known as active biosorption, is a metabolism-driven process that occurs only in living organisms. In this mechanism, hexavalent chromium ions are initially bound to cell surface ligands, forming a metal–ligand complex. This complex is then transported into the cell via transporter proteins. Inside the cell, metal-binding proteins such as metallothioneins and phytochelatins interact with the metal, facilitating processes like methylation and precipitation.

Both living and non-living microbial biomass (bacteria, fungi, and algae) can aid in removing toxic chromium through biosorption, bioaccumulation, and biotransformation (Chandra et al., 1997; Losi et al. 1994b). During microbial remediation, various microorganisms are employed to rapidly degrade hazardous pollutants in soil, water, sludge, and other media into less harmful or environmentally benign forms. This process harnesses microbial defense mechanisms developed over millions of years, such as metal ion reduction, intracellular sequestration, active transport, and extracellular barriers (Ma et al., 2019). Depending on microbial metabolism, metal removal can occur via biosorption (a passive process not requiring metabolic energy) or bioaccumulation (an active, energy-dependent process). In bioaccumulation, higher membrane permeability in living cells allows heavy metals to be absorbed into the cell, where they accumulate (Ramli et al., 2023). Biosorption, in contrast, is a quick, passive mechanism where dead or inactive biomass binds heavy metals to cell walls through extracellular interactions. In this process, metals do not enter the cell but attach to functional groups such as carboxyl, sulfhydryl, amino, phosphate, phenol, hydroxyl, and others found on the biosorbents' surfaces (Ayele & Godeto, 2021).

Microorganisms can interact with metal ions in several ways, including:

- Binding to cell walls
- Accumulating metals intracellularly
- Producing metal-chelating compounds (siderophores)
- Facilitating extracellular transformations
- Mobilizing or immobilizing metals
- Volatilizing metal ions (Bhattacharya et al., 2019; Wani et al., 2019).

4.1 Biosorption and Bioaccumulation

One of the most common methods of removing heavy metals from contaminated sites using fungi is the biosorption. Variety of groups like carboxyl, hydroxyl, and amine groups are abundant in the cell walls of fungi which can bind metal ions via chelation, complexation, and ionic exchange (Jeyakumar et al., 2023). *Aspergillus niger* and *Penicillium chrysogenum* have high ability for the sorption of lead, mercury, and cadmium. Bioaccumulation entails the uptake and internalization of metals into fungal cells, where they may attach to intracellular proteins or be trapped and stored in vacuoles (Ozdemir et al., 2004; Ozer et al., 2012).

4.2 Chelation and Precipitation

Heavy metals are bound and precipitated by organic acids like citric and oxalic acid, which are released by fungi like Aspergillus niger and other phosphate-solubilizing species. Citric acid decreases the toxicity and mobility of hazardous metals like lead, cadmium, and aluminum by forming stable complexes with them. Aspergillus niger, for instance, can produce up to 15 g/L of citric acid, which chelates metals like cadmium and helps remove them from the environment and immobilize them. The production of insoluble metal oxalates is facilitated by fungi that release oxalic acid. By filtering or sedimentation, these oxalates can be readily extracted from soil and water after precipitating out of solution. To efficiently reduce lead concentrations in contaminated water, for example, it has been demonstrated that A. niger precipitates lead as lead oxalate.

4.3 Enzymatic Degradation

Numerous extracellular enzymes produced by fungi aid in the breakdown of intricate organic contaminants. These contaminants are oxidized by enzymes such as laccases, manganese peroxidases, and lignin peroxidases, which reduce them to less dangerous and simpler forms. For instance, ligninolytic enzymes released by white-rot fungi like Phanerochaete chrysosporium can break down lignin and other complex compounds, such as synthetic colors and polycyclic aromatic hydrocarbons (PAHs). Laccases, in particular, oxidize phenolic compounds, reducing their toxicity and breaking down complex organics into smaller molecules (Srinath et al., 2002).

4.4 Redox Reactions

Additionally, fungi aid in redox processes, which are essential for detoxification and pollutant transformation. Fungal cells and pollutants exchange electrons during these processes, which might result in the contaminant being reduced or oxidized. The detoxification of the extremely poisonous hexavalent form of chromium [Cr (VI)] into the less deadly trivalent form [Cr (III)] is one prominent example. *Aspergillus niger* has demonstrated the ability to reduce Cr (VI) to Cr (III) in both oxygen-rich and oxygen-poor environments, making it a promising candidate for chromium bioremediation.

4.5 Ion Exchange

Fungal cell walls, rich in polysaccharides, can exchange ions with bivalent metal ions present in polluted soils. This ion exchange occurs via two mechanisms: non-metabolic binding and metabolic binding at hyphal sites. The cation exchange capacity (CEC) of fungal cell walls is determined by negatively charged functional groups and varies with both pH and fungal species. For arbuscular mycorrhizal (AM) fungi, higher pH levels can enhance the sorption of metal cations on the fungal cell wall.

4.6 Complexation

The rhizosphere which is the zone around plant roots plays a critical role in metal detoxification due to the release of organic acids, amino acids, sugars, and other compounds that bind metals. AM fungal hyphae, through cystine ligands, are involved in metal complexation near the root zone, helping to reduce the concentration of free metals. Both intracellular and extracellular sequestration mechanisms contribute to the detoxification process. While intracellular sequestration occurs within fungal cells and is often applied in wastewater treatment, extracellular complexation typically happens in the periplasmic space, where it aids in reducing metal availability in the environment.

4.7 Precipitation

Another mechanism that resulted from extracellular sequestration is precipitation. Geobacter species like *G. sulfurreducens* and *G. metallireducens* use this mechanism to convert hexavalent chromium to less toxic trivalent chromium. Higher concentrations of hydrogen sulfide cause metal cations to precipitate in sulfate-reducing bacteria. This mechanism is also used by AM fungi to recover soil from Cr-containing tannery effluents. The metallic cations were intracellularly precipitated in the hyphae of AM fungi. Furthermore, the precipitation of heavy metals is caused by glomalin, a protein that AM fungi produce in large quantities from their extraradical hyphae. Fungal structures like vesicles and intracellular spaces had higher metal precipitation.

4.8 Solubilization

Metals in metalliferous soils bind firmly to the soil particles, making it necessary to take specific steps to remove them. This is where solubilization comes in, and it's aided by the rhizosphere becoming more acidic due to the production of organic acids. Heavy metal solubility varies; Pb2 + and Cr are less soluble than Zn, Ni, and Cu, which dissolve readily. Relatively insoluble metal ligands are dissolved by the organic acids secreted by many bacterial genera. Phosphobacteria are one of the significant bacterial groups involved in this process. By introducing these bacteria that produce biosurfactants, which further liberate metals from the soil particles, the bioavailability of metals can be further enhanced. AM fungi's hyphae generate organic acids and dissolve the minerals and their metal ligands (Li et al., 2019).

4.9 Biotransformation of Cr (VI)

The biotransformation of chromium element is thought to be a detoxification mechanism. According to some recent research, the supernatant of microbial cultures contains extracellular chromate reductase activity. Some bacteria such as *Pseudomonas*, *Bacillus* and *Arthrobacter* are involved in chromium reduction either through insoluble cell membrane enzymes or soluble cytosolic proteins. Another alternative method for chromium reduction involved various metabolites like amino acids, nucleotides, sugars, vitamins, organic acids especially citrate (fungal metabolites) and oxalate, either by Mn alone or by the photocatalytic action of Fe (III) have been studied by scientists (Jobby et al., 2018).

5 Scale-Up Strategy

Microbial-based Cr (VI) removal has been explored using various process modes such as batch, fed-batch, and continuous systems (Sanjay et al., 2020). When scaling up, factors like fluid dynamics, mass transfer, and microbial growth conditions must be considered when choosing an appropriate bioreactor. Compared to in situ or solid-phase systems, bioreactors offer higher Cr (VI) removal efficiency, are easier to manage, and provide more predictable performance in controlled environments (Saeed et al., 2022; Verma & Kuila, 2019).

5.1 Stirred Tank Reactors (STR)

These reactors use mechanical stirring and can function in either batch or continuous operation. They are typically applied to treat smaller volumes of industrial wastewater. In continuous mode, the Cr (VI)-contaminated solution is steadily fed into the system. Multiple reactors can be used in series, adjusting hydraulic retention times based on microbial growth rates. Studies comparing bioreactor types under varying conditions tested aerobic suspended, aerobic attached, and anoxic attached growth systems. A strain of *Arthrobacter rhombi*, isolated from chromium-contaminated soil, achieved a 98% Cr (VI) reduction in aerobic attached systems, outperforming the other configurations.

5.2 Fixed-Bed Reactors

These systems pass contaminated liquid through a packed column containing immobilized biosorbent materials. Immobilization was achieved using techniques like entrapment, encapsulation, or bonding to carriers such as polymers, textiles, or inorganic substances. A two-stage anaerobic packed bed reactor using immobilized *Pseudomonas aeruginosa* and *Bacillus subtilis* on agar was effective in treating effluent from electroplating industries.

5.3 Fluidized-Bed Bioreactors (FBRs)

FBRs rely on biofilms forming on suspended solid particles. A counter-flow liquid stream keeps these particles in a constant fluidized motion within the reactor column. When using chemically modified and polysulphone immobilized *Rhizopus nigricans* biomass, STRs showed the highest Cr (VI) removal, followed by fluidized and then packed bed reactors under identical operating conditions.

5.4 Airlift Reactors

These gas-driven reactors, like airlift and bubble columns, are ideal for fungi due to their low shear stress environments. They reduce pellet formation and encourage hyphal dispersion, improving biosorption. Fungi like *Trichoderma viridae* and *Hypocrea tawa* performed well in airlift systems. In contrast, STRs caused mycelium fragmentation, which reduced biosorption efficiency.

6 Factors Influencing Cr (VI) Remediation

Microbial biosorbents with multiple functional groups have greater capacity to bind contaminants. Functional groups involved in chromium biosorption include carboxyl, phosphate, amine, alcohols/phenols, aldehydes, esters, and others (Asira, 2013; Naik & Duraphe, 2012; Sharma et al., 2022). These are mainly found on microbial cell walls and specific in organisms.

6.1 PH Influence

pH affects both metal ion solubility and the charge of functional groups on biosorbents. At lower pH, Cr (VI) reduction is favored due to proton consumption. Maximum Cr (VI) adsorption by *Pedigrum boryanum* occurred at pH 2.0 and efficiency decreased as pH increased, showing that low pH enhances biosorption.

6.2 Temperature Effect

Temperature plays a crucial role in microbial enzymatic activity and biosorption. High temperatures can damage microbial cells by affecting DNA, proteins and membranes. *Desulfovibrio CR-1* showed increased Cr (VI) removal with rising temperatures from 30 to 40 °C, indicating optimal removal at moderate warmth.

6.3 Contact Time

Contact time directly influences removal efficiency. As exposure time increases, removal improves until it reaches at equilibrium. *Pantoea sp. TEM18* rapidly adsorbed Cr (VI) in the first five minutes, stabilizing after 15 min, indicating fast biosorption kinetics.

6.4 Initial Metal Concentration

Initial Cr (VI) concentration affects removal performance. While higher concentrations increase biosorption capacity, removal efficiency drops due to limited binding sites. For example, removal dropped from 99.88% to 83.69% as concentration increased from 100 to 500 mg/L, but uptake capacity increased from 6.27 to 28.95 mg/g.

6.5 Biosorbent Dosage

Increasing biosorbent dosage improves metal removal due to more available surface area. In one study, raising the *Scenedesmus sp.* inoculum from 2 to 10% (w/v) significantly improved Cr (VI) removal from 45.9 to 92.9%, with a slight increase to 96 at 30% (w/v).

6.6 Agitation Speed

Agitation enhances oxygen transfer and microbial activity, boosting Cr (VI) reduction. However, excessive agitation can damage cells. Key parameters influencing Cr (VI) removal include pH, temperature, contact time, agitation speed, and biosorbent dosage all of which must be optimized based on the microbial strain used.

6.7 Nutrient Availability

Microbial activity and bioremediation depend on nutrient balance, especially carbon, nitrogen, and phosphorus. Adjusting C: N: P ratios improve degradation performance. In nutrient-poor environments, supplementation is essential to enhance microbial metabolism and pollutant breakdown especially in cold or aquatic systems.

6.8 Environmental Conditions

Environmental variables such as temperature, pH, moisture, soil structure, oxygen, redox potential, and contaminant characteristics affect microbial growth and bioremediation success. Bioremediation efficiency hinges on the bioavailability of pollutants and the surrounding physical and chemical environment.

References

Akhtar, N., & Mannan, M. A. U. (2020). Mycoremediation: Expunging environmental pollutants. *Biotechnol. Rep., 26*, Article e00452.

Al-Battashi, H., Joshi, S. J., Pracejus, B., & Al-Ansari, A. (2016). Geomicrobiology of chromium (VI) pollution: Microbial diversity and its bioremediation potential. *Open Biotechnology Journal, 10*(1), 379–389.

Al-homaidan, A. A., Al-qahtani, H. S., Al-ghanayem, A. A., Ameen, F., & Ibraheem, I. B. M. (2018). Potential use of green algae as a biosorbent for hexavalent chromium removal from aqueous solutions. *Saudi Journal of Biological Sciences, 25*(8), 1733–1738.

Alimba, C. G., Dhillon, V., Bakare, A. A., & Fenech, M. (2016). Genotoxicity and cytotoxicity of chromium, copper, manganese and lead and their mixture in WIL2-NS human B lymphoblastoid cells is enhanced by folate depletion. *Mutation Research, Genetic Toxicology and Environmental Mutagenesis, 798*, 35–47.

Anah, L., & Astrini, N. (2017). Influence of pH on Cr (VI) ions removal from aqueous solutions using carboxymethyl cellulose-based hydrogel as adsorbent. *IOP Conference Series: Earth and Environmental Science, 60*, Article 012010.

Aryal, M. (2020). A comprehensive study on the bacterial biosorption of heavy metals: Materials, performances, mechanisms, and mathematical modellings. *Reviews in Chemical Engineering, 36*, 1–40.

Asira, E. E. (2013). Factors that determine bioremediation of organic compounds in the soil. *Academic Journal of Interdisciplinary Studies, 2*(13), 125.

Ayele, A., & Godeto, Y. G. (2021). Bioremediation of chromium by microorganisms and its mechanisms related to functional groups. *Hindawi Journal of Chemistry, 7694157*, 21.

Banu, S. K., Stanley, J. A., Sivakumar, K. K., Arosh, J. A., Taylor, R. J., & Burghardt, R. C. (2017). Chromium VI- induced developmental toxicity of Placenta is mediated through spatiotemporal dysregulation of cell survival and apoptotic proteins. *Reproductive Toxicology, 68*, 171–190.

Beaver, L. M., Stemmy, E. J., Constant, S. L., Schwartz, A., Little, L. G., Gigley, J. P., et al. (2009). Lung injury, inflammation and Akt signalling following inhalation of particulate hexavalent chromium. *Toxicology and Applied Pharmacology, 235*, 47–56.

Benila Smily, J. R. M., & Sumithra, P. A. (2017). Optimization of chromium biosorption by fungal adsorbent, *Trichoderma sp.* BSCR02 and its desorption studies. *Hayati Journal of Biosciences, 24*(2), 65–71.

Bhattacharya, A., Gupta, A., Kaur, A., & Malik, D. (2019). Alleviation of hexavalent chromium by using microorganisms: Insight into the strategies and complications. *Water Science and Technology, 79*(3), 411–424.

Carolin, C. F., Kumar, P. S., Saravanan, A., Joshiba, G. J., & Naushad, M. (2017). Efficient techniques for the removal of toxic heavy metals from aquatic environment: A review. *Journal of Environmental Chemical Engineering, 5*, 2782–2799.

Chandra, P., Sinha, S., & Rai, U. N. (1997). Bioremediation of Cr from water and soil by vascular aquatic plants. In E. L. Kruger, T. A. Anderson, & J. R. Coats (Eds.), *Phytoremediation of soil and water contaminants, ACS symposium series 664* (pp. 274–282). American Chemical Society.

Chen, Q. Y., Murphy, A., Sun, H., & Costa, M. (2019). Molecular and epigenetic mechanisms of Cr (VI)-induced carcinogenesis. *Toxicology and Applied Pharmacology, 377*, Article 114636.

Dinakarkumar, Y., Ramakrishnan, G., Gujjula, K. R., Vasu, V., Balamurugan, P., & Murali, G. (2024). Fungal bioremediation: An overview of the mechanisms, applications and future perspectives. *Environmental Chemistry and Ecotoxicology, 6*, 293–302.

Duarte, I., Mendonça, R. F., Korkes, K. L., Lazzarini, R., & Hafner, M. D. (2018). Nickel, chromium and cobalt: The relevant allergens in allergic contact dermatitis. Comparative study between two periods:1995–2002 and 2003–2015. *Anais Brasileiros De Dermatologia, 93*, 59–62.

El-Sikaily, A., El Nemr, A., Khaled, A., & Abdelwehab, O. (2007). Removal of toxic chromium from wastewater using green alga *Ulva lactuca* and its activated carbon. *Journal of Hazardous Materials, 148*(1–2), 216–228.

Garg, S. K., Tripathi, M., & Srinath, T. (2012). Strategies for chromium bioremediation of tannery effluent. In D. Whitacre (Ed.), *Reviews of environmental contamination and toxicology* (Vol. 217). Springer.

Gomathy, M., Sabarinathan, K. G., Subramaian, K. S., Sivashankari Devi, T., Ananthi, K., Kalaiselvi, P., & Jeyshree, M. (2022). Microbial remediation of chromium, chapter 10. Advances in environmental microbiology. In C. J. Hurst (Ed.), *Book: Microbial metabolism of metals and metalloids* (pp. 255–278). Springer Nature.

Huvinen, M., Uitti, J., Oksa, P., Palmroos, P., & Laippala, P. (2002). Respiratory health effects of long-term exposure to different chromium species in stainless steel production. *Occupational Medicine, 52*(4), 203–212.

Jeyakumar, P., Debnath, C., Vijayaraghavan, R., & Muthuraj, M. (2023). Trends in bioremediation of heavy metal contaminations. *Environment Engineering Research, 28*(4), Article 220631.

Jobby, R., Jha, P., Yadav, A. K., & Desai, N. (2018). Biosorption and biotransformation of hexavalent chromium [Cr (VI)]: A comprehensive review. *Chemosphere, 207*, 255–266.

Li, X., Fan, M., Liu, L., Chang, J., & Zhang, J. (2019). Treatment of High-concentration chromium-containing wastewater by sulfate-reducing bacteria acclimated with ethanol. *Water Science and Technology, 80*(12), 2362–2372.

Losi, M. E., Amrhein, C., & Frankenberger, W. T. (1994). Factors affecting chemical and biological reduction of hexavalent chromium in soil. *Environment Toxicology Chemistry, 13*(11), 1727–1735.

de Lourdes Pereira, M., das Neves, R. P., Oliveira, H., Santos, T. M., & De Jesus, J. P. (2005). Effect of Cr (V) on reproductive organ morphology and sperm parameters: an experimental study in mice. *Journal Environment Health, 4*, 1–6.

Ma, L., Xu, J., Chen, N., Li, M., & Feng, C. (2019). Microbial reduction fate of Chromium (Cr) in aqueous solution by mixed bacterial Consortium. *Ecotoxicology and Environmental Safety, 170*, 763–770.

Malaviya, P., & Singh, A. (2016). Bioremediation of chromium solutions and chromium containing wastewaters. *Critical Reviews in Microbiology, 42*(4), 607–633.

Mamyrbaev, A. A., Dzharkenov, T. A., Imangazina, Z. A., & Satybaldieva, U. A. (2015). Mutagenic and carcinogenic actions of chromium and its compounds. *Environmental Health and Preventive Medicine, 20*, 159–167.

Mishra, S., & Bharagava, R. N. (2016). Toxic and genotoxic effects of hexavalent chromium in environment and its bioremediation strategies. *Journal of Environmental Science and Health Part C Environment Carcinogenesis and Ecotoxicology Reviews, 34*(1), 1–32.

Murthy, M. K., Khandayataray, P., Padhiary, S., & Samal, D. (2022). A review on chromium health hazards and molecular mechanism of chromium bioremediation. *Reviews on Environmental Health, 38*(3), 461–478.

Naik, M. G., & Duraphe, M. D. (2012). Review paper on-parameters affecting bioremediation. *International Journal Life Science Pharma Research, 2*(3), L77–L80.

Nath, K., & Kumar, N. (1988). Hexavalent chromium: Toxicity and its impact on certain aspects of carbohydrate metabolism of the freshwater teleost *Colisa fasciatus*. *Science of the Total Environment, 72*, 175–181.

Nickens, K. P., Patierno, S. R., & Ceryak, S. (2010). Chromium genotoxicity: A double-edged sword. *Chemico-Biological Interactions, 188*, 276–288.

Nigam, H., Das, M., Chauhan, S., et al. (2015). Effect of chromium generated by solid waste of tannery and microbial degradation of chromium to reduce its toxicity: A review. *Advances in Applied Science Research, 6*(3), 129–136.

Ozdemir, G., Ceyhan, N., Ozturk, T., Akirmak, F., & Cosar, T. (2004). Biosorption of Cr (VI), Cd (II) and Cu (II) by *Pantoea sp.* TEM18. *Chemical Engineering Journal, 102*(3), 249–253.

Ozer, T. B., Erkaya, I. A., Udoh, A. U., et al. (2012). Biosorption of Cr (VI) by free and immobilized *Pediastrum boryanum* biomass: Equilibrium, kinetic, and thermodynamic studies. *Environmental Science and Pollution Research, 19*(7), 2983–2993.

Pagano, G., Esposito, A., Bove, P., de Angelis, M., Rota, A., & Giordano, G. G. (1983). The effects of hexavalent and trivalent chromium on fertilization and development in sea urchins. *Environmental Research, 30*, 442–452.

Pakshirajan, K., Worku, A. N., Acheampong, M. A., Lubberding, H. J., & Lens, P. N. L. (2013). Cr (III) and Cr (VI) removal from aqueous solutions by cheaply available fruit waste and algal biomass. *Applied Biochemistry and Biotechnology, 170*(3), 498–513.

Pan, X., Hu, J., Xia, W., Zhang, B., Liu, W., Zhang, C., et al. (2017). Prenatal chromium exposure and risk of preterm birth: A cohort study in Hubei. *China Science Report, 7*, 1–8.

Ramli, N. N., Othman, A. R., Kurniawan, S. B., Abdullah, R. S., & Hasan, H. A. (2023). Metabolic pathway of Cr (VI) reduction by bacteria: A review. *Microbiological Research, 268*, Article 127288.

Ray, R. R. (2016). Adverse hematological effects of hexavalent chromium: An overview. *Interdisciplinary Toxicology, 9*(2), 55–65.

Remy, L. L., Byers, V., & Clay, T. (2017). Reproductive outcomes after non-occupational exposure to hexavalent chromium, Willits California. *Environmental Health, 16*, 1–15.

Saeed, M. U., Hussain, N., Sumrin, A., Shahbaz, A., Noor, S., Bilal, M., Aleya, L., & Iqbal, H. M. (2022). Microbial bioremediation strategies with wastewater treatment Potentialities–A review. *Science Total Environment, 818*, 151754.

Sanjay, M. S., Sudarsanam, D., Raj, G. A., & Baskar, K. (2020). Isolation and identification of chromium reducing bacteria from tannery effluent. *Journal King Saud University Science, 32*, 265–271.

Sharma, P., Singh, S. P., Parakh, S. K., & Tong, Y. W. (2022). Health hazards of hexavalent chromium (Cr (VI)) and its microbial reduction: A Review. *Bioengineered, 13*(3), 4923–4938.

Shelnutt, S. R., Goad, P., & Belsito, D. V. (2007). Dermatological toxicity of hexavalent chromium. *Critical Reviews in Toxicology, 37*, 375–387.

Srinath, T., Verma, T., Ramteke, P. W., & Garg, S. K. (2002). Chromium (VI) biosorption and bioaccumulation by chromate resistant Bacteria. *Chemosphere, 48*, 427–435.

Srivastava, S., & Thakur, I. S. (2006). Isolation and process parameter optimization of *Aspergillus sp.* for removal of Chromium from tannery effluent. *Bioresource Technology, 97*(10), 1167–1173.

Tripathi, P. P., & Sumathy, J. H. (2018). A comparative study on the application of algae for bioremediation of cooum water. *International Journal of Engineering and Techniques, 4*(1), 426–442.

Trzeciak, A., Kowalik, J., Malecka-Panas, E., Drzewoski, J., Wojewódzka, M., Iwanenko, T., et al. (2000). Genotoxicity of chromium in human gastric mucosa cells and peripheral blood lymphocytes evaluated by the single cell gel electrophoresis (comet assay). *Medical Science Monitor, an International Journal Publishing Experimental and Clinical Medicine, 6*, 24–29.

Tumolo, M., Ancona, V., De Paola, D., et al. (2020). Chromium pollution in European water, sources, health risk, and remediation strategies: An overview. *International Journal of Environmental Research and Public Health, 17*(15), 5438.

Verma, S., & Kuila, A. (2019). Bioremediation of heavy metals by microbial process. *Environmental Technology and Innovation, 14*, Article 100369.

Wani, P. A., Wahid, S., Khan, M. S. A., Rafi, N., & Wahid, N. (2019). Investigation of the role of chromium reductase for Cr (VI) reduction by *Pseudomonas sp.* isolated from Cr (VI)contaminated effluent. *Biotechnology Research and Innovation, 3*(1), 38–46.

Wu, G., Xiao, X., Feng, P., Xie, F., Yu, Z., Yuan, W., Liu, P., & Li, X. (2017). Gut remediation: A potential approach to reducing chromium accumulation using *Lactobacillus plantarum* TW1-1. *Science and Reports, 7*(1), 15000.

Yen, H. W., Chen, P. W., Hsu, C. Y., & Lee, L. (2017). Use of autotrophic *Chlorella vulgaris* in chromium (VI) reduction under different reduction conditions. *Journal of the Taiwan Institute of Chemical Engineers, 74*, 1–6.

Zayed, A. M., & Terry, N. (2003). Chromium in the environment: Factors affecting biological remediation. *Plant and Soil, 249*, 139–156.

Integrated Strategies for Effective Remediation of Chromium-Contaminated Soils

Prasann Kumar

Abstract Chromium contamination of soils presents a significant environmental and health risk due to its toxicity, mobility, and persistence. Traditional remediation techniques often concentrate only on physical or chemical treatments, which may not fully address the complexities of chromium-contaminated sites. This chapter explores integrated strategies that combine biological, chemical, and physical methods for the effective remediation of chromium-contaminated soils. It provides a comprehensive overview of current technologies, emphasizing synergistic approaches that enhance contaminant removal, reduce secondary pollution, and restore soil functionality. The chapter discusses phytoremediation using chromium-accumulating plants, microbial-assisted bioremediation, chemical stabilization with organic and inorganic amendments, and electrokinetic remediation techniques. Furthermore, it highlights the importance of site-specific assessments, risk analysis, and sustainability considerations in designing remediation plans. Emerging technologies and future directions, such as nanotechnology and genetic engineering, are also reviewed. By integrating multiple remediation approaches, this chapter aims to provide a roadmap for practitioners and researchers seeking efficient and eco-friendly solutions for managing chromium-polluted soils.

Keywords Chromium · Environment · Toxicity · Remediation · Organic · Plants · Soil · Bioaccumulation

1 Introduction

The two most environmentally significant types of chromium compounds are trivalent chromium (Cr(III)), found in formaldehyde, and hexavalent chromium (Cr(VI)). Cr(III), which represents the standard valence of chromium, is an essential trace element for the human body, participating in glucose metabolism, lipid metabolism,

P. Kumar (✉)
Department of Agronomy, School of Agriculture, Lovely Professional University, Phagwara, Punjab, India
e-mail: prasann0659@gmail.com

N. Kumar (ed.), *Chromium Toxicity*, https://doi.org/10.1007/978-3-032-14323-5_11

and protein biosynthesis. In contrast, Cr(VI) is extremely toxic, carcinogenic, and poses serious risks to both ecosystems and human health (Xing et al., 2025; Zhao et al., 2024). Chromium exhibits bioaccumulation and persistence properties, failing to decompose naturally, thus requiring eventual elimination. It is recognized as a contaminant of soils and waters globally and as a pollutant in the atmosphere. Cr(VI) originates from both human and natural sources. For example, rocks that contain significant amounts of chromium can weather, releasing chromite and ferrochromite, or volcanic eruptions can occur. However, most pollution arises from industrial processing methods such as electroplating, leather tanning, paint manufacturing, stainless steel treatment, pigment production, wood antiseptics, and hydrosement hardeners (Yuan et al., 2024; Zhang et al., 2025a). These industries emit heavy pollutants into the environment through liquid effluents and solid waste containing chromium; when untreated, these materials contaminate everything in the vicinity. Moreover, improper handling of waste materials containing chromium and emissions from vehicle combustion significantly contribute to the widespread presence of chromium in various environmental sinks. Once released into the environment, chromium undergoes complex geochemical transformations that determine its mobility, concentration, and toxicity. The success of strategies to prevent Cr pollution often hinges on whether a well-informed pollution control technician understands the exact state of chromium in the soil and acts accordingly (Kumar & Choudhury, 2025; Kumar & Singh, 2024a). Key environmental parameters influencing whether Cr(III) can become bioavailable include the presence or absence of organic matter, pH, redox potential, cation exchange capacity, and the amount of organic material in the soil. While mono-atomic salts derived from nitric acid may enhance the solubility of Cr(III) in acidic soils, elevated levels of ions such as carbonate and sulfite can often increase the availability of Cr(VI). The spread of Cr within soil profiles may lead to groundwater contamination, endangering drinking water supplies. Surface waters near contaminated sites can also be affected by leaching and erosion, which spread contamination to organisms inhabiting the area (Deoli et al., 2025; Shen et al., 2025). Regarding farming and crops, Cr storage in soils presents a serious problem. From carbon uptake by plants to digestion in humans, this metal introduces multiple points of potential entry into the food chain. Plants consistently grown in a medium containing chromium show reduced seedling survival rates, inhibited growth, chlorosis that may progress from dark yellow to white, metabolic alterations due to oxidative stress, and inhibited enzymatic functions within their interconnected control systems. For bacteria residing in the spaces between roots and soil particles, the presence of chromium in irrigation water undoubtedly impacts soil ecology. Aquatic communities are also at risk of mutations, including cancer, leukemia, and skin disorders, from Cr pollution (Kumar & Choudhury, 2025; Kumar & Singh, 2025a, 2025b; Kumar & Singh, 2025, 2024a). One way people can be exposed to Cr(VI) is by inhaling dust or fumes containing it, both of which can irritate the respiratory tract and may lead to lung cancer. They can also cause mucosal ulcers, such as those in the nasal septum, and trigger asthma (Maiti et al., 2023; Shaghaleh et al., 2025). If ingested through contaminated drinking water or food, Cr(VI) can result in stomach disorders, liver and kidney damage, and general poisoning. Skin contact can cause dermatitis. In workplaces such as leather tanning and electroplating factories, workers are consistently at risk of deteriorating health over time. Extended exposure, even at low levels, may lead to lung or gastrointestinal cancers. Children and individuals with weakened immune systems are especially vulnerable to long-term chromium exposure. Regulatory agencies such as the United States Environmental Protection Agency (USEPA), the World Health Organization (WHO), and the European Union have established limits for chromium levels in drinking water, soil, and air. For instance, the WHO has set a guideline of 0.05 mg/L for Cr(VI) in drinking water (Ajao et al., 2025; Zheng et al., 2025). Depending on land usage by residents in surrounding areas, soil remediation values in several countries range from 1 to 5 mg/km. While many regulations are still in development, chromium pollution remains a significant issue in numerous areas, particularly those without strict laws. These regulations apply to both urban and industrial regions. Hence, there is an urgent need for coordinated strategies addressing the entire science and technology of Cr remediation, considering both its chemical and biological aspects. Management decisions should not only focus on reducing toxic species; they must also assess recovery opportunities at the soil level and devise plans for ecological impacts, particularly frameworks to ensure sustainability (Pradhan & Nambi, 2025; Santalunai et al., 2024). This will necessitate a multidisciplinary approach, involving collaboration among environmental scientists, engineers, toxicologists, policymakers, and community stakeholders to develop site-specific, integrated solutions. Given the challenges in predicting the next sites of large-scale pollution, understanding the origins and fate of chromium in the environment is crucial for designing effective monitoring systems and intervention strategies. As the industrial age advances rapidly, controlling and mitigating existing Cr pollution, along with reducing emissions of this metal, is increasingly important for maintaining environmental balance and safeguarding human health (Fig. 1).

Fig. 1 Environmental and health impacts of chromium compounds

2 Challenges in Chromium Soil Remediation

Chromium-contaminated soil presents a significant challenge involving a combination of biological, physicochemical, and environmental factors that influence how chromium behaves within the soil system. The inherent complexity of soil matrices remains one of the major obstacles. Several key factors are crucial: the ratio of mineral particles to organic matter, texture and structure, organic matter content, mineral composition, pH value, and redox potential. Cation exchange capacity (CEC) is another important variable with implications for both the root causes and consequences of environmental safety at a contaminated site. These factors significantly affect the speciation, bioavailability, mobility, and resilience of chromium in contaminated soils (Cheng et al., 2025; Hou et al., 2025; Wang et al., 2024). Soil is a complex medium composed of mineral particles, organic materials, water, air, and microorganisms. Each component contributes to the dynamic exchange processes between contaminants, which typically have a slight electrical charge at their origin, and the surrounding medium. In contrast, soils with high CEC, primarily consisting of clay particles, effectively adsorb cations like Cr(III), thus decreasing their mobility. Conversely, soils rich in iron and aluminum oxides may form strong complexes with Cr(VI), enhancing its retention and making removal challenging, akin to chiseling off hard limestone cement. Organic matter can either promote or inhibit the movement of chromium, depending on its chemical nature and the prevailing situation. For instance, humus compounds may play dual roles; they can assist in reduction via a Cr(III) pathway or contribute to the dissolution of Cr into the soil solution by complexing it with ligands. In the presence of competitive ions such as carbonate, sulfate, and phosphate, the fate of chromium becomes increasingly complex, as they alter its speciation, inducing desorption or precipitation reactions (Ahmad et al., 2024; Ali et al., 2024; Suanon et al., 2025). This variability necessitates site-specific solutions; methods that may be effective in one soil type could fail in another. Additionally, the environmental parameters of each soil type must be considered, as water content, soil depth, and biological activity create irregularities in three dimensions, both horizontally and vertically. Without a precise initial definition coupled with ongoing monitoring, effective management becomes nearly impossible. Even if matrix composition is fixed, chromium species consistently emerge through various chemical reaction pathways, directly influencing the overall environmental behavior of materials and their potential liability for remediation. Chromium exists in several oxidation states, but the most significant in an ecological context are Cr(III) and Cr(VI). Its extreme mobility and relatively lower toxicity under typical environmental conditions lead to challenges that single-treatment methods struggle to overcome if we aim to remediate chromium-contaminated soil effectively and sustainably. Each remediation method—including soil washing, chemical stabilization, and phytoremediation—presents distinct strengths and weaknesses. For instance, while chelating agents or acids used in soil washing can effectively remove heavy metals such as chromium from the soil, this process is typically costly and generates significant amounts of waste byproducts. Moreover, it may strip essential nutrients from the land, consequently reducing soil fertility. In chemical stabilization, substances such as lime and zero-valent iron immobilize chromium, decreasing its bioavailability and leaching risks (Liu et al., 2023; Taneja et al., 2024; Zhang et al., 2025b). However, this process does not effectively remove chromium from the environment, and its success relies on specific geochemical conditions. For example, alterations in pH or redox status may lead to the remobilization of the material. Phytoremediation, though environmentally friendly and relatively low-cost, is generally slow and has limitations. Factors such as plant growth rate, metal uptake efficiency, and the tolerance levels of various plant species to high Cr(VI) levels hinder its effectiveness. Electrokinetic remediation utilizes electric fields to transport charged species to electrodes for collection, demonstrating considerable promise in leaching Cr(VI) from fine-grained soil. Nevertheless, it is energy-intensive and may produce undesirable side effects, including soil acidification and nutrient loss. Bioremediation particularly the use of microorganisms to reduce Cr(VI) to Cr(III)—offers an eco-friendly alternative. However, its efficiency depends on the presence of suitable microbial populations, the prevailing environmental conditions, and the absence of inhibitory substances such as other heavy metals or organic pollutants (Kumar et al., 2025; Sarankumar et al., 2025; Zhang et al., 2024). Additionally, these biological processes are often slow and challenging to control in the field, particularly in heterogeneous soil types with varying moisture and temperature regimes. Due to the ineffectiveness of independent approaches, integrating physical, chemical, and biotic methodologies is essential. The limitations of individual methods can be overcome while enhancing overall efficiency. Improving remediation effectiveness through multistep treatments is feasible; for example, Cr(VI) could be reduced to Cr(III) locally using reducing agents, followed by stabilization or encapsulation. Adding electrokinetic remediation to biostimulation or bioaugmentation can also enhance the local activity of anaerobic bacteria around the cathode material, resulting in greater in situ reduction of Cr(VI). Recycling wash solutions from soil washing processes alongside subsequent chemical or biological remediation can effectively manage pollution streams while recovering valuable materials (Kumar, 2024a; Kumar et al., 2024g; Kumar & Singh, 2024b; Sarker et al., 2023). However, integrating these systems requires a deep understanding of chromium chemistry, soil properties, microbial ecology, engineering principles, thorough risk assessments, and compliance with relevant regulations. Additionally, public perception, stakeholder involvement, and aligning costs with benefits significantly impact the long-term success of any remediation project. Currently, no recognized protocol or standard exists for this area of work regarding chromium treatment methods. Although laboratory studies can provide deeper insights into the mechanisms and kinetics, translating those findings to field conditions remains challenging due to the diverse nature of contamination sites, which vary widely in both spatial and temporal dimensions (Kumar, 2024b; Kumar & Choudhury, 2024a; Kumar & Dey, 2024; Kumar & Singh, 2024c). Monitoring and evaluating the success of remediation efforts necessitate robust analytical techniques capable of distinguishing among different chromium species and quantifying their transformations. Advanced spectroscopic methods, including X-ray absorption spectroscopy (XAS), electron paramagnetic resonance (EPR), and synchrotron-based systems, provide precise levels of speciation information; yet, they are often costly and technically demanding to implement, limiting their practical application in environmental monitoring. Furthermore, regulatory standards for chromium remediation differ widely across regions, with varying acceptable land-use thresholds based on risk models, exposure pathways, and land-use histories. The absence of uniform standards complicates the determination of a single remediation method and restricts the adoption of new technologies that do not meet detailed regulatory requirements. Additionally, climate change and environmental changes add an extra layer of uncertainty for future chromium remediation planning and implementation. Variations in temperature, rainfall patterns, and extreme weather events can influence soil moisture content, redox status, and microbial activity, all of which impact the transformation and transport of chromium (Kumar, 2024a; Kumar et al., 2024b; Kumar & Choudhury, 2024c; Kumar & Singh, 2024a). In cases of elevated rainfall, Cr(VI) pollution may leach readily into groundwater. Meanwhile, during droughts, heavy metal contamination levels can increase in the topsoil, where dust can be easily blown into the air by wind. Increased temperatures may accelerate microbial processes, but they could also trigger the re-release of previously immobilized chromium into the soil through changes in chemical conditions. Consequently, future remediation strategies must incorporate climate resilience and adaptability to ensure long-term protection for ecosystems and public health. Addressing the multitude of issues associated with chromium-contaminated soils necessitates a sustained focus on research, the development of innovative technologies, multidisciplinary cooperation, and the formulation of policies that facilitate the translation of advances in our understanding of chromium behavior into more efficient and sustainable options (Fig. 2).

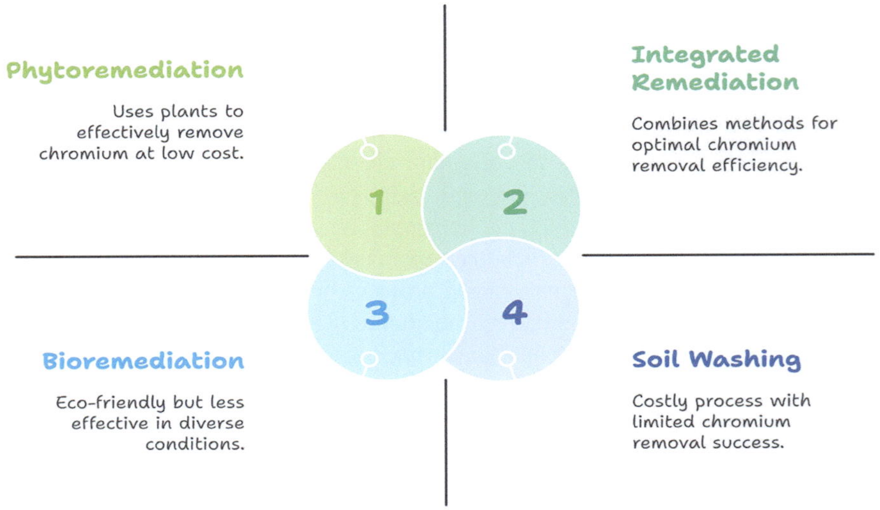

Fig. 2 Chromium soil remediation process

3 Physical Remediation Techniques

This method has emerged due to the growing need to remediate soil contaminated by chromium. In many cases, it quickly removes or isolates the affected area. Soil washing, thermal desorption, and electrokinetic remediation are all effective methods for remediating contaminated land. Each has its advantages and disadvantages in specific situations, depending on the characteristics of the contamination site (Kumar, 2024b; Kumar & Singh, 2024b, 2024c, 2024d). These techniques mainly rely on physical mechanisms, such as solubilization, volatilization, and electromigration, to lower the levels of toxic chromium species in soils, particularly the highly mobile hexavalent form Cr(VI) found in industrial chemical production waste sites. This presents both an environmental hazard and a risk to public health.

Soil washing is among the most widely used ex-situ physical methods for addressing soil contamination, particularly in areas where the soil primarily consists of sand or silt, and the pollutants are either loosely bound or have a readily extractable chemical structure. This method utilizes aqueous solutions, often combined with chelating agents, acids, alkali agents, surfactants, or ligands that bind to and dissolve heavy metals, including chromium, transforming them from a solid to a liquid phase. Common washing agents include hydrochloric acid (HCl), sulfuric acid (H_2SO_4), EDTA, citric acid, and various organic and inorganic ligands. Acid solutions are particularly effective at recovering Cr(III) by dissolving fractions obtained from carbonates and oxides. The addition of chelating agents also enhances extraction effectiveness, leading to increased recoveries of both Cr(III) and Cr(VI) from fully mobilized fractions by forming stable chelates with metal ions (Kumar, 2024c; Kumar et al., 2024a; Kumar & Choudhury, 2024b; Sharma & Kumar, 2024). Typically, soil washing involves several steps: excavating and cleaning the contaminated soil, mixing it with a washing solvent, and then separating the liquid and solid phases through sedimentation or filtration before treating the wash liquor to reclaim or safely dispose of metals. Moreover, soil washing provides a significant advantage over traditional methods of digging and disposing of hazardous waste: it can considerably reduce the volume of contaminated material awaiting further burial or incineration, thereby lowering long-term liability and costs. However, there are notable limitations to using soil washing as a treatment method, particularly if the target soil is fine-grained clay, which easily adsorbs metals due to its high cation exchange capacity and has substantial potential for expanding the practical range of exposure factors. Additionally, excessive use of strong acids can damage the soil structure, essential nutrients may be lost through erosion, and the pH level of the treated soil may not align with the natural conditions of both agricultural and ecological restoration sites (Kumar, 2024a, 2024b; Kumar & Choudhury, 2024c; Kumar & Dwivedi, 2024; Kumar & Raut, 2024; Kumar & Singh, 2024d). Therefore, the soil must undergo a series of post-treatment amendments before it is suitable for use in agro- or eco-restoration projects. Furthermore, the large volumes of secondary wastewater produced from soil washing, laden with heavy metal ions and chemical reagents, necessitate an expensive treatment process before discharge into the environment or reuse. Despite these limitations, soil washing remains a viable option in regions with loose soils, such as coarse sand or gravel, and very high concentrations of mobile chromium species, particularly when used in conjunction with other methods to address residual pollution.

It can also be used in soils contaminated with non-volatile polar compounds as well as semi-volatile organics. Thermal desorption is another essential remedial strategy based on physical methods and is particularly suitable for soils contaminated with semi-volatile or non-volatile organics. However, it has also shown promise for treating specific inorganic pollutants, such as chromium. This technique involves heating the contaminated soil media to elevated temperatures (typically between 300 and 600 °C for low-temperature thermal desorption and up to 1000 °C for high-temperature thermal destruction) to volatilize and remove contaminants from their solid matrix. Volatilized contaminants are then collected and managed using off-gas systems, such as activated carbon filters, scrubbers, or thermal oxidizers (Kumar et al., 2024b; Kumar, 2024a; Kumar & Choudhury, 2024c; Kumar & Dey, 2024). However, spirits have a relatively high boiling point of approximately 2671 °C, and their behavior under thermal conditions may be influenced by their interaction with organic matter or volatile compounds. Notably, Cr(VI) may be reduced to Cr(III) at moderate temperatures due to the presence of reducing agents, such as organic matter or carbonaceous materials, in the soil. Furthermore, some studies have shown that thermal treatment can induce mineralogical transformations in soil matrices, promoting the incorporation of chromium into more stable crystalline forms, such as spinel or silicates, thereby significantly reducing its bioavailability and leachability. High-temperature thermal desorption (also known as vitrification) melts the soil into an unreactive glass-like material, effectively containing the pollution as a permanent solid. However, this method is extremely energy-intensive and is typically limited to highly contaminated soils that require destruction or permanent melting. It offers several advantages over other methods of bioremediation; for example, it can treat many pollutants simultaneously, its removal efficiency can be very high, and the final products are usually safe and reusable. Nevertheless, the technique remains an energy-dependent operation and is associated with considerable energy consumption (Kumar & Choudhury, 2024b; Kumar & Dey, 2024; Kumar & Singh, 2024d, 2024a). Additionally, it may produce volatile contaminants or other unusable products that could be harmful if not adequately controlled. In many cases, thermal treatment can alter the physicochemical properties of soil, such as texture, porosity, and nutrient composition, often necessitating post-treatment rehabilitation measures, including soil conditioning and revegetation. Moreover, the feasibility of thermal desorption significantly depends on the depth and extent of contamination, the site's accessibility, and whether regulatory authorities will permit the material left behind after treatment. As such, it is primarily employed at industrial sites or other locations where severe contamination is involved, despite substantial economic and environmental costs.

Electrokinetic remediation (EKR) is a less destructive and invasive technology for remediation. It maintains everything in situ by using direct electric current to mobilize charged contaminants in the soil pore water, which opens highly variable oxymy. EKR is particularly suited for fine-grained soils, such as clays and silts, where traditional flushing or washing methods are impractical due to their low permeability. In an EKR setup, electrodes inserted into the Earth's crust are connected to a DC power source, generating an electric field that promotes electromigration, electrophoresis, and electro-osmosis—three primary mechanisms through which pollutants move. Positively charged species, such as Cr(III) cations, travel as far as the cathode, while negatively charged Cr(VI) oxyanions also migrate in that direction (Hou et al., 2025; Kumar & Singh, 2024b; Santalunai et al., 2024; Suanon et al., 2025). Electro-osmotic flow drives water from the anode to the cathode. Locally, this movement increases brilliance and impurity via pore water. After the water reaches the cathode, its reduction generates hydroxide ions (OH-), which create a high pH level near the cathode, potentially resulting in metals precipitating as hydroxides or oxides. However, near the anode, the water undergoes a similar process, producing hydrogen ions (H+), which can increase the solubility of some metals and accelerate their movement. Modifications have been made to enhance the efficiency of EKR for chromium-contaminated soils, including the installation of pH control systems, the addition of enhancers (such as acids, bases, chelating agents, and surfactants), and integration with bioremediation or redox-chemical oxidation processes (Hou et al., 2025; Shaghaleh et al., 2025; Shen et al., 2025; Zheng et al., 2025). An example is the injection of iron or citric acid into the anode pool, which increases the solubility of Cr(III). If you connect the cathode with reducing agents, such as sodium dithionite and zero-valent iron, conditions are created around the cathode where Cr(VI) is reduced to Cr(III), a less toxic form, thus lowering toxicity but also reducing the metal's reactivity. This process is further advanced by combining these agents with electrodes for direct reduction, a method known as bioelectroremediation. As a result, microorganisms inserted in the soil or indigenous organisms capable of performing redox reactions are encouraged to facilitate this in situ. The EKR approach has several operational limitations: it is challenging to scale up, remediation is slow, energy is consumed, and the electrodes corrode (Kumar & Choudhury, 2024a; Kumar & Singh, 2024b; Kumar & Singh, 2024c, 2024b). EKR's effects are significantly influenced by soil properties, such as moisture content, clay mineralogy, and buffering capacity, all of which relate to the conductivity and mobility of ions. Additionally, preferential pathways can develop, and differences may arise in the pH profile, which may imply that not all contaminants are removed from a polluted region. For the next step, the post-treatment recovery of chromium from electrode interiors and the management of generated waste streams must be addressed sensibly and skillfully. Nonetheless, EKR still holds promise as a method for cleaning up chromium-contaminated soils (Ajao et al., 2025; Cheng et al., 2025; Liu et al., 2023; Pradhan & Nambi, 2025). This potential is particularly valuable in situations where minimal soil disturbance is desired. Combining it with other methods can enhance the process's efficiency.

Physical remediation techniques, such as soil washing, thermal desorption, and electrokinetic remediation, provide valuable methods for managing chromium-contaminated soil, each with unique mechanisms, functional characteristics, and limitations. Soil washing offers an effective ex-situ solution for loamy soils; however, it is susceptible to secondary pollution and soil degradation. Thermal desorption allows for the safe removal and immobilization of pollutants but demands high levels of both energy and financial resources, making it most suitable for sites focused on cleaning up deeply contaminated areas (Ali et al., 2024; Cheng et al., 2025; Maiti et al., 2023; Pradhan & Nambi, 2025; Yuan et al., 2024). Electrokinetic remediation, as an in-situ alternative, creates less disturbance to the soil and has potential for integration with chemical or biological treatments. However, its effectiveness is constrained by soil heterogeneity and operational challenges. The options, whether applied individually or in combination, depend on various factors specific to each site, including soil type, pollution levels, chromium speciation, regulatory standards, and available resources. To maximize the potential of future physical remediation efforts, research should concentrate on optimizing existing methods, such as exploring ways to merge multiple remedial mechanisms into hybrid technologies and aligning environmental footprints with sustainability principles. Furthermore, sensor applications, real-time monitoring, and predictive models will enhance the accuracy and predictability of physical remediation strategies. This will provide decision-makers with more information, leading to more effective management of chromium-contaminated sites. Ultimately, the key to comprehensive, low-cost, and long-term remediation lies in integrating physical techniques with complementary biological and chemical approaches (Fig. 3).

Fig. 3 Chromium physical remediation process

4 Chemical Remediation Approaches

Due to their capability to affect speciation, mobility, and bioavailability of chromium through controlled chemical reactions, chemical remediation techniques are now a popular approach for addressing moderately to severely contaminated soil. They specifically aim to reduce the highly toxic hexavalent chromium (Cr(VI)) into less mobile and less harmful trivalent chromium (Cr(III)), or to immobilize chromium forms within the soil matrix to prevent their migration to groundwater and potential absorption by living organisms (Deoli et al., 2025; Hou et al., 2025; Pradhan & Nambi, 2025; Suanon et al., 2025; Zheng et al., 2025). The most widely used chemical strategies include chemical oxidation and reduction, stabilization and solidification, and the application of various amendments such as lime, biochar, and zero-valent iron (ZVI). Each method has its advantages and limitations, depending on the characteristics of the soil being treated, the degree of contamination, and the localized environmental conditions specific to each site.

We can change the redox state of chromium in contaminated soils primarily through chemical oxidation and reduction. For example, since Cr(VI) is the most hazardous form, reductions are used much less frequently than reductive steps. The hexavalent state of chromium exists mainly as chromate (CrO_4^{2-}) or dichromate ($Cr_2O_7^{2-}$), both of which are highly mobile, soluble, and cancer-causing substances. Reductive treatments aim to transform Cr(VI) into its Cr(III) form, which typically precipitates as hydroxides or oxyhydroxides at neutral to alkaline pH levels. This significantly reduces its solubility and toxicity (Kumar et al., 2024b; Kumar, 2024b; Kumar et al., 2024g; Kumar & Choudhury, 2024c). Commonly used reductants include ferrous sulfate ($FeSO_4$), hydrogen sulfide, sodium dithionite ($Na_2S_2O_4$), zero-valent iron (ZVI), and organic matter. Indeed, ferrous iron plays two roles: it serves as a reductant and acts as a co-precipitant, producing stable mixed Fe-Cr hydroxides that effectively help sequester chromium. In addition to chemical reductants, organic matter, such as humic substances formed through the microbial decomposition of plant material, appears capable of mediating Cr(VI) reductive processes. Lignin, a substance that is intermediate between cellulose and resin, as well as other plant-derived materials, are additional examples of this type. However, the success of chemical reduction largely depends on maintaining appropriate redox conditions. If subjected to oxidizing agents or allowed to absorb a large amount of air, Cr(III) may revert to its hazardous Cr(VI) form, particularly in environments rich in manganese oxides and other minerals that act as oxidizers (Ahmad et al., 2024; Cheng et al., 2025; Liu et al., 2023; Wang et al., 2024; Zhang et al., 2025b). Therefore, long-term monitoring is necessary, regardless of the reductive conditions. Only by keeping them stable can redox chromium species be maintained.

Compared to environmental reduction, however, chemical oxidation has not been as widely used to treat chromium pollutants. It can still be suitable in certain situations where Cr(III) forms organic complexes or where there is a need to convert one metal species to another for ultimate treatment. Oxidants like permanganate ions (MnO_4^-), peroxides (H_2O_2), and ozone (O_3) may change the valence of chromium.

However, caution must be exercised when using them, ensuring that the mobilization of other metals is prevented and that soil texture is not compromised. Regarding chromium-contaminated sites, applications involving oxidation are not universally ideal. If mismanaged, they could produce more toxic or mobile compounds rather than reducing them (Kumar, 2024b, 2024a; Kumar & Dey, 2024; Kumar & Singh, 2024a; Kumar & Singh, 2024c). Thus, while oxidation has its specific applications, it is not usually the first method employed to address chromium pollution through chemical means. The preferred approach remains reduction coupled with detoxification.

Stabilization and solidification (S/S) technologies represent a key direction in controlling chromium-contaminated soils chemically through physical encapsulation, which confines contaminants within a stable matrix or binds them chemically, thereby reducing their bioavailability. Common solidifying agents include Portland cement, fly ash, slag, or polymers, which are used in the solidification process to prevent the concentration of contaminants. The soil reaction with acid can result from adding lime for stabilization or from the transformation of pollutants into other forms, indicating that stabilizing agents often play a pivotal role. In the case of chromium, stabilization generally involves introducing phosphates, silicates, and compounds high in calcium, leading to the formation of less soluble or reactive Cr phases. Several factors, including the initial chemical form of chromium, soil pH, soil mineralogy, and competing ions, influence the effectiveness of S/S. For example, acidic soils will be affected when lime or cement additives are applied to increase pH, encouraging the precipitation of $Cr(OH)_3$ and thus decreasing its solubility (Ahmad et al., 2024; Cheng et al., 2025; Kumar et al., 2025; Taneja et al., 2024; Zhang et al., 2024; Zheng et al., 2025). However, at higher pH levels, some anionic species may become more soluble, or the performance of binding agents may be influenced, necessitating careful control over the amounts and conditions of amendment dosage required. A significant advantage of S/S methods is that they allow for treatment either on-site or off-site with minimal disruption to the surrounding area, making these technologies particularly suitable for large-scale contamination issues. However, a limitation of this technology is that it does not remove pollutants; instead, it immobilizes them. At some unspecified point in the future, under yet-to-be-specified geochemical conditions, their remobilization could pose a significant risk. Therefore, long-term performance monitoring and modeling of prediction scenarios are essential components of any S/S remediation plan to ensure that immobilized pollutants do not become active again (Ajao et al., 2025; Deoli et al., 2025; Hou et al., 2025; Liu et al., 2023; Yuan et al., 2024).

Amendments play a vital role in enhancing the effectiveness of remediation through chemical conditioning, such as modifying soil properties to inhibit the activation or transformation of contaminants, thereby reducing their persistence. Lime, biochar, and zero-valent iron (ZVI) are the most widely used amendments for chromium remediation. Each of these offers unique mechanisms of action and suitability for different soil types or contamination scenarios, providing tools to address this problem. Lime is primarily used to adjust soil pH, with a high pH being preferred.

Such conditions lead to the precipitation of Cr(III) as $Cr(OH)_3$, while Cr(VI) oxyanions have low solubility due to competing adsorption by carbonate and phosphate ions. Additionally, lime promotes the formation of stable Fe-Cr or Al-Cr coprecipitates, which are even less soluble, further reducing chromium mobility. However, excessive liming can cause nutrient imbalances in the soil, inhibit microbial activity, and alter its structure, resulting in modified fertility that makes the land less usable after remediation. Therefore, lime should be administered cautiously, considering the soil's buffering capacity and existing pH levels to minimize potential side effects (Ajao et al., 2025; Pradhan & Nambi, 2025; Shaghaleh et al., 2025; Zhang et al., 2025a).

Biochar, a carbon-rich material produced by pyrolyzing organic biomass, has emerged as an effective solution for removing toxic heavy metals. Its effectiveness primarily stems from its high surface area, porosity, cation exchange capacity, and the presence of various functional groups that can adsorb or chelate metal ions. Biochar facilitates electron transfer, leading to the disassembly of organic matter (Ajao et al., 2025; Hou et al., 2025; Pradhan & Nambi, 2025; Shen et al., 2025). Regarding chromium pollution, biochar acts as an inert substance onto which Cr6+ can be firmly adsorbed on its surface and within its pores; it also functions as an electron donor. Some studies indicate that biochar with high concentrations of oxygen-containing functional groups, such as –OH and –COOH, can enhance Cr(VI) reduction, particularly under slightly reductive conditions. Moreover, biochar can improve both the physical and chemical properties of soil, stimulate microbial activity, and serve as a long-term carbon sink. It also contributes to the sustainable recovery of land. However, the effectiveness of biochar is influenced by factors such as the type of feedstock, processing temperature, and method of activation, emphasizing the need to tailor its selection to local conditions. Additionally, concerns exist regarding the potential desorption of adsorbed metals under varying environmental conditions, along with the necessity for production and application technologies that can be scaled economically (Kumar, 2024a; Kumar & Singh, 2024b; Kumar et al., 2025; Sarankumar et al., 2025).

Environmental remediation has extensively studied zero-valent iron (ZVI) due to its strong reduction capability, particularly in reducing Cr(VI) to Cr(III). In the presence of water and dissolved oxygen, ZVI generates Fe^{2+} and Fe^{3+} ions, which act as electron donors that can reduce Cr(VI) to Cr(III). Alongside this process, which partially contributes to ionic bond formation but does not encompass all chemical occurrences, we also observe mixed Fe—Cr precipitation, such as $FeCrO_4$ or $FeCr_2O_4$. This effectively traps chromium in the soil (Ali et al., 2024; Maiti et al., 2023; Shen et al., 2025; Wang et al., 2024; Zheng et al., 2025). Nanoscale zero-valent iron (nZVI) particles show greater reactivity and efficiency than microscale ZVI due to their larger surface area and enhanced reducing power. However, nZVI particles are susceptible to agglomeration and may lose their reactivity over time, leading to research on modified ZVI formulations that include stabilizers, surfactants, or substrates like biochar and silica to enhance their dispersion lifespan. While ZVI-based methods can be effective at small laboratory scales, scaling them up for use in varying field conditions presents challenges due to issues with delivery, even distribution, and long-term stability. Additionally, secondary products, such as iron oxides and hydroxides, are produced when ZVI corrodes. These products may change the soil's chemistry and influence the fate of other contaminants, making thorough investigation necessary before implementation (Cheng et al., 2025; Kumar & Choudhury, 2024a; Shaghaleh et al., 2025; Wang et al., 2024).

Their overall efficiency can be significantly improved by coupling different mechanisms in a way that synergizes their effects, combining, for instance, reductive amendments and synthesis-enhanced stabilization methods, or biochar with zero-valent iron (ZVI) and other low-valent metals. This approach utilizes both the sorption capacity of carbonaceous matter and the role of these metals in transformation (Hou et al., 2025; Pradhan & Nambi, 2025). When zero-valent iron (ZVI) is followed by lime or phosphate amendments, it serves a dual purpose: the reduction of chromium is followed by its precipitation. Omitting this second step has been found, through practice, to significantly enhance the effectiveness of immobilization methods when biochar is integrated into stabilization and solidification systems. The introduction of superior physical properties, such as increased porosity and a rich habitat for microorganisms, is one such example. Such integrated or hybrid techniques are increasingly recognized as a means of overcoming the limitations of single-treatment approaches and achieving more lasting environmental soundness (Kumar & Choudhury, 2024b; Kumar & Singh, 2024c).

However, despite the potential of various chemical remediation methods, several issues remain that need to be addressed for their technical success to be realized. Key among these is the possible re-mobilization of immobilized chromium under altered environmental conditions, such as those caused by acid rain, fluctuations in groundwater levels, or redox changes. Household hardware and car rental are not enough; any chemical fix worth its salt also requires post-treatment monitoring and prognosis modeling to estimate how long your treatment will last and identify any potential failure modes (Kumar, 2024b, 2024a; Kumar et al., 2024a; Kumar & Choudhury, 2024c; Kumar & Singh, 2024d). The variability of soil properties across contaminated sites presents another challenge, requiring close procedural attention in both pre-treatment characterization and on-site post-treatment adjustment. Clay content, organic matter, mineralogy, and the presence of co-contaminants are influenced by these factors, which can significantly determine the effectiveness of chemical amendments or their failure, as well as the overall success rates of remediation efforts.

Additionally, the regulatory environment and policies will determine how chemical remediation technologies evolve and whether they will expand. Many jurisdictions have reduced the allowable concentration of residual contaminants that can remain on site for certification as "remediated" and extended stability data to 30 years. Today, there is a need for standardized testing protocols and risk assessments that can accurately predict the environmental safety and sustainability of chemically treated soils. Another variable is public opinion. Whether people view chemical cleanup as socially acceptable is influenced by the stakeholders involved and their specific concerns. The use of on-site methods for cleaning up polluted land continues to pose another obstacle to achieving public acceptance of chemical remediation. Chemically customized computational simulations are promoting recognition of the contributions that contaminant science makes (Kumar, 2024a; Kumar & Choudhury, 2024b; Kumar & Singh, 2024b). With this new emphasis, future trends in the chemical remediation of chromium-contaminated soils are likely to focus on enhancing selectivity, efficiency, and environmental friendliness in existing technologies. Frontier areas of nanotechnology, such as tailor-made nanoparticles and newly developed functional composites, hold potential for minimal-impact remediation. In addition, the field of green chemistry, characterized by biodegradable additives and environmentally friendly reagents (with nature serving as the ultimate model), may enhance the ecological acceptability of chemical treatments. Moreover, combining chemical remediation with biological therapies, such as strengthening immobilized compounds followed by phytostabilization or leaving them in a form that causes no more ecological disturbance to people than planting a garden, might provide synergistic economic and social benefits (Kumar & Choudhury, 2024b, 2024c; Kumar & Singh, 2024d).

To manage chromium-contaminated soils, both to degrade the pollutant into non-toxic forms and limit its potential mobility, chemical remedial measures have become essential to this toolkit. Chemical oxidation and reduction methods can directly detoxify Cr(VI). Additionally, solidification techniques can stabilize the metal in soils; options such as lime, biochar, and zero-valent iron serve as flexible tools that can be adapted to site-specific conditions (Kumar et al., 2024g; Kumar & Choudhury, 2024a; Kumar & Singh, 2024c; Kumar et al., 2025). While each method has its characteristics and limitations, strategically combining these approaches is crucial for overcoming individual shortcomings and achieving comprehensive, sustainable remediation. Research, technological innovation, and interdisciplinary collaboration are vital for refining these methods and addressing emerging challenges. This will ensure that the world's chromium-contaminated soils are cleaned of this toxic metal safely for many generations to come (Fig. 4).

5 Biological Remediation Strategies

Among phytoremediation strategies, which include both phytoextraction and phytostabilization, as well as microbial bioremediation, the reduction of toxic hexavalent chromium (Cr(VI)) to less toxic trivalent chromium (Cr(III)), along with the reduction of chlorine and the application of acid at elevated temperatures and pressures, is the most promising alternative. Strategies such as bioaugmentation and biostimulation are increasingly being utilized to enhance the performance of indigenous or introduced microbial populations in contaminated soils (Liu et al., 2023; Suanon et al., 2025; Taneja et al., 2024; Wang et al., 2024). These biological techniques not only offer potential for in-situ treatment with minimal ecological disturbance, but also aid in soil restoration by promoting microbial activity and enhancing nutrient

Fig. 4 Achieving sustainable chromium remediation

cycling, rather than simply relocating problems elsewhere, which is inevitable if we continue to produce more waste than we can handle responsibly.

Phytoremediation, a plant-based approach, involves the ability of specific species to accumulate, tolerate, or transform pollutants through various mechanisms. It is divided into two main categories related to chromium contamination: phytoextraction and phytostabilization. Phytoextraction relies on the uptake and translocation of heavy metals from soil into harvestable plant tissues, which can then be removed and processed for the contaminants they contain or the elements extracted from them. A key factor in its success is the selected plant variety, particularly hyperaccumulator and high-biomass species that are capable of absorbing large amounts of chromium without exhibiting phytotoxic effects (Kumar, 2024b; Kumar et al., 2024b; Kumar, 2024a; Kumar & Singh, 2024d). Several studies have identified plant species that can accumulate significant quantities of Cr, including Brassica juncea (Indian mustard), Helianthus annuus (sunflower), Atriplex patula (orach), and Sedum alfredii, among others, which excel in this process. Moreover, these plants often possess specialized physiological traits absent in other plants, such as enhanced root exudation and metal transporter proteins. Their root cell walls contain areas capable of storing substantial amounts of chromium. However, the effectiveness of phytoextraction is influenced by several factors, including complexation, soil pH, redox conditions, and competition from other heavy metal ions, which can alter chromium levels in the soil. Since Cr(VI) is highly mobile and toxic, converting it into the less toxic Cr(III) is essential to reduce its toxicity and enhance translocation, either before or during plant uptake (Ali et al., 2024; Kumar et al., 2024g; Liu et al., 2023; Zhang et al., 2025b, 2024). Unfortunately, the relatively slow growth rate of many hyperaccumulators, combined with the need for multiple crop cycles to extract a significant percentage of contaminants, limits the practical use of this technology. Additionally, the post-treatment and disposal of Cr-bearing biomass present environmental challenges, necessitating safe disposal methods such as incineration or pyrolysis, which can also lead to secondary pollution.

In contrast, phytostabilization aims to use plants to localize contaminants in the root zone. Without intervention, contaminants can alter their mobility and bioavailability, not necessarily permanently, at a given site. Instead, utilizing resistant plants and shrubs is one potential strategy; it is particularly promising when removing all chromium from a contaminated area does not make sense or cannot currently be achieved. The primary goal is to protect humans and other living creatures from potential exposure to airborne particles, as this could involve the off-site migration of contaminants across both large and small distances (Deoli et al., 2025; Maiti et al., 2023; Santalunai et al., 2024; Yuan et al., 2024). Any chromate (VI) absorbed by grass roots quickly oxidizes to hydroxides, meaning chromate presence can be expected in any of these environments. Once the soil pH reaches 10.5 or higher, the chromate ion becomes strongly complexed and associates with soluble iron(III) and aluminum during the precipitation process, while also being immobilized in situ by plants. Plants used in phytostabilization typically have aggressive root systems that can penetrate deeply into the earth, such as Populus spp. and alfalfa (Medicago sativa), which exhibit high tolerance to heavy metals. Grasses like Poa annua are also employed. You may be familiar with alfalfa, which is also known as Medicago sativa in Latin. Its rooting depth is excellent. Some mechanisms involved in phytostabilization include the root exudation of organic acids and amino acids from the cortex. This alters the soil pH and redox potential, allowing for the precipitation and adsorption of Cr, thereby stabilizing it without posing health hazards to humans, plants, or other organisms (Kumar et al., 2025; Liu et al., 2023; Suanon et al., 2025; Zhang et al., 2025b). Additionally, specific plant species can support rhizospheric microorganisms that reduce or co-precipitate Cr(VI) with iron or aluminum oxides. Phytostabilization is superior to phytoextraction due to its lower risk of contaminant dispersion, as the pollutant remains in situ and does not accumulate above ground. Furthermore, this approach promotes long-term stabilization and erosion control of the soil, as well as habitat restoration, making it well-suited for large-scale or diffuse contamination issues, among other extensive challenges. Although phytostabilization is effective, it heavily relies on maintaining a suitable environment for root growth, ensuring the ground is continuously covered by plants year-round during growing seasons and throughout harsh weather. In practice, this may require ongoing maintenance, in addition to routine mowing, as a preventive measure (Deoli et al., 2025; Pradhan & Nambi, 2025; Zhao et al., 2024; Zheng et al., 2025). Occasionally, contamination can become widespread again in areas where plants have died back after a dispersive event; thus, timely intervention is essential to prevent the re-mobilization of contaminants under changing environmental conditions or to address recovery failures of vegetation due to a lack of active community support. Unknown text.

Another key aspect of biological remediation is microbial bioremediation, particularly in detoxifying Cr(VI) through reductive transformation. Various species of bacteria, fungi, and archaea can biochemically reduce Cr(VI) to Cr(III), which decreases the water solubility, mobility, and acute toxicity of Cr(VI). This reduction process typically occurs via intracellular or extracellular enzyme pathways, involving chromate reductases that transfer electrons from Cr(VI) to Cr(III) at each stage of reduction, assisted by electron donors such as NADH, flavins, or quinones. Many well-documented bacteria can effectively facilitate this biochemical reduction, including Pseudomonas fluorescens, Shewanella confuciana, Bacillus subtilis, and Escherichia coli. Similarly, fungi such as Aspergillus niger, Penicillium janthinellum, and Fusarium oxysporum have shown comparable abilities in vitro. The various microbes employ diverse metabolic strategies to achieve these results, such as respiratory reduction under anaerobic conditions and metabolic reduction when alternative electron acceptors are available. Several physical, chemical, and biological factors influence the efficiency of Cr(VI) reduction to Cr(III) by microbes. These environmental factors encompass pH, temperature, salinity, oxygen availability, and the presence or absence of organic carbon sources, which provide energy for microbial growth and function (Cheng et al., 2025; Pradhan & Nambi, 2025; Shaghaleh et al., 2025; Zhang et al., 2025a). Under anaerobic conditions, some anaerobic or facultatively anaerobic species can simultaneously reduce Cr(VI) while utilizing organic substrates as electron donors, recovering energy while cleansing the environment of the contaminant. Conversely, aerobic species may partly mitigate their chromium load through non-respiratory mechanisms such as enzyme detoxication or passive adsorption onto cell surfaces.

In bioaugmentation, specific strains of microbes known for their ability to reduce Cr(VI) are deliberately added to contaminated environments to either replace or supplement the existing population. Laboratory studies and field tests have shown that using genetically altered strains sourced from nature, which are resistant to high levels of Cr, has proven very promising with this approach. However, this success relies on the survival of the bioaugmented organisms, natural competition with indigenous microbes, and various environmental factors that often hinder the establishment and thriving of these strains (Kumar, 2024a; Kumar et al., 2024b; Kumar et al., 2024g; Kumar & Choudhury, 2024a; Kumar & Singh, 2024c). To address these challenges, researchers are experimenting with encapsulation techniques, biofilm formation, and collaborative approaches that combine multiple beneficial organisms to create functionally redundant and resilient units. Conversely, biostimulation aims to enhance the growth and activity of existing microbial populations by providing nutrients, electron donors, or Cr(VI)-reducing substances, thereby encouraging the development of these organisms. Molasses, glycerol, lactate, acetate, and vegetable oils are commonly used to supply carbon and energy sources, thus promoting microbial metabolism while increasing Cr(VI) reduction rates. Generally, biostimulation is more sustainable than bioaugmentation and less intrusive. It works with existing microbial communities in local environments that have evolved over many years to adapt to local conditions. However, it must be managed carefully to prevent issues such as promoting undesirable microbial processes or depleting essential nutrients over the long term (Kumar et al., 2024a; Kumar & Choudhury, 2024c; Kumar & Dwivedi, 2024; Kumar & Singh, 2024c).

Other biological or physicochemical techniques create even more powerful methods to maximize the advantages of microbial bioremediation. At the same time, by combining plant microbial reduction with phytoremediation, we establish a dual-nature system. Although the roots of the two technologies lie in different areas, phores are also excellent locations for microbe-reducing rhizosphere species. Growing plants in contaminated Cr(VI)-rich soil helps absorb more of the contaminant and prevents its escape into the environment. Likewise, biostimulation combined with electrokinetic remediation or chemical amendments can enhance microbial activity near electrode areas, making Cr(VI) more accessible for microbial attack. With tools like these, a broader vision emerged that focusing on bacteria with chromate reductive enzymes might provide a novel biological remediation strategy capable of addressing agricultural and soil environmental pollution problems (Kumar, 2024a; Kumar & Choudhury, 2024b; Kumar & Singh, 2024d, 2024b). However, scaling up these technologies for field applications is not easy. Especially in soils with heterogeneous quality, soil moisture, temperature, and nutrients vary widely throughout different times of the year. Similarly, the long-term stability of reduced species and conditions that may promote reoxidation should be carefully considered when assessing an entity's potential for returning to normal after a specific period has elapsed.

Meanwhile, fungi are recognized as hyperaccumulators of Cr(III) or Cr(VI), although their effectiveness in removing chromium from contaminated sewage remains poorly understood. Biogeochemical remediation technology utilizes the natural abilities of plants and microorganisms to transform, sequester, or degrade pollutants. By employing this strategy to clean up contaminated sites, various receptors can act as alternative sinks for heavy metals in leachate. However, with current treatment methods, such as plant bioreactors where pollutants are absorbed by the plant and ultimately destroyed through oxidation, metals that do not volatilize and return to a regular cycle will instead accumulate in sediments. These sediments cannot remineralize due to their relatively low oxygen levels (Hou et al., 2025; Liu et al., 2023; Sarankumar et al., 2025; Suanon et al., 2025). Compared to phytoextraction, phytostabilization exhibits lower plant growth rates, metal uptake rates, and post-harvest handling burdens. Microbial bioremediation, particularly Cr(VI) reduction, presents an effective detoxification method when combined with bioaugmentation or biostimulation, which boosts both microbial activity and resilience in contaminated environments. As the roles of plant-microbe interactions and enzyme function are being uncovered, the regulation of metal resistance genes becomes clearer, and new insights continue to emerge in these areas (Kumar, 2024a; Kumar et al., 2024a; Kumar & Dwivedi, 2024; Kumar & Singh, 2024b). The potential for developing even more efficient and resilient biological remediation systems is likely to expand. Yet,

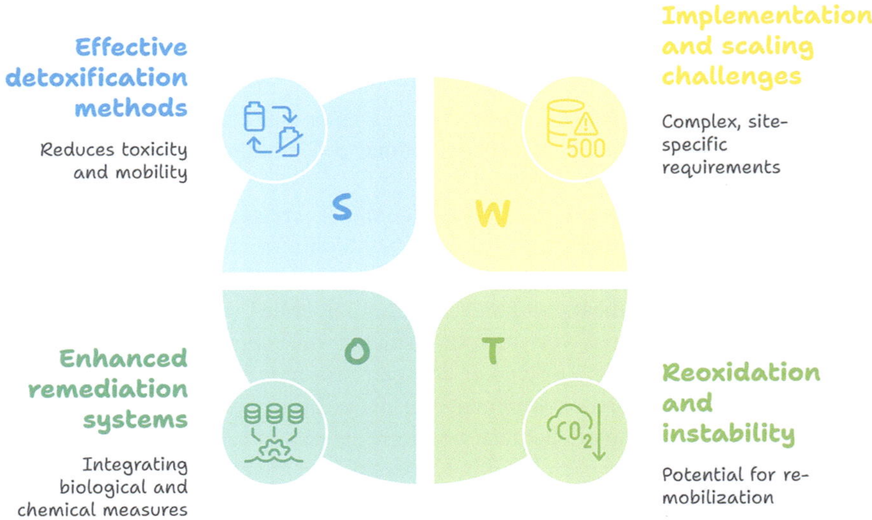

Fig. 5 Biological remediation strategies for chromium contamination

success in implementing these strategies will hinge on integrating them with complementary physical and chemical measures, aligning interventions with specific site conditions, and ensuring thorough monitoring and risk assessment throughout every stage of the remediation process. Through multidisciplinary research and collaboration across the scientific, engineering, and policy domains, biological remediation offers a pathway to transforming contaminated landscapes into both ecologically viable and socially accepted habitats (Fig. 5).

6 Integration of Remediation Approaches

The entire process of matching different remediation methodologies represents a revolutionary shift from traditional single-method approaches to integrated technologies in the treatment of chromium-contaminated soil, with beneficial effects extending beyond our immediate project budget. This includes the synthesis of basic means such as physical treatment, chemical treatment, or biological remediation, combined with physical, chemical, and biological techniques. These not only improve the efficiency of remediating chromium-contaminated soil but also accelerate the process over time while ensuring natural restitution for decades to come (Ali et al., 2024; Liu et al., 2023; Sarker et al., 2023; Zhang et al., 2025b, 2024). Stand-alone remediation technologies (e.g., soil washing, thermal desorption, chemical stabilization, phytoremediation, or microbial bioremediation) have all been successful to some extent in both laboratory and field settings. However, in their application to actual sites, they often prove ineffectual individually due to constraints such as partial recovery; this also involves high fees, and aftercare work is usually necessary to prevent secondary contamination, as well as to restore the ecological balance that pollutants have damaged. These constraints are pretty standard for any single method (Kumar, 2024c, 2024b; Kumar & Choudhury, 2024c; Kumar & Singh, 2024d). Therefore, the integration of several different remediation means, whether sequential or synchronous in their preparation phase, whether one continuous operation follows another, or various types run concurrently, now addresses these challenges and combines strengths at different points. This not only counteracts each method's respective weaknesses in remediation efficacy but also enhances local environmental compatibility, reduces treatment time, and aids in restoring soil ecological functions (Kumar, 2024a; Kumar, 2024b; Kumar et al., 2024g; Kumar & Dey, 2024).

While previous studies have focused on detailing and summarizing these efforts, this paper is unique in that it consolidates them into three success stories, examining commonalities and general features. If necessary, contaminants transported from Here to There must undergo a process of redistribution (migration), as some of the pollution may infiltrate people's land. At the same time, it is more likely to end up in the groundwater around Volta. This model is especially effective for addressing complex and mixed pollution scenarios (Kumar, 2024b; Kumar et al., 2024b; Sarker et al., 2023; Suanon et al., 2025; Zhang et al., 2024). A commonly used sequential strategy begins with chemical reduction to convert the toxic Cr(VI) that injections or slammers would leave in place into harmless Cr(III). Next, stabilization procedures follow, including adding lime to the soil or treating amendments such as phosphate or biochar, which are in direct contact with and within (through diffusion) reduced forms of chromium that are irreversibly trapped by crystalline mineral surfaces. Here, the first reductive step significantly reduces toxicity and aids in the future immobilization of Cr. For instance, soil washing can serve as a preliminary treatment to remove loosely held or highly mobile chromium species before implementing more targeted biological or electrokinetic countermeasures against the remaining residues. The sooner the entire system filters out pollutants, the more manageable the subsequent steps will be.

Consider electrokinetic remediation (EKR) as an alternative method in microbial bioremediation. The first step in this process is to use EKR to drive the low-perm soils, where high levels of Crions exist, toward the cathode region through electromigration and electro-osmosis. After establishing the Cr near the cathode, the reduced zone created by water electrolysis enhances indigenous or added Cr-reducing microorganisms (Ahmad et al., 2024; Kumar et al., 2024b; Kumar & Choudhury, 2024c; Kumar & Singh, 2024b; Sarankumar et al., 2025). This process improves the in situ biotransformation of Cr to Cr(III). The combination of these two methods leverages their strengths: EKR addresses microbial access issues in compacted soils, while bioremediation detoxifies Chromium harmlessly without producing hazardous byproducts. Field test results indicate that such systems achieve higher Cr removal rates than either stand-alone EKR or microbial methods, especially in clay-rich soils where conventional flushing techniques are less effective. Alternatively, simultaneous integrated approaches typically involve the concurrent use of two or more remediation types within a single system, if feasible. These models are beneficial when rapid transformation or removal of contaminants is required, or when different mechanisms work together advantageously. A prime example is the combination of zero-valent iron (ZVI) nanoparticles with microbial stimulation. ZVI particles act as electron donors, facilitating the reduction of Cr(VI) to Cr(III). This process includes adding organic substrates (e.g., lactate, molasses) to co-stimulate microbial metabolism and maintain a stable reducing environment over time. This dual approach not only accelerates Cr(VI) reduction but also stabilizes the end product as a co-precipitant with Fe oxides, minimizing the risk of reoxidation. Furthermore, the presence of ZVI can shield microbial populations from oxidative stress, thereby increasing their survival and ability to thrive in contaminated environments (Ahmad et al., 2024; Ali et al., 2024; Pradhan & Nambi, 2025; Zhang et al., 2025b).

Additionally, combining phytoremediation with chemical oxidation or reduction shows promise in hybrid treatments. For instance, when organic amendments or chelating agents are used in conjunction with phytoextraction, the bioavailability of Cr(III) in the rhizosphere can be significantly enhanced, leading to increased plant uptake rates and translocation. Some plant species demonstrate considerable chromium accumulation when grown in soils treated with citric acid or EDTA, both of which improve metal solubility and root absorption to varying degrees in different plant species. However, it is equally essential to be cautious when selecting the type and quantity of chelating agent; otherwise, there may be an overloading effect on groundwater due to excessive leaching of metals by plants that do not require them (Deoli et al., 2025; Hou et al., 2025; Maiti et al., 2023; Santalunai et al., 2024; Yuan et al., 2024). Alternatively, the chosen chelating agent may hinder root absorption of essential metals, preventing their incorporation into plant tissues altogether, or worse, turn all earthworms below ground level into pillars of salt as it once did in ancient Egypt! In contrast, phytoextraction can be conducted at a lower cost by combining phytostabilization with chemical stabilization, and the plants themselves offer a safeguard that works in two ways. Without this approach, we would have to invest a substantial sum surrounding the mines to keep contaminants out, and even then, we would return to square one because groundwater levels quickly rebound to their previous states after heavy rain or snowfall.

Another example is electrokinetic-assisted bioremediation, where electric fields are utilized to enhance the effectiveness of microbial organisms in subsurface soil. By inducing electro-osmotic flow and electrophoretic movement, EKR facilitates the movement of nutrients throughout the contaminated zone. The closest thing to ecstasy in field studies, electrokinetics combined with inoculating bacteria, has been shown to significantly enhance the rates at which presumably toxic Cr(VI) salts are reduced by naturally occurring or added Cr-resistant strains of bacteria-themselves no small performers. Moreover, local acidity is created at the anode, while alkalinity is generated at the cathode. The parsimony principle dictates where to meet these equilibrium points: these are managed strategically, not only to reduce pH but also to ensure sustained detoxification of chromium over time (Kumar, 2025a, 2025b, 2024a, 2024b).

The split remediation strategy offers numerous benefits, including improved pollutant removal efficiency, reduced treatment time, energy and material savings, and habitat rehabilitation. By targeting only one aspect of chromium action at a time, such as mobility or solubility, single-approach methodologies remain incomplete and ineffective. Today, it is possible to address many facets of the problem simultaneously, allowing solution efforts to be as thorough and enduring as any other endeavor one might undertake (Dey et al., 2024; Kumar, 2024c). For instance, the compounds extracted from the soil through physical methods will not only decrease the total amount of toxic contaminants in a specific area but also ensure that residual levels remain within regulatory standards over an extended period. In this manner, long-term liability is minimized even while relatively small amounts are removed at any given time. Furthermore, coupling microbial reduction with phytoremediation creates a closed-loop system that facilitates continuous remediation. In this system, plant root exudates support microbial growth, and vice versa; thus, both biological processes are mutually reinforcing.

Because of this, these integrated techniques allow for much more flexibility in matching a particular site's peculiarities and needs than any single procedure applied uniformly. A sandy ground can be flushed with chemicals and microorganisms introduced into it, since it has good drainage—this is, for example, the case with soil that Water will readily leach through. Coincidentally, clayey soils may require special treatment to overcome the fact that the charged particles in them (clays) are very slow to penetrate barriers. Urban brownfields will tend towards in-situ solutions, which minimize diurnal variation and avoid secondary pollution of soil disturbed from its original location; here, one might hope for a combination of phytostabilization, chemical amendments, and microbial encouragement. On the other hand, in agriculture, emphasis falls on the land's fertility. Therefore, biochar, compost, or green manures must be used in conjunction with other methods to restore productivity to affected soils (Devi et al., 2024a; Devi and Kumar, 2024; Kumar and Dwivedi, 2024; Kumar and Singh, 2024c; Saini et al., 2024; Sharma et al., 2024).

However, despite the apparent benefits of this approach and in light of increasingly strict environmental regulations, its application is fraught with challenges that must be carefully addressed to ensure successful deployment. One of these challenges lies in the multifaceted nature of designing and managing ecohydraulics. This field requires a wide range of expertise, including soil chemistry, microbiology, environmental modeling, and engineering. Each component of such an integrated system must be consistent regarding its operating conditions, reaction kinetics, and spatial distribution. This is essential to prevent antagonism between the elements, which can result in inefficiency at best. For example, if levels of chemical oxidants are too high, they may inhibit microbial activity; if electrodes are awkwardly positioned within the trajectory of electrokinetic processes, they may damage standard plant root systems in a phytoremediation scheme (Kumar, 2025a, 2025b, 2025c; Kumar & Raj, 2025; Kumar & Singh, 2025c; Raj & Kumar, 2025).

A second significant concern revolves around the unintended consequences arising from interactions among various remediation mechanisms. On one hand, while microbial stimulation can improve Cr(VI) reduction, it may also create conditions that promote anaerobic sulfate reduction, producing hydrogen sulfide and potentially mobilizing other heavy metals, such as arsenic or cadmium. Similarly, introducing exotic microflora through bioaugmentation may disrupt native microbial communities and may not thrive under local environmental conditions. Therefore, careful pilot-scale tests, sophisticated predictive models, and adaptive management techniques are all essential for anticipating and mitigating these risks.

Regulatory frameworks and policy considerations play a crucial role in adopting integrated problem-solving strategies. Many of today's regulations remain rooted in prescriptive standards that encourage single-technology solutions, which are already well-proven, rather than hybrid approaches. To enable broader adoption, new guidelines are needed that acknowledge the value of integrated systems, conduct life-cycle evaluations, and allow for flexible, performance-based cleanup targets. Furthermore, the social acceptability of these complex methods in sensitive situations—such as residential areas or ecologically sensitive sites—will be significantly influenced by public perception. Transparent communication, public participation, and demonstration projects can enhance confidence among regulators and scientists regarding the deployment of complex remediation schemes. Additionally, industry participants may assume a more active role in collaboration with affected communities as this development unfolds (Kumar et al., 2024f, 2024a, 2024d; Kumar & Singh, 2025, 2024d).

The future of remediating chromium-contaminated soil lies in integrating technological innovations with ecological principles in a coherent manner. Combining hybrid remediation strategies with precision technologies, such as synthetic biology and intelligent monitoring systems, will enable even greater remedies on a larger scale than before. Nanotechnology, advanced gene-editing microbes, and online sensor networks will be utilized to create dynamic, responsive remediation systems that adapt in real-time to new environmental conditions. Additionally, by incorporating related digital tools into projects, such as decision support systems utilizing GIS mapping software for modern IT supervisory duties on a project basis, they will become more efficient and data-driven. The integration of remediation approaches provides a pathway to address chromium-contaminated soils by combining the strengths of physical, chemical, and biological methods to achieve better recovery results at lower costs. These integration models, whether for sequential or simultaneous treatment settings, provide clear guidelines for optimizing the treatment sequence or introducing new improvements. In contrast, hybrids conduct experiments based on those tested in practice, without artificial constraints (Devi et al., 2024b; Dhiman et al., 2023; Kumar, 2025c; Kumar et al., 2024b, 2024a; Kumari et al., 2024). The benefits of hybrids are clear: improved efficiency, reduced environmental impact, and long-term sustainable development. Our future lies beyond isolated remediation strategies and moves toward a more integrated, adaptive dialogue on how best to solve the problem, grounded in ecological principles combined with technological innovation. As our scientific understanding deepens and technological capabilities expand, integrated remediation will play an increasingly central role in restoring contaminated lands and ensuring both environmental and human health worldwide (Fig. 6).

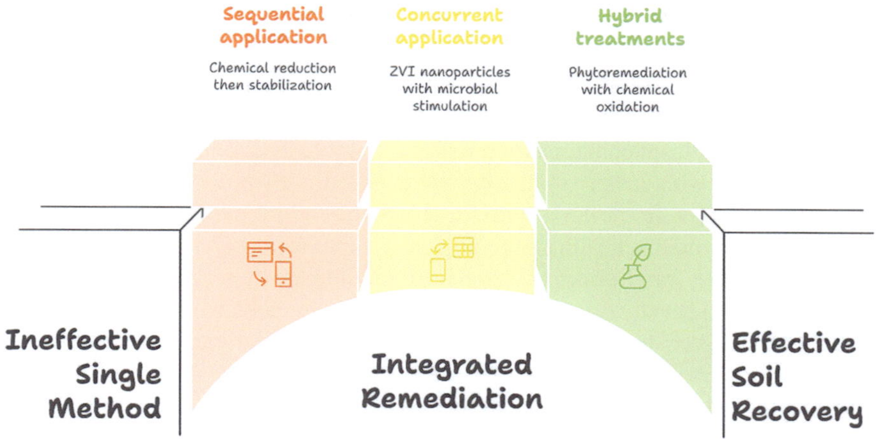

Fig. 6 Integrated remediation for chromium-contaminated soil

7 Site Assessment and Risk Management

Preliminary site characterization is a critical step in environmental site assessment and management. During this stage, existing data is collected and analyzed. The goal is to gain a comprehensive understanding of the historical and current site conditions, which include potential contaminants, land use history, and hydrogeological characteristics. This initial investigation aims to identify possible sources of pollution at the site, understand its movement, and determine who or what may be affected by it. A thorough preliminary characterization often begins by examining historical records, such as black-and-white aerial photographs. These fire insurance maps show the types of buildings constructed next to one another, transaction histories from county assessors' offices related to recent property sales, and archives maintained by regulatory agencies. This documentary evidence offers valuable insights into past uses of the site that may have involved hazardous chemicals. For instance, former factories, gas stations, dry cleaners, or agricultural areas might be flagged as potential sources unless an Environmental Site Assessment has been performed. In addition to historical research, a site visit is conducted to observe the current construction or structures in the area. Signs such as stained soil and unusual odors that could indicate pollution are also sought. Information is also gathered on nearby water bodies, residential areas, and ecologically sensitive sites to assess potential effects that may spill over to neighboring environments if further investigation of hazardous sites is pursued. Finally, the data collected during the preliminary characterization helps determine whether a detailed Phase II Environmental Site Assessment (ESA) is necessary. If significant risks are identified, the next step is typically to conduct intrusive investigations, including soil and groundwater sampling, to quantify the levels of contaminants and evaluate potential risks to human health and the environment.

This early stage is vital for laying the groundwork for all subsequent risk management efforts. It enables stakeholders to allocate resources more effectively, avoid unnecessary tests, and make informed decisions about site remediation or redevelopment. Without proper preliminary characterization, there is a considerable risk of overlooking significant contamination issues that could lead to long-term liabilities, non-compliance with laws or standards, and even harm to public health.

Risk assessment frameworks provide a structured approach to evaluating the potential harm to human health and ecological systems resulting from environmental pollutants. Robust frameworks are crucial for guiding the systematic identification, quantification, and evaluation of risks associated with contaminated sites. They serve as essential decision-making tools for managing these locations. Today, the most recognized framework is the USEPA four-step risk assessment model, which includes the following basic components: (1) Hazard Identification: In this stage, we identify the contaminants present at the site and assess their potential to cause harm. We may refer to toxicological profiles from organizations such as the Agency for Toxic Substances and Disease Registry (ATSDR) or the World Health Organization (WHO) to understand the health effects linked to each chemical. (2) Exposure Assessment: This step focuses on how individuals or ecosystems may come into contact with contaminants. Key elements include defining exposure pathways (e.g., ingestion, inhalation, dermal contact), the duration and frequency of each exposure, and identifying at-risk populations or species. (3) Dose-Response Assessment: In this phase, we quantify the relationship between the level of exposure to a contaminant and the likelihood of adverse effects occurring. Dose-response models may estimate toxic values such as reference doses (RfDs) for non-carcinogens and cancer slope factors for carcinogens. (4) Risk Characterization: This final step integrates information from the previous three steps to estimate probable risk and its associated likelihood. It also conducts uncertainty analysis and provides a concise summary for decision-makers. Besides the USEPA framework, there are region-specific approaches, such as the Canadian Environmental Quality Guidelines (CEQGs) and soil screening values established by the European Union. These alternatives are tailored to local environmental conditions and regulations. In some writing frameworks, particularly the most rigorous, the Best International Practices protocol also incorporates probabilistic techniques that tackle variability and uncertainty in exposure parameters and toxicological data.

USEPA guidance for ecological risk assessments (ERAG), for example, provides interpretive frameworks. These frameworks primarily focus on species, excluding humans, and ecosystems, addressing endpoints such as population viability, community structure, and functionality to monitor their impact. The framework for risk assessment has been dynamic and evolving, incorporating advances in science, technology, and policy. Tools such as emerging biomonitoring and bioavailability assessments, along with the application of machine learning algorithms, are gradually being integrated into modern risk assessments to enhance both accuracy and relevance. Regulatory standards and guidelines are crucial in ensuring that site assessments and risk management efforts comply with legal requirements and public health goals. These standards define acceptable contamination levels, establish cleanup targets, and outline procedures for site investigation and remediation practices. At the international level, organizations such as the World Health Organization (WHO), the International Agency for Research on Cancer (IARC), and the Organisation for Economic Co-operation and Development (OECD) provide scientific guidance. These global standards often influence domestic and regional policies. The Comprehensive Environmental Response, Compensation, and Liability Act (CERCLA) encompasses Superfund, the response system for cleaning up toxic waste sites. The CERCLA Hazard Ranking System (HRS) is used to prioritize sites as candidates for the National Priorities List (NPL). The Resource Conservation and Recovery Act (RCRA) regulates the management of hazardous waste throughout its entire life cycle.

State regulations at the federal level supplement federal standards but may be more stringent or equivalent to local standards, depending on local environmental conditions and policy priorities. Many states have developed risk-based cleanup protocols that allow for site-specific risk assessments to guide cleanup decisions instead of relying on universal cleanup levels. This approach is especially beneficial for redevelopment projects on brownfields where removing all pollution is impossible or unnecessary. Countries in the Asia-Pacific region differ significantly in their regulatory maturity. Japan has long had laws in place to protect soil and groundwater. In contrast, China has recently issued the "Soil Pollution Prevention and Control Action Plan" to substantially strengthen its regulatory regime. Australia employs a combination of national guidelines and state legislation for managing contaminated sites. Public participation in the risk assessment and decision-making processes is now regarded as a vital part of effective site management. The interpretation of combined meanings is highly encouraged for both the general public and ecologists alike. Practical site assessment and risk management rely on preliminary site characterization, a solid risk assessment framework, and compliance with regulations. To adhere to ever-changing rules and effectively protect human health and the environment, environmental scientists, engineers, and those concerned with environmental policy must apply the knowledge gained from the legislation that governs their actions (Fig. 7).

8 Sustainability and Cost Considerations

One of the essential tasks of life cycle assessment is to remediate contaminated sites. Typically, a life cycle assessment examines the environmental impacts over the duration that a remediation strategy operates to clean up a site, encompassing everything from material extraction and energy consumption associated with construction and operation to its eventual shutdown.

Each of the conventional remedies, including pump-and-treat, soil vapor extraction, thermal desorption, bioremediation, and in situ chemical oxidation technology, exhibits a distinct environmental impact. Take pump-and-treat systems as an example. The concept is to use this method for treating groundwater. However, these are not merely large-scale operations requiring significant energy input; each

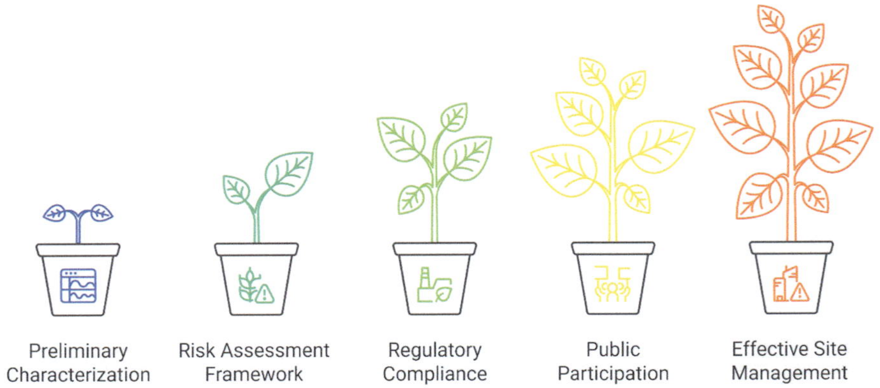

Fig. 7 Steps to effective site management

necessitates its own extensive and long-term monitoring. Such requirements lead to substantial total carbon dioxide emissions. Bioremediation, on the other hand, often has a smaller carbon footprint and causes less disruption to surrounding areas when indigenous microbial groups thrive. Thus, it may offer local communities a more sustainable alternative at specific sites. Occasionally, the aspheric regime is applied in LCA projects that involve remediation. Depending on how the form or mass of all soils removed is defined as equivalent to one functional unit, the next step may aim to demonstrate effectiveness in any treated area. This study also promotes a proper understanding of elements in the environmental analysis, whether greenhouse gas (GHG) emissions, energy consumption, water use, and nutrient contamination are environmentally significant, or where hazardous chemicals might enter the food chain. The key point to note is that an LCA comparing formulas may reveal a discrepancy if a tree were blown away in one day. In contrast, a slowly growing type of vegetation might have similar effects but emit far more CO_2 per megawatt-hour. Additionally, LCA can show how to use recycled materials in construction, recycle waste products from remediation, or power remediation equipment using alternative energy. By incorporating LCA into their decision-making processes, developers can direct people and activities toward solutions that realistically align with broader initiatives for climate change intervention as well as global environmental efforts.

Public technology is essential for maintaining the economy of environmental faith, agreed Mulberry. In addition to completing necessary legal paperwork and conducting ecological inspections, engineering comparisons were also performed, where possible, alongside public attorneys. On the other hand, ex-situ methods like excavation and off-site disposal yield quick results but incur higher upfront costs, including transportation fees, landfill charges, and site restoration expenses, among others. Public funding, insurance, and private sector partnerships significantly influence economic feasibility. Many governments have sustainability programs that offer grants or provide low-interest loans to assist in developing and cleaning up brownfield sites. Risk-based corrective action (RBCA) methods allow for cost-effective solutions tailored to various site conditions, instead of relying on overly conservative or one-size-fits-all standards for environmental cleanup. Additionally, it is crucial to consider the opportunity costs, which reflect what other uses these funds could support. Critically investing in a technically aggressive cleanup that yields only marginal benefits could divert resources from other pressing environmental or community needs. For economic evaluation, a balance must be struck between environmental protection and welfare to maximize outcomes.

Turning to the global Anthropocene, we observe an increasing movement within the environmental sustainability literature. The detection of greater detail about ecological phenomena has led to extensive process breakdowns, but it also reveals possible, albeit sparse, trends in advance. With growing global awareness of environmental sustainability, there is a heightened focus on eco-friendly practices across all stages of site assessment and remediation. From a long-term perspective, every effort is made to minimize disturbance to local habitats by using alternative methods. A low carbon footprint, along with intensity and energy output, are key priorities for today's remediation operations. Renewable energy sources, such as wind, solar, or bioenergy, are employed to power environmental cleanup equipment. This practice reduces dependence on fossil fuels and consequently contributes less to overall greenhouse gas emissions. Phytoremediation, bioaugmentation, and natural attenuation are some examples of low-impact cleanup methods that rely entirely on natural processes to break down or contain pollutants. Material efficiency and waste minimization strategies such as using clean soil in situ for on-site backfilling, recycling concrete and asphalt, or reducing packaging contribute to a model grounded in recycling and ecological economics. To preserve freshwater resources, measures like implementing water-efficient treatment systems or utilizing treated groundwater for irrigation and industrial processes can be enacted. Biodiversity enhancement post-cleanups, through native vegetation restoration and wildlife habitat creation, further entrenches local ecological diversity and ecosystem services. Thus, it makes logical sense to establish green procurement policies designed to promote environmentally friendly practices. By selecting non-toxic reagents, biodegradable solvents, and energy-efficient equipment, we can further minimize our environmental impact. The strategy of eco-friendly remediation increasingly considers community engagement and environmental justice issues. For instance, involving nearby communities, especially those historically affected by pollution, in the decision-making process fosters trust, enhances transparency, and helps achieve more equitable outcomes. New concepts like nanotechnology, electrokinetic remediation, and microbial fuel cells are also shaping eco-friendly practices. These promising developments require less energy and provide greater decontamination capability than established methods. Finally, standardized assessment tools and regulatory mechanisms, such as the ITRC Green and Sustainable Remediation Metrics or LEED for Sustainable Sites, offer common frameworks for organizations to measure the environmental performance of their remediation projects. These widely used conventions enable corporations to compare their efforts with those of other businesses, signal their goodwill and eco-friendly practices early on, and meet continually evolving ecological expectations in a verifiable manner.

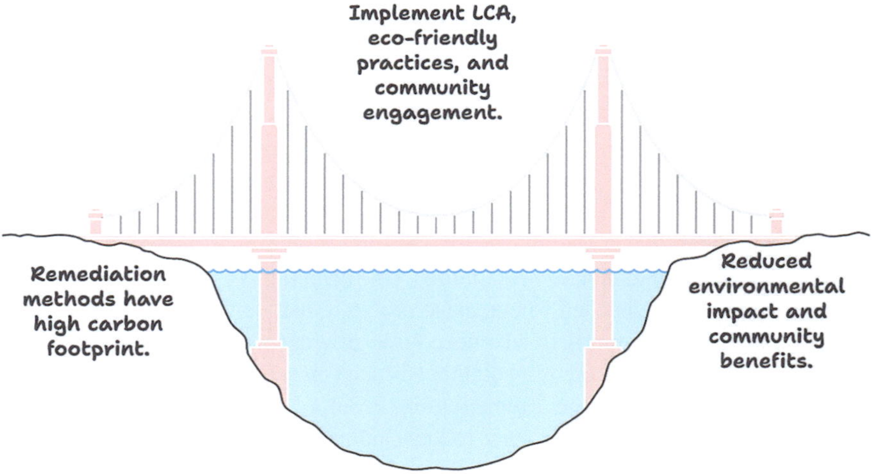

Fig. 8 Transitioning to sustainable site remediation through LCA and Eco-friendly practices

Life cycle analysis, economic feasibility evaluation, and the incorporation of eco-friendly measures are the three pillars of modern site assessment and risk management. As increasingly stringent environmental regulations are implemented and public expectations for sustainability rise, professionals across all disciplines must collaborate to ensure that remedial actions are not only effective in mitigating risk but also align with broader ecological and economic objectives. By combining innovative technologies with sustainable design concepts, transparent methods for cost evaluation can keep stakeholders engaged in long-term environmental management responsibilities while also protecting human health and ensuring that land is reused responsibly. Ultimately, the future of contaminated site management relies on a solid scientific foundation paired with prudent financial management and ethically driven practices that produce resilient and sustainable solutions for future generations (Fig. 8).

9 Emerging Technologies and Innovations

Environmental remediation is rapidly evolving with the introduction of new technologies that offer more cost-effective and sustainable approaches to treating contaminated sites. These advancements include nanomaterials for chromium removal, genetic engineering of plants and microorganisms, and machine learning (ML) combined with remote sensing, which transforms how contaminants are detected, monitored, and treated. These advancements not only enhance remediation performance but also provide mechanisms for data-driven decision-making, greater accuracy in risk assessment, and a reduced environmental footprint.

One of the most interesting and appealing achievements in addressing environmental pollutants is the use of engineered nanomaterials for removing heavy metals, particularly hexavalent chromium (Cr(VI)), a hazardous and mobile pollutant found in industrial wastewater, groundwater, and soils. Cr(VI) has adverse effects on human health due to its carcinogenicity and persistence in the environment. Nanotechnology has emerged as an exciting new field for developing materials with high surface areas, reactivity, and adsorption capabilities, including zero-valent iron nanoparticles (nZVI), titanium dioxide nanoparticles (TiO_2 NPs), carbon nanotubes (CNTs), and graphene-based composites. These materials can act as effective reducing agents to convert Cr(VI) to Cr(III) or immobilize the hexavalent species through strong adsorption. Moreover, modifying nanomaterials with species-specific ligands or polymers enhances selectivity and stability in complex environmental matrices. The small size and high mobility of nanoparticles allow them to penetrate deeply into soil and groundwater systems, reaching contaminated areas that standard remediation techniques cannot access. Challenges related to the long-term environmental fate, potential toxicity, and scaling up of nanomaterials remain. However, investments in product development primarily focus on optimizing delivery systems, enhancing biocompatibility, and developing cost-effective production methods to make these solutions viable in the field.

Concomitant with progress in materials science, the use of genetic engineering in manipulating plants and microbes is advancing into new frontiers of phytoremediation and bioremediation. Through genetic engineering of the organisms, their natural uptake, sequestration, degradation, or transformation of degradation products can be improved. For instance, transgenic mustard has been engineered to tolerate and accumulate higher amounts of heavy metals, such as cadmium, arsenic, and chromium, from the soil. The expression of genes encoding metal transporters, chelating agents (such as phytochelatins or metallothioneins), or detoxifying enzymes that promote the accumulation and storage of these pollutants in the soil is characteristic of these plants. In the same vein, genetically modified microorganisms, such as bacteria, fungi, and algae, are being developed to degrade organic pollutants (e.g., PCBs, petroleum hydrocarbons, pesticides) or convert inorganic pollutants into benign forms. Some lines have a biosensor function and can monitor water pollution levels in real-time using detectable reporter genes. In addition to being more effective at cleanup, these bioengineered organisms offer a greener approach than traditional, energy-heavy methods of remediation. However, risks related to gene transfer, ecological consequences, and public acceptance necessitate stringent regulation and thorough environmental risk evaluation before such activities can be carried out on a laboratory scale.

Advancements in biology and chemistry, along with machine learning (ML) and remote sensing, are transforming how we acquire, manage, and interpret environmental data. These digital tools provide practical solutions for characterizing sites, detecting contaminants, modeling risks, and making predictions, significantly enhancing the accuracy and speed of decision-making. Additionally, remote sensing, supported by satellite imagery, aerial drones, and LiDAR (Light Detection and Ranging), facilitates non-invasive monitoring of key environmental variables, including land cover changes, vegetation stress, soil moisture, and surface water quality. HSI and MSI deliver multi-wavelength information that distinguishes subtle differences in spectral signatures linked to individual contaminants, allowing for the identification of polluted areas in the field without extensive laboratory-based physical sampling efforts. When combined with GIS, remote sensing data aids in the spatial mapping and visualization of risks, helping prioritize clean-up projects.

Machine learning adds an additional layer of analysis by enabling the automatic processing of data and the discovery of patterns in extensive, multidimensional datasets. Various machine learning models, including random forests, support vector machines, neural networks, and deep learning models, are being used to predict contaminant transport, simulate groundwater flow, assess potential exposure, and inform the design of remediation techniques. These models can take in a wide range of inputs, such as historical site data, sensor readings, lab results, and climate variables, providing dynamic analysis suitable for an adaptive management approach. Furthermore, ML-based digital twins of polluted sites are being developed to create virtual replicas that simulate various remediation options across different environmental scenarios. This facilitates cost-effective virtual trial and error of interventions, leading to their implementation and yielding superior results. As artificial intelligence matures, a stronger integration with IoT sensor networks is expected, enhancing real-time monitoring and autonomous response systems, which will support new responsive environmental management strategies.

Collectively, these new technologies are redefining the roles of environmental remediation and risk management. Not only do they enhance technical performance, but they also contribute to broader sustainability goals by reducing energy consumption, minimizing waste, and promoting environmentally friendly practices. However, this progress depends on interdisciplinary collaboration among environmental scientists, engineers, data scientists, policymakers, and residents. Addressing technical, ethical, and regulatory challenges (e.g., ensuring data privacy, validating model predictions, mitigating unintended ecological effects, and fostering public trust) is essential to the responsible scaling of these innovations.

In the future, the combination of nanotechnology, synthetic biology, and artificial intelligence offers significant potential for environmental solutions. We may observe self-healing biofilms that can respond automatically to pollution, nano-biohybrid systems that integrate the functions of microorganisms and nanoparticle-based catalysis, and artificial intelligence platforms that consolidate site assessment, risk analysis, and remediation planning into a cohesive, intelligent workflow. As we enhance our understanding of the natural world and develop new tools for manipulating it, the advancement and application of such sophisticated technology will likely be essential for addressing the complex issues related to chemical pollution and for designing ecosystems that are resilient against increasing environmental stress.

10 Future Directions and Research Needs

Now, with the evolution of environmental remediation, the prospects demand the design of green technologies to support sustainability and high-efficiency processes for pollutant removal. Recent academic research is trending toward scalable, low-energy technologies such as filtration units using biochar, electrochemical treatment, and solar-operated treatment systems that reduce dependence on non-renewable energy sources. Developments in green technology, like biodegradable surfactants, plant-based extractants, and enzyme-aided degradation, have garnered increased attention for their ability to lessen secondary contamination and enhance remediation outcomes. Furthermore, advancements in nanotechnology and biotechnology, including bioengineered microorganisms and nano-enabled delivery systems, are being optimized for the selective degradation of pollutants without disturbing the environment. However, further studies are still needed to evaluate the long-term fate and toxicity of these novel materials, especially in complex soil and groundwater systems. Additionally, future work should focus on developing cost-effective manufacturing processes and environmentally friendly recycling methods to ensure that green technologies are both sustainable and affordable for large-scale applications.

The adoption of these technologies in routine remediation practices necessitates significant policy implications and regulatory support. Government institutions and ecological organizations should upgrade existing mechanisms to accommodate new methods that preserve public health and protect the environment. For instance, current laws are outdated, and control technology is so old that it becomes challenging to deploy new remediation technologies. Policymakers should establish clear guidelines for the application of nanomaterials, genetically modified organisms, and AI-supported monitoring systems, alongside regulations for risk assessment, environmental monitoring, and liability. Incentive-based policies, such as tax credits, grants, and public-private partnerships, could encourage the adoption of green technologies by industry and expedite the shift to sustainable remediation. Furthermore, reconciling international standards will foster international cooperation and technology transfer across borders, addressing transboundary contamination, including groundwater pollution, and the migration of industrial waste. Regulatory agencies should also integrate the principles of adaptive management into their programs, ensuring that decisions remain flexible and can evolve in response to new scientific research and site-specific circumstances, based on data.

Equally important as advancements in technology and policy is the role of public involvement and our level of awareness regarding the future of managing contaminated sites. Environmental cleanup is not just a technical issue; it's deeply connected to social equity, cultural sensitivity, and public trust. Engaging local communities from the outset ensures that the remediation effort aligns with the best interests of affected populations, particularly those from underserved or vulnerable communities. Open dialogue about the risks, timing, and potential health impacts of contamination fosters greater responsibility and a willingness to support such sites. Educational

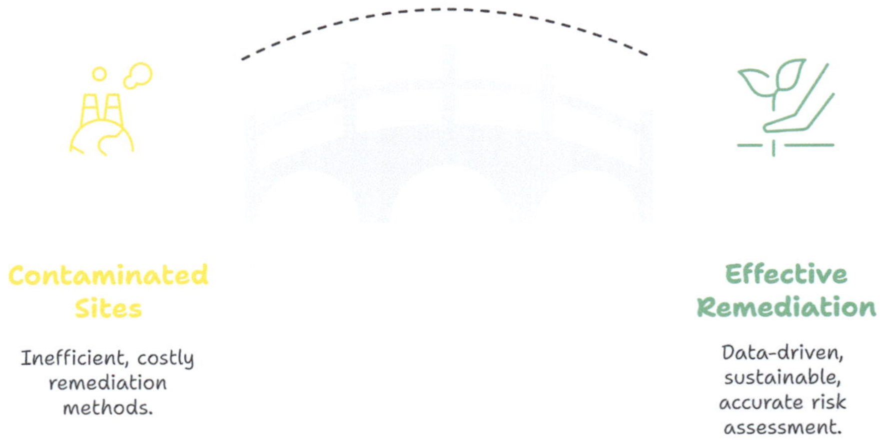

Fig. 9 Remediation involves technology, improving accuracy, and sustainability

initiatives that raise public awareness of environmental health risks and remediation science empower communities to participate more actively in the decision-making process. Similarly, incorporating citizen science platforms, where individuals engage in ecological monitoring using affordable sensors or mobile reporting applications, can improve data collection and strengthen community bonds. Best practices for involving diverse stakeholders, including culturally sensitive outreach strategies and methods for addressing environmental justice issues in the management of contaminated sites, also merit further investigation.

Innovative green technologies, progressive policies, and strong local engagement form the foundation for the future of site assessment and risk management through interdisciplinary cooperation and funding that marries science, regulation, and public engagement. Furthermore, stakeholders can create more durable and sustainable cleanup systems to address the increasingly complex contaminations we face today. This integrated strategy will not only enhance environmental performance but also support long-term public health, economic stability, and social equity as the pressures of environmental change continue to grow (Fig. 9).

11 Conclusion

Together, the real-world SSARP approach to site evaluation and risk management demonstrates that cleaning up polluted spaces involves much more than just science or regulation; it also includes infrastructure, economics, and politics at local, regional, and national levels, along with other elements unique to each location. Main conclusions: Preliminary site characterization is essential for identifying the source of contamination, determining exposure pathways, and assessing the potential risks to human health and the ecosystem. Risk-assessment models, such as the USEPA's four-step model, provide structured approaches to analyzing hazards. Compliance with evolving regulations is also a fundamental law that protects the environment. Integrating the LCA process into remedial planning highlights sustainability, encouraging stakeholders toward solutions that reduce ecological footprints while maintaining the desirability of specific outcomes. Cost-effectiveness is a vital aspect to consider, and cost-benefit analysis is a critical factor in determining the most effective remediation technology to implement. Furthermore, newly developed and emerging technologies, including nanoparticles, bioengineered organisms, and machine learning, promise to qualitatively enhance the robustness, efficacy, and precision of remediation.

Based on these findings, several recommendations are presented as best practices for managing contaminated sites in the future. First, the RBCA concept of 'taking a site-by-site approach' to remediation would alleviate the 'one-size-fits-all' mindset of regulators. This flexible system enables effective resource allocation while protecting public health and the environment. Second, the adoption of green and sustainable remediation (GSR) principles, including the use of more renewable energy, waste reduction, and biodiversity enhancement, should be emphasized throughout all stages of the remediation cycle. Third, rigorous stakeholder engagement and transparency are essential elements that must be incorporated into every stage of the process, ultimately ensuring that communities at risk are always informed, heard, and included in the decision-making process. Public involvement promotes trust and enhances project acceptance, as well as long-term environmental justice benefits. Fourth, it is critical to maintain investment in new technologies and appropriately deploy these technologies through a clear regulatory framework to address complex contamination issues. Finally, refresher courses and advanced training programs should be developed to equip environmental professionals with the most current tools and information for applying contemporary remediation techniques.

If all these elements work together, they will succeed; ultimately, only an integrated approach that combines science, policy, economics, and social responsibility can achieve this. Environmental remediation is more than just cleaning up pollution and discarded waste; it involves revitalizing natural habitats, communities, and promoting long-term sustainability. As human-induced pressures on global ecosystems rise (through industrialization, climate change, and altered land use), the urgency for resilient, adaptable, and inclusive remediation becomes increasingly urgent. By embracing emerging advancements, forging cross-sector collaborations, and aligning cleanup objectives with broader sustainability priorities, stakeholders can move beyond traditional remediation practices and work toward a vision where polluted areas are transformed into valuable resources for both nature and society. We cannot only repair the damage we have caused, but also build healthier, fairer, and more resilient communities for future generations.

Acknowledgements The author would like to express their sincere gratitude to Lovely Professional University for providing all kinds of support.

References

Ahmad, S., Liu, X., Liu, L., Waqas, M., Zhang, J., Hassan, M. A., Zhang, S., Pan, B., & Tang, J. (2024). Remediation of chromium contaminated water and soil by nitrogen and iron doped biochars. *Science of the Total Environment, 954*, 176435. https://doi.org/0.1016/j.scitotenv.2024.176435

Ajao, A. T., Awe, S., Aborisade, W. T., Jimoh, F. A., & Ajao, F. J. (2025). Bioremediation of soil contaminated with cassava mill effluents—a structural, functional, and metagenomics approach. *Next Research, 2*, 100439. https://doi.org/10.1016/j.nexres.2025.100439

Ali, J., Abid, R., Zheng, C., Waseem, H., Khan, K., Wang, L., & Yang, Y. (2024). Electro-bioremediation of hexavalent chromium and bioenergy extraction from contaminated water: Past to future perspectives. *Next Research, 1*, 100074. https://doi.org/10.1016/j.nexres.2024.100074

Cheng, Z., He, Y., Wang, N., Wu, L., Xu, J., & Shi, J. (2025). Uncovering soil amendment-induced genomic and functional divergence in soybean rhizosphere microbiomes during cadmium-contaminated soil remediation: Novel insights from field multi-omics. *Environmental Pollution, 368*, 125787. https://doi.org/10.1016/j.envpol.2025.125787

Deoli, V., Gupta, P. K., & Malik, A. (2025). Chapter 10—Field scale remediation technologies for vadose zone contaminated with chromium contaminants. In D. Kumar, P. K. Gupta, B. Singh, & S. Verma (Eds.), *Decontamination of subsurface water resources system using contemporary technologies* (pp. 131–141). Elsevier. https://doi.org/10.1016/B978-0-443-26639-3.00010-7

Devi, P., & Kumar, P. (2024). Alleviation of different climatic conditions by foliar application of salicylic acid and sodium nitroprusside and their interactive effects on pigments and sugar content of maize under different sowing dates. *Nature, Environment and Pollution Technology, 23*, 1161–1166. https://doi.org/10.46488/NEPT.2024.v23i02.050

Devi, P., Kumar, P., & Singh, J. (2024a). Application of potential microbes in bioremediation. In *Development in waste water treatment research and processes: Role of environmental microbiology in industrial wastewater research* (pp. 201–225). Elsevier. https://doi.org/10.1016/B978-0-443-13609-2.00021-5

Devi, P., Kumar, P., & Singh, J. (2024b). Role of global climate change in crop yield reductions. *Environ. Sci. Eng. Part, F3742*, 465–488. https://doi.org/10.1007/978-3-031-63296-9_18

Dey, S. R., Sharma, M., & Kumar, P. (2024). Chapter 16—Phytoremediation of phenolic compounds from soil. In P. Kumar, A. L. Srivastav, V. Chaudhary, E. D. van Hullebusch, & R. Busquets (Eds.), *Bioremediation of emerging contaminants from soils* (pp. 335–358). Elsevier. https://doi.org/10.1016/B978-0-443-13993-2.00016-5

Dhiman, H., Kumar, P., & Kundathil, C. (2023). Enhancing rapeseed yield and quality via growth regulators and microelement optimization: Implications for food production and security. *Journal Of Food Chemistry And Nanotechnology, 9*, S453–S460. https://doi.org/10.17756/jfcn.2023-s1-057

Hou, G., Xu, J., Guo, L., Han, T., Wang, Z., Zuo, D., Lv, J., Shao, G., & Lu, H. (2025). Chemically consolidated Cr(VI) contaminated soil based on granulation of blast furnace slag and Phosphogypsum. *KSCE Journal of Civil Engineering, 100285*. https://doi.org/10.1016/j.kscej.2025.100285

Kumar, P. (2024a). Bioremediation for emerging organic pollutants (pesticides) present and future, Biotechnologies for Wastewater Treatment and Resource Recovery. In *Current Trends and Future Scope*. Elsevier. https://doi.org/10.1016/B978-0-443-27376-6.00005-0

Kumar, P. (2024b). Biochar production from marine algae and its application in the treatment of industrial effluent. In *Marine biomass: biorefinery, bioproducts and environmental bioremediation* (pp. 57–93). De Gruyter. https://doi.org/10.1515/9783111353951-004

Kumar, P. (2024c). Advances in phytoremediation approach for resource recovery from wastewater. In *Biotechnologies for wastewater treatment and resource recovery: current trends and future scope*. Elsevier. https://doi.org/10.1016/B978-0-443-27376-6.00013-X

Kumar, P. (2025a). Nanofertilizers affecting soilless agricultural system. In *Nanofertilizers for sustainable agriculture: assessing impacts on health, environment, and economy*. Springer Science+Business Media. https://doi.org/10.1007/978-3-031-78649-5_13

Kumar, P. (2025b). Intersections of heavy metal toxicity, protein misfolding, and neurodegenerative disorders in humans. In *Protein misfolding in neurodegenerative diseases* (pp. 413–461). Elsevier. https://doi.org/10.1016/B978-0-443-18716-2.00012-0

Kumar, P. (2025c). Nanofertilizers, Nanopesticides for individual's (farmer's) economy. In *Nanofertilizers for sustainable agriculture: assessing impacts on health, environment, and economy*. Springer Science+Business Media. https://doi.org/10.1007/978-3-031-78649-5_16

Kumar, P., & Choudhury, D. (2024a). Hydroponic removal of organic contaminants from water. In *Springer water* (pp. 143–164). Springer Nature. https://doi.org/10.1007/978-3-031-53258-0_6

Kumar, P., & Choudhury, D. (2024b). Pollutants from agricultural activities in the global south. In *Biomonitoring of pollutants in the global South* (pp. 89–126). Springer Nature. https://doi.org/10.1007/978-981-97-1658-6_3

Kumar, P., & Choudhury, D. (2024c). Toxicity of Rhizospheric cadmium contaminated soil and its phytoremediation. In *Springer water* (pp. 227–264). Springer Nature. https://doi.org/10.1007/978-3-031-54005-9_10

Kumar, P., & Dey, S. R. (2024). Hydroponics phytoremediation: An overview. In *Springer water* (pp. 361–396). Springer Nature. https://doi.org/10.1007/978-3-031-53258-0_13

Kumar, P., & Dwivedi, P. (2024). Recent advances in the role of nitric oxide and cross-talk with signalling molecules under environmental stress in plants. In *Plant growth regulators: resilience for sustainable agriculture* (pp. 55–78). Springer Nature. https://doi.org/10.1007/978-981-97-2918-0_4

Kumar, P., & Raut, A. M. (2024). Microbes-assisted bioaugmentation process in the reduction of emerging industrial pollutants from soil. In *Bioremediation of emerging contaminants from soils: soil health conservation for improved ecology and food security* (pp. 519–540). Elsevier. https://doi.org/10.1016/B978-0-443-13993-2.00023-2

Kumar, P., & Singh, J. (2024a). Microbes in xenobiotics biodegradation. In *Development in waste water treatment research and processes: applied technologies for clean up of environmental contaminants*. Elsevier. https://doi.org/10.1016/B978-0-443-13615-3.00012-1

Kumar, P., & Singh, J. (2024b). Emerging contaminants and ways to reduce them. In *Trends in biological processes in industrial wastewater treatment* (pp. 4.1–4.34). Institute of Physics Publishing. https://doi.org/10.1088/978-0-7503-5678-7ch4

Kumar, P., & Singh, J. (2024c). Microbes in bioremediation of heavy metals. In *Development in waste water treatment research and processes: applied technologies for clean up of environmental contaminants*. Elsevier. https://doi.org/10.1016/B978-0-443-13615-3.00017-0

Kumar, P., & Singh, J. (2024d). Bioprocesses in industrial wastewater treatment: Trends and prospects. In *Trends in biological processes in industrial wastewater treatment* (pp. 15.1–15.50). Institute of Physics Publishing. https://doi.org/10.1088/978-0-7503-5678-7ch15

Kumari, P., Kumar, P., & Siddique, A. (2024). Biostimulants and calcium-mediated optimization in morpho-phenological growth of mung bean (Vigna radiata L.) in the summer season. *Journal of Applied and Natural Science, 16*, 1544–1551. https://doi.org/10.31018/jans.v16i4.5898

Kumar, P., Choudhury, D., Devi, M. I., & Sahoo, J. (2024a). Impact of radiation on Parasitoid ecology: Unveiling complex interactions. *Biopestic. Int., 20*, 165–176. https://doi.org/10.59467/BI.2024.20.165

Kumar, P., Choudhury, D., & Singh, J. (2024b). Performance of nitrification-denitrification processes for wastewater treatment. In *Microbial applications of nitrifiers and denitrifiers in industrial wastewater treatment*. Elsevier. https://doi.org/10.1016/B978-0-443-13929-1.00005-2

Kumar, P., Dwivedi, P., & Upadhyay, S. K. (2024c). Optimization of polyamine and mycorrhiza in sorghum plant for removal of hazardous cadmium. *Plant Physiology and Biochemistry, 214*. https://doi.org/10.1016/j.plaphy.2024.108846

Kumar, P., Dey, S. R., & Choudhury, D. (2024d). Effectiveness of cadmium on biochemical shift of pea plant treated with mycorrhiza and putrescine. *Nature, Environment and Pollution Technology, 23*, 275–285. https://doi.org/10.46488/NEPT.2024.v23i01.023

Kumar, P., Devi, P., & Singh, J. (2024e). Microbiology and application of the nitrification-denitrification process. In *Microbial applications of nitrifiers and denitrifiers in industrial wastewater treatment*. Elsevier. https://doi.org/10.1016/B978-0-443-13929-1.00004-0

Kumar, P., Dey, S. R., Sharma, M., & Singh, J. (2024f). Microbes in bioremediation of pesticides. In *Development in waste water treatment research and processes: applied technologies for clean up of environmental contaminants*. Elsevier. https://doi.org/10.1016/B978-0-443-13615-3.00011-X

Kumar, P., Raj, A., & Kumar, V. A. (2024g). Approach to reduce agricultural waste via sustainable agricultural practices. In *Valorization of biomass wastes for environmental sustainability: green practices for the rural circular economy*. Springer. https://doi.org/10.1007/978-3-031-52485-1_2

Kumar, P., & Choudhury, D. (2025). Advancements in precision agriculture: Integrating machine learning techniques for crop monitoring and management. *Microorganisms for Sustainablity, 45*, 29–57. https://doi.org/10.1007/978-981-96-3448-4_3

Kumar, P., & Raj, A. (2025). Degradation of misfolded proteins by ubiquitin-proteasome system and autophagy lysosomal proteolytic pathways. In *Protein misfolding in neurodegenerative diseases* (pp. 253–289). Elsevier. https://doi.org/10.1016/B978-0-443-18716-2.00007-7

Kumar, P., & Singh, J. (2025). 3D printed materials in environmental remediation. In *Nanomaterials in environmental remediation: developments and challenges*. CRC Press. https://doi.org/10.1201/9781003484189-15

Kumar, P., & Singh, J. (2025a). Nanobiosensors: Versatile tool for diagnosis of infectious diseases. In *Nano-biosensor technologies for diagnosis of infectious diseases* (pp. 121–172). Wiley. https://doi.org/10.1002/9781394287697.ch4

Kumar, P., & Singh, J. (2025b). Design and synthesis of novel nanomaterials emphasizing infectious diseases. In *Nano-biosensor technologies for diagnosis of infectious diseases* (pp. 23–73). Wiley. https://doi.org/10.1002/9781394287697.ch2

Kumar, P., & Singh, J. (2025c). Perspective and challenges of synergistic removal of toxic contaminants from effluent using different treatment techniques. In *Environmental science and engineering* (pp. 419–451). Department of Agronomy, School of Agriculture, Lovely Professional University, Punjab, Phagwara, India. https://doi.org/10.1007/978-3-031-62660-9_17

Kumar, V., Singh, V., & Pandit, S. (2025). Advanced omics approach and sustainable strategies for heavy metal microbial remediation in contaminated environments. *Bioresource Technology Reports, 29*, 102040. https://doi.org/10.1016/j.biteb.2025.102040

Liu, J., Sun, S., Zhang, H., Kong, Q., Li, Q., & Yao, X. (2023). Remediation materials for the immobilization of hexavalent chromium in contaminated soil: Preparation, applications, and mechanisms. *Environmental Research, 237*, 116918. https://doi.org/10.1016/j.envres.2023.116918

Maiti, S., Neogi, S., & Dutta, B. K. (2023). Remediation and immobilization of Cr(VI)-contaminated soil using stabilized nanoscale iron sulfide and ecological impact. *Heliyon, 9*, e15009. https://doi.org/10.1016/j.heliyon.2023.e15009

Pradhan, P., & Nambi, I. M. (2025). Sustainable remediation method for hexavalent chromium contaminated site. *Process Safety and Environment Protection, 199*, 107268. https://doi.org/10.1016/j.psep.2025.107268

Raj, A., & Kumar, P. (2025). Advances in methods to detect and study protein aggregation. In *Protein misfolding in neurodegenerative diseases* (pp. 51–65). Elsevier. https://doi.org/10.1016/B978-0-443-18716-2.00003-X

Saini, L., Kumar, P., & Upadhyay, H. (2024). Zinc and boron foliar application effects on primed mung bean (Vigna radiata L.) growth and productivity. *Nature, Environment and Pollution Technology, 23*, 1407–1418. https://doi.org/10.46488/NEPT.2024.v23i03.012

Santalunai, S., Pakprom, J., Charoensiri, W., Thongsopa, C., Thosdeekoraphat, T., Janpangngern, P., Yodrot, T., Santalunai, N., & Santalunai, S. (2024). Determination of dielectric properties of lead-contaminated soils: Potential application to soil remediation. *Heliyon, 10*, e35787. https://doi.org/10.1016/j.heliyon.2024.e35787

Sarankumar, R. K., Sun, S., Sathishkumar, K., Devanesan, S., AlSalhi, M. S., Mangalaraja, R. V., Arulraj, A., Han, F., & Wang, X. (2025). Enhanced bio-electrokinetic remediation of cesium-contaminated soil through an optimized Cr(VI)-resistant alkalophilic bacterial degradation approach. *Journal of Environmental Chemical Engineering, 13*, 115251. https://doi.org/10.1016/j.jece.2024.115251

Sarker, A., Al Masud, M. A., Deepo, D. M., Das, K., Nandi, R., Ansary, M. W. R., Islam, A. R. M. T., & Islam, T. (2023). Biological and green remediation of heavy metal contaminated water and soils: A state-of-the-art review. *Chemosphere, 332*, 138861. https://doi.org/10.1016/j.chemosphere.2023.138861

Shaghaleh, H., Alhaj Hamoud, Y., Sun, Q., Sheteiwy, M. S., & AbdElgawad, H. (2025). Soil flushing coupled with aminated-nanocellulose/MOF hydrogel nanocomposites adsorbents: A novel sustainable remediation strategy for Cr(VI)-contaminated agricultural soils. *Separation and Purification Technology, 353*, 128440. https://doi.org/10.1016/j.seppur.2024.128440

Sharma, K., & Kumar, P. (2024). Chapter 9—Environmental threats posed by xenobiotics. In P. Kumar, A. L. Srivastav, V. Chaudhary, E. D. van Hullebusch, & R. Busquets (Eds.), *Bioremediation of emerging contaminants from soils* (pp. 183–201). Elsevier. https://doi.org/10.1016/B978-0-443-13993-2.00009-8

Sharma, K., Kumar, P., & Singh, J. (2024). Ecology and diversity of microbial communities involved in the removal of priority contaminants and micropollutants in waste. In *Development in waste water treatment research and processes: role of environmental microbiology in industrial wastewater research* (pp. 181–199). Elsevier. https://doi.org/10.1016/B978-0-443-13609-2.00014-8

Shen, Y., Qi, W., Li, S., Yang, Z., & Chen, K. (2025). Numerical simulation of copper-contaminated sediment consolidation and remediation through vacuum electro-osmosis. *Geomechanics for Energy and the Environment, 41*, 100655. https://doi.org/10.1016/j.gete.2025.100655

Suanon, F., Tomètin, L. A. S., Zveushe, O. K., de Dios, V. R., Han, Y., Ifon, B. E., Atakpa, E. O., Yete, P., Sesu, F., Li, J., Omer, A. M., & Dong, F. (2025). Electrokinetic remediation of chromium-contaminated soils: The potential for advanced materials in three-dimensional EKR approaches. *Journal of Environmental Chemical Engineering, 13*, 116774. https://doi.org/10.1016/j.jece.2025.116774

Taneja, S., Karaca, O., & Haritash, A. K. (2024). Electrokinetic remediation: Past experiences and future roadmap for sustainable remediation of metal-contaminated soils. *Journal of Geochemical Exploration, 259*, 107437. https://doi.org/10.1016/j.gexplo.2024.107437

Wang, J., Guan, Y., Guo, M., Gao, J., Yang, M., Liu, Y., Liu, X., Wang, W., Jin, Y., & Qu, J. (2024). New insights into the remediation of chromium-contaminated industrial electroplating wastewater by an innovative nano-modified biochar derived from spent mushroom substrate: Mechanisms, batch study, stability and application. *Chemosphere, 367*, 143621. https://doi.org/10.1016/j.chemosphere.2024.143621

Xing, Y., Zheng, Y., & Wang, X. (2025). Integrated strategies for effective remediation of chromium-contaminated soils: Advancements, challenges, and sustainability implications. *Environmental Advances, 19*, 100614. https://doi.org/10.1016/j.envadv.2025.100614

Yuan, Z., Peng, A., Chu, Z., Zhang, X., Huang, H., Mi, Y., Xia, D., Wu, X., Ye, Z., Tao, Y., & Yan, X. (2024). Sustainable remediation of Cr(VI)-contaminated soil by soil washing and subsequent recovery of washing agents using biochar supported nanoscale zero-valent iron. *Science of the Total Environment, 921*, 171107. https://doi.org/10.1016/j.scitotenv.2024.171107

Zhang, C., Deng, Z., Zhang, H., Wang, J., Tong, J., Lu, F., Ruan, C., & Shi, J. (2025a). Biochar-enhanced remediation of Cr(VI)-contaminated soil by Penicillium oxalicum SL2: Enhanced reduction based on extracellular electron transfer. *Journal of Environmental Science*. https://doi.org/10.1016/j.jes.2025.04.079

Zhang, C., Wang, J., Zhang, H., Lu, F., Ruan, C., Tong, J., Ren, J., & Shi, J. (2025b). Biochar enhances the simultaneous stabilization of chromium and lead in contaminated soil by Penicillium oxalicum SL2. *Chemical Engineering Journal, 509*, 161522. https://doi.org/10.1016/j.cej.2025.161522

Zhang, Y., Zhang, Y., & Wu, A. (2024). Design and construction of magnetic nanomaterials and their remediation mechanisms for heavy metal contaminated soil. *Science of the Total Environment, 951*, 175369. https://doi.org/10.1016/j.scitotenv.2024.175369

Zhao, X., Liu, X., Zhang, Z., Ren, W., Lin, C., He, M., & Ouyang, W. (2024). Mechanochemical remediation of contaminated soil: A review. *Science of the Total Environment, 946*, 174117. https://doi.org/10.1016/j.scitotenv.2024.174117

Zheng, X., Wu, S., Tan, Q., Hu, C., & Sun, X. (2025). Molybdenum tolerance of strains screened from molybdenum tailings and their potential application in molybdenum-contaminated soils. *Environmental Pollution, 377*, 126433. https://doi.org/10.1016/j.envpol.2025.126433

Artificial Intelligence and IoT-Driven Technologies for Environmental Chromium Toxicity Monitoring and Management

Prasann Kumar

Abstract Chromium (Cr) toxicity, primarily from hexavalent Cr (Cr(VI)), poses a significant and widespread environmental issue resulting from heavy industrial activities such as tanning, electroplating, and mining. Traditional monitoring methods are often labor-intensive, slow, and limited in coverage, which hinders effective and real-time management of contamination. The adoption of Artificial Intelligence (AI) and the Internet of Things (IoT) marks a significant change, shifting from passive observation to proactive, intelligent environmental oversight. This chapter thoroughly examines how AI and IoT-based technologies are transforming the tracking and control of ecological chromium toxicity. It details the environmental chemistry and toxicological effects of chromium, underscoring the urgent need for advanced solutions. The chapter also explores the design of IoT sensor networks, emphasizing sophisticated sensing platforms for in-situ, continuous, and multiplexed detection of chromium compounds in soil, water, and air. It primarily discusses how AI and machine learning (ML) algorithms can be applied, including supervised learning for predicting concentration, unsupervised learning for identifying contamination sources, and deep learning for analyzing complex spectral data from sensors. Additionally, it examines how these technologies can be integrated into innovative decision-support systems for predictive pollution modeling, automated remediation, and risk evaluation. The chapter also highlights key challenges, including sensor calibration, data security, energy efficiency, and model transparency. Finally, it considers prospects, including the potential of digital twins for ecosystems affected by chromium and how these digital tools can promote a circular economy. This work aims to offer researchers, environmental engineers, and policymakers a solid foundation and guidance for leveraging advanced digital technologies to address the global issue of chromium pollution.

Keywords Zero hunger · No poverty · Global challenge · Circular economy

P. Kumar (✉)
Department of Agronomy, School of Agriculture, Lovely Professional University, Phagwara, Punjab, India
e-mail: prasann0659@gmail.com

N. Kumar (ed.), *Chromium Toxicity*, https://doi.org/10.1007/978-3-032-14323-5_12

1 Introduction

The pervasive and cunningly nefarious nature of Cr pollution makes it a serious environmental problem resulting from a wide range of human activities, mainly industrial processes such as leather tanning, electroplating, chromate production, textile manufacturing, and informal electronic waste recycling. These activities release millions of tons of Cr-contaminated effluents and solid waste into the environment each year, creating a legacy of contamination that threatens ecosystems and human health worldwide (Safeer et al., 2022). This issue is especially urgent in rapidly industrializing regions with weaker regulatory systems. The toxicity of chromium heavily depends on its chemical form, a key aspect of its environmental chemistry that exists mainly in two stable oxidation states: trivalent chromium (Cr(III)) and hexavalent chromium (Cr(VI)). These forms differ significantly in mobility and toxicity: Cr(III), which is less soluble and less mobile in soils, is also an essential nutrient involved in glucose and lipid metabolism in humans and animals. Conversely, Cr(VI) is highly mutagenic and carcinogenic due to its high solubility and mobility in water, as well as its ability to cross cell membranes. This paradoxical behaviour means that Cr(VI) levels primarily determine the environmental impact of chromium release. However, the fate of Cr(VI) is complex, as redox transformations driven by organic matter, ligands, minerals, pH, and Eh levels in soils can interconvert Cr species, complicating monitoring and mitigation efforts. The toxic effects of Cr(VI) on ecosystems and human health are well-documented (Hakeem Anwer et al., 2024; Moretti et al., 2024). It bioaccumulates in plants and animals when it enters the food chain through contaminated water or soil. It can cause phytotoxicity by inhibiting seed germination, root growth, and photosynthesis in plants. At the same time, in aquatic environments, it induces genotoxic and cytotoxic effects on fish and invertebrates, disrupting entire communities. For humans, exposure occurs through inhalation of airborne particulates in industrial settings and consumption of contaminated water, with Cr(VI) linked to higher risks of nasopharyngeal and lung cancers, severe skin conditions such as dermatitis and ulcers, and other non-cancerous health issues, including liver and kidney damage, respiratory problems, and weakened immunity. Due to these risks, agencies such as the WHO and the EPA classify Cr(VI) as a priority pollutant. Despite the apparent threat, current methods for monitoring and treating chromium toxicity are limited. Traditional approaches, which rely on infrequent field sampling and laboratory analyses, such as AAS and ICP-MS, are slow, costly, and unable to capture the dynamic spatial and temporal variations of contamination plumes (Islam & Ali, 2025; Yadav & Sood, 2025). Cleanup operations, such as pump-and-treat, chemical reduction, and excavation, are similarly slow, expensive, and disruptive, and they lack real-time feedback, which hampers effective management. Amid this growing global challenge and technological gap, a paradigm shift is underway: integrating Artificial Intelligence (AI) and the Internet of Things (IoT) into environmental management. This innovative approach involves networks of wireless, intelligent sensor nodes that provide continuous, high-resolution, real-time data on chromium speciation and concentrations. These data are wirelessly transmitted via low-power wide-area networks to cloud or edge computing platforms where advanced AI and machine learning tools—such as regression models for predictive mapping, anomaly detection for pollution events, and deep learning networks for spectral data analysis—extract valuable insights from the large data streams. The result is actionable intelligence that enables predictions of contaminant fate and transport, optimizes real-time remediation strategies, and supports automated decision-making systems for proactive environmental stewardship. This chapter aims to guide readers through this revolutionary technology, beginning with an overview of chromium's ecological chemistry and toxicology, which motivates the need for these advancements (Moulahoum & Ghorbanizamani, 2024; Malode et al., 2025; Robert & Kalpathy, 2025). It will detail the design and components of IoT sensor networks used for chromium monitoring, explain how AI and machine learning models convert sensor data into predictive insights and actionable management, and present illustrative case studies demonstrating the practical benefits. Additionally, it will critically examine ongoing challenges and scientific needs for future development. Ultimately, this chapter seeks to serve as a comprehensive resource for researchers, environmental engineers, and policymakers interested in leveraging these cutting-edge digital tools for the global control and remediation of chromium pollution (Fig. 1).

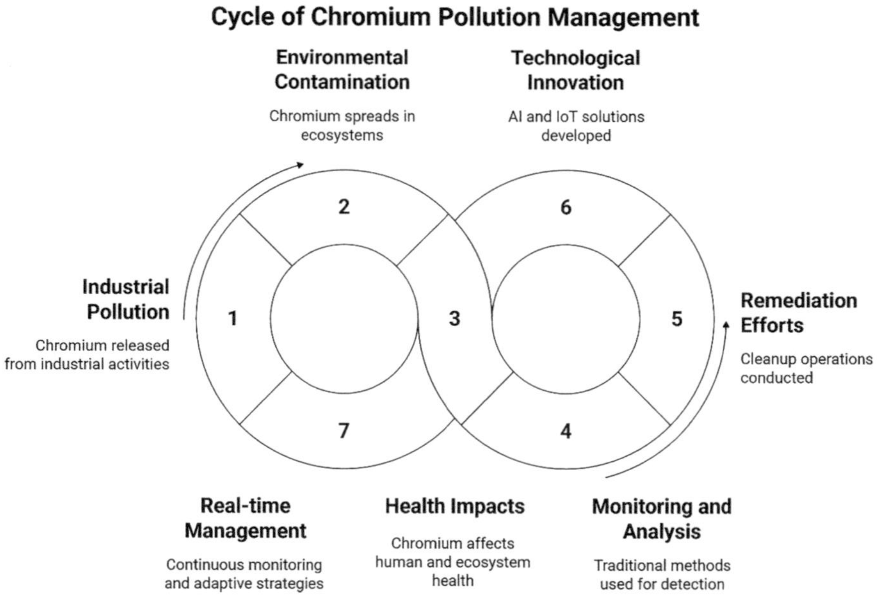

Fig. 1 Cycle of chromium pollution management

2 IoT-Based Framework for Chromium Sensing and Data Acquisition

The rise in industrial activities has created a need for robust environmental monitoring systems, particularly for toxic heavy metals such as chromium. Chromium exists mainly in two states: the hazardous hexavalent (Cr(VI)) and the less toxic trivalent (Cr(III)). This necessity has led to the development of an IoT-based framework for chromium sensing and data collection a complex system designed for real-time, detailed evaluation of toxicity levels across different environments. It is based on the core framework outlined in "Architecture of an Environmental IoT System for Toxicity Monitoring," which is organized into separate but interconnected layers (Ghanbarlou et al., 2025; Mohd Adnan et al., 2025; Saha et al., 2025).

The first layer is the Sensing Layer, which includes Chromium-specific sensors. This layer forms the physical connection to the environment, comprising sensors placed at key measurement sites. These sensors use selective recognition elements, such as ion-selective membranes or functionalized surfaces, to detect chromium ions and convert their presence and concentration into electrical or optical signals. These signals are distinguishable from background noise and generate initial data points that serve as catalysts for further analysis and interpretation.

Next, the data flows into the Network Layer: Communication Protocols (LoRaWAN, 5G, NB-IoT). This layer functions as the circulatory system of the IoT architecture, where the chosen protocol balances range, bandwidth, and power consumption. For example, LoRaWAN is suitable for low-power, wide-area applications, such as agriculture or remote environments, where it transmits small data packets over long distances. (Jayaraman et al., 2024; Giwa et al., 2025; Khoshvaght & Khiadani, 2025). NB-IoT is better suited for urban or underground coverage, such as water systems, while 5G provides reliable, low-latency communication necessary for high-density sensor arrays near industrial sites. This ensures the seamless transfer of data from physical sensors to the digital realm.

Once transmitted, data is received and processed at the Cloud/Edge Layer: Data Storage and Preprocessing. A hybrid architecture is typically employed, with edge computing nodes located near sensors, performing immediate preprocessing tasks such as analog-to-digital conversion, noise filtering, and temporary caching. These initial processes enable quick local alerts if chromium levels exceed thresholds. Meanwhile, cloud platforms aggregate large datasets for more complex operations, such as advanced signal processing, long-term trend analysis through machine learning, and correlation with environmental parameters like pH, temperature, and dissolved oxygen. The refined information is then presented via user-friendly dashboards for stakeholders and regulatory agencies. (Kumar et al., 2024; Silvanir et al., 2024; Li et al., 2025).

The success of this framework heavily relies on advancements in ASPC, particularly electrochemical sensors, which are promising for in situ monitoring due to their portability, high sensitivity, and affordability. Techniques such as anodic stripping voltammetry (ASV) are employed, where Cr ions are electroplated onto a working electrode and then stripped off, producing a current proportional to the concentration of Cr ions. Recent improvements involve nanostructured electrodes made of graphene or carbon nanotubes, which increase the surface area and electron transfer rate, thereby lowering detection limits.

Additionally, optical and spectroscopic sensors offer benefits like minimal sample preparation and non-destructive measurement. Colorimetric sensors, which use functionalized nanoparticles such as gold and silver that change colour upon detecting Cr(VI), can be visually quantified. More advanced fiber optic sensors, based on surface plasmon resonance (SPR) or fluorescence quenching, modify their optical properties upon binding with chromium ions, enabling continuous water monitoring (Bharti et al., 2024; Bin Abu Sofian et al., 2025; Mahapatra et al., 2025).

State-of-the-art sensors include biosensors and nanomaterial-based sensors that leverage biological recognition elements and nanomaterials. Biosensors utilize biomaterials, such as enzymes, antibodies, or bacteria, that specifically interact with chromium, resulting in detectable metabolic changes or electrical signals. Nanomaterial sensors utilize engineered nanoparticles, quantum dots, or molecularly imprinted polymers (MIPs) as artificial antibodies, offering exceptional selectivity and stability in harsh environmental conditions. These sensors are key for next-generation, self-calibrating, and self-regenerating systems. (Kundathil et al., 2023; Singh et al., 2023; Sharma et al., 2025).

Deployment strategies for these sensors are tailored to each medium. In water, sensors are often deployed on floating platforms, integrated into existing infrastructure, or attached to autonomous underwater vehicles to create real-time maps of contamination in rivers, lakes, and outflow channels. For soil, durable probe arrays are inserted at specific depths to monitor leaching and plume movement, often combined with additional sensors for soil moisture and pH. Atmospheric monitoring near industrial facilities, such as electroplating or tanneries, tracks particulate matter (PM2.5 and PM10) to assess total chromium exposure, enabling a comprehensive evaluation (Karam et al., 2023; Kumar et al., 2023a; Kumar, 2023).

However, several challenges remain. Sensor calibration and drift are significant issues. Manual calibration against standard solutions can be impractical at large scales, and environmental factors such as biofouling and weathering cause signal drift, necessitating robust correction routines and periodic maintenance. Power management is critical, especially in remote sites, where energy-harvesting solutions, such as solar, wind, or biomass, are utilized in conjunction with ultra-low-power electronics. Advanced sleep-wake schedules extend operational lifetime (Dey et al., 2023b; Kumar & Kumar, 2023; Sharma & Kumar, 2023). Lastly, ruggedization is vital to ensure sensor durability against extreme environmental conditions, employing waterproof, dustproof (IP67/IP68 rated) casings resistant to chemicals, mechanical stresses, wind, water currents, and vandalism. Proper housing is as important as the sensing element in acquiring high-quality, time-coherent data, which is essential for protecting the environment and public health. IoT-based chromium toxicity monitoring is illustrated in Fig. 2.

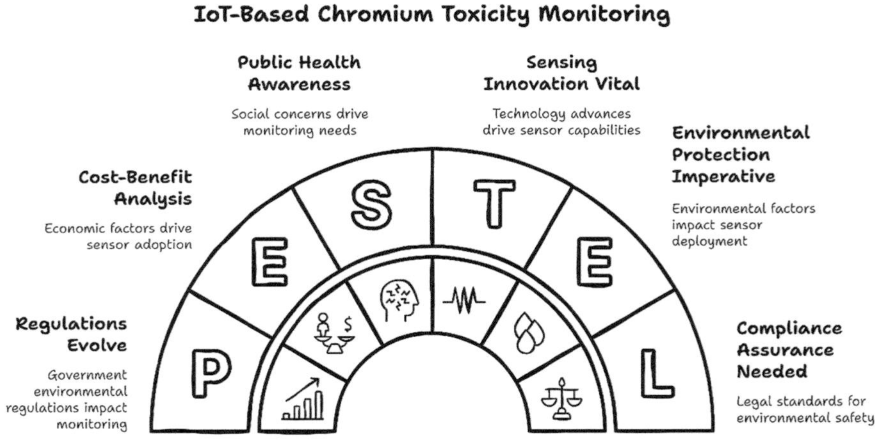

Fig. 2 IoT-based chromium toxicity monitoring

3 The Role of Artificial Intelligence and Machine Learning

The transformational capability of an IoT framework to chromium sensing is realized when one moves beyond a naive IoT (i.e., that of a glorified data logger) to a system that becomes intelligent, predictive, and even prescriptive. It is in the intelligent application of AI and ML technologies that an IoT framework evolves from merely serving as a data logger to a data analyst, focusing on the Fundamentals of AI/ML Relevant to Environmental Data Analysis, which involves applying the core paradigms of learning to the complex, noisy, multidimensional nature of environmental data, taken up in terms of both the theoretical understanding of the algorithms as well as the practical considerations in terms of model selection, training, validation, and deployment so that the resulting insights are both statistically robust as well as environmentally significant (Devi et al., 2023c; Devi & Kumar, 2023a; Kumari et al., 2023). The first and most important step is Data Preprocessing and Fusion from Heterogeneous IoT Sources, which is a daunting task as the raw data streams from distributed sensors are inherently sparse, noisy, and nonuniform in scale; this includes careful data cleaning to process the missing data values through imputation techniques or intelligent interpolation, outlier detection to identify and mitigate outliers arising from sensor failures and burst or environmental interference, and normalization to map disparate data layers (such as chromium level in ppb, temperature level in °C, pH level, turbidity in NTU) to a common scale, followed by advanced sensor fusion algorithms that combine these multimodal data to produce an enriched, more contextual, and high-resolution information product than any single data stream could deliver, to reduce uncertainty and increase the overall signal-to-noise ratio (Devi et al., 2023a; Devi & Kumar, 2023b; Sharma et al., 2023a; Sinam et al., 2023). Once a clean and merged dataset is established, the heavy lifting is done by Machine Learning Models for Chromium Prediction and Analysis, which come from two families of learning following the traditional ML approach: Supervised Learning: Regression Models for Concentration Prediction (e.g., Random Forest, Gradient Boosting) that use labelled historical data with known chromium concentration to train models to predict the levels for unknown data given the value of the input features captured by the rest of the sensors, for example a Gradient Boosting Machine (GBM) like XGBoost would model non-linear and complex relationships between water quality parameters such as, pH, oxidation-reduction potential, electrical conductivity, to give a few, and chromium concentration, learning that some combinations of water quality parameters are highly predictive of having a chromium plume and then being able to give real-time inferred concentration values through a specific chromium sensor being temporarily offline or needing calibration:. In addition to effortless prediction, Unsupervised Learning: Clustering for Source Identification and Anomaly Detection is of particular use in exploratory data analysis where labelled data is limited, algorithms like K-Means or DBSCAN can independently rediscover latent structures and group similar observations in clusters, which can help identify specific pollution fingerprints characteristic of a location (e.g., discharges in an illegal dump or industrial discharges) or naturally occurring contamination (e.g., mineral leaching), as well as, to vigilantly monitoring the latest data points and comparing them to the identified populations to detect divergent values that point to pollution events, equipment failures, or illicit discharges in substantially small datasets most of which could fly under the radar in the immense time series (Dey et al., 2023c; Kumar et al., 2023c; Saini et al., 2023). For vibrant data structures, Convolutional Neural Networks (CNNs) for Spectral Image Anaysis, Recurrent Neural Networks (RNNs) for Time-Series Forecasting have unrivaled power—CNNs can be trained to analyze complex spectral data from advanced optical sensors, and to automatically extract features from absorbance or fluorescence spectra to identify and quantify chromium species, with high precision, while RNNs, and in particular Long Short-Term Memory (LSTM) networks, excel at modelling temporal dependencies in time-series data, learning from weeks or months of historical sensor readings to forecast future trends in contamination, predict the migration of a groundwater plume, and to give early warnings about an impending violations of threshold conditions based on the evolving sensor readings. Finally, these capabilities are in-context spatialized with Geospatial AI: Integrating GIS with ML for Contamination Hotspot Mapping, where beyond-treatment points integrate to form a continuous, dynamic spatial understanding of risk; with the temporal outputs derived from the sensor network matched with geographic information system (GIS) data layers such as topography, hydrology, soil type, land use categories (e.g. industrial zones, agricultural areas), and transportation networks, ML models such as geostatistical kriging or spatially-weighted regression can interpolate between sensors to generate high-resolution predictive contamination maps, visually revealing hotspots of pollution, modelling potential contaminant pathways across watersheds, and attributing likely sources of contamination based on spatial correlation, providing decision-makers and environmental agencies with a powerful, evidence-based tool for focused remediation, resource allocation, and long-term environmental planning and, ultimately,

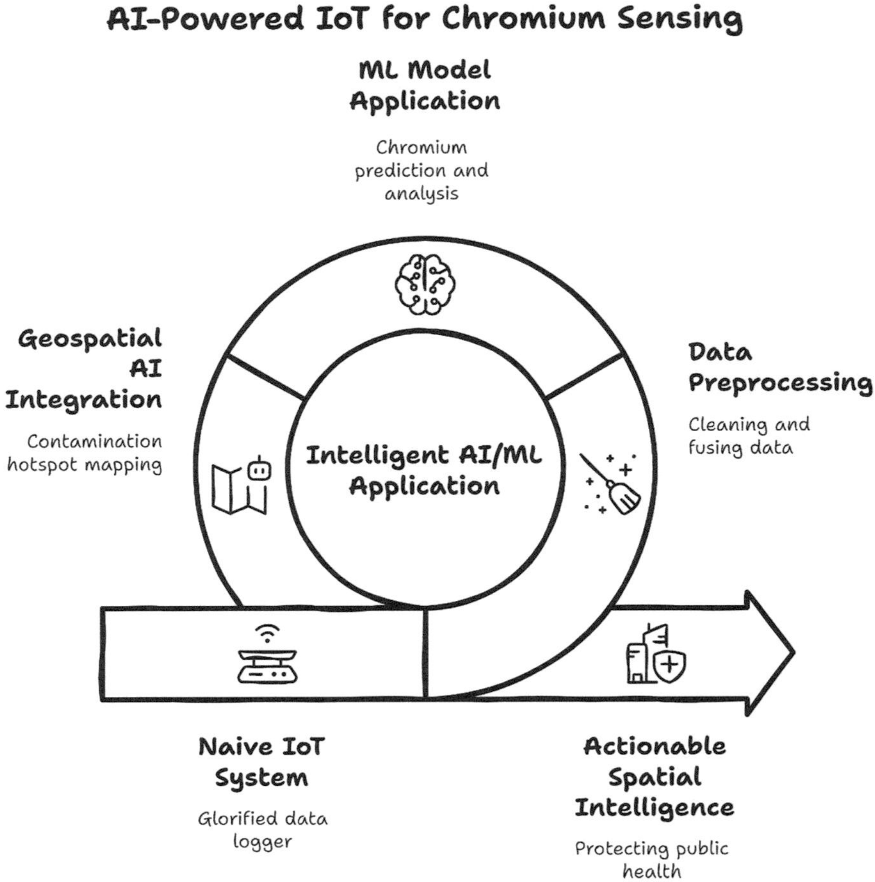

Fig. 3 AI-powered IoT for chromium sensing

turning raw IoT data into actionable spatial intelligence to protect public health and ecosystems (Fig. 3).

4 AI and IoT for Chromium Management and Remediation

The intersection of AI and IoT is more than simply monitoring—it represents a new paradigm for Active Chromium Management and Remediation, involving proactive, precise, and continuously self-improving algorithms that will radically change how we prevent pollution risks and rehabilitate the environment, beginning with Predictive Analytics for Pollution Plume Modeling and Early Warning Systems, where the federation of real-time IoT sensor data and sophisticated AI models, including physics-informed neural networks (PINNs) and ensemble forecasting methods, will produce dynamic simulations of contaminant transport in four dimensions spanning advection, dispersion, adsorption, and chemical conversion of chromium species in complex geological and hydrological settings, thereby anticipating the future trajectory, presence, and dilution of pollution plumes with an unprecedented level of accuracy, empowering authorities to alert downstream communities, strategically place barriers like interception wells or permeable reactive barriers long before the plumes spread, and simulate a variety of intervention scenarios for their potential efficacy, shifting the response paradigm away from reactive and towards truly anticipatory (Devi et al., 2023b; Dey et al., 2023a; Hidangmayum et al., 2023; Sharma et al., 2023d). This predictive intelligence is fluidly translated into Intelligent Decision Support Systems (DSS) for Policymakers that serve as the central cognitive gateway, seamlessly integrating AI-generated predictions with multi-criteria decision analysis (MCDA) that take into consideration environmental, economic, and social dimensions remediation cost, potential human health impact, ecological sensitivity, and land-use priorities informing ranked, actionable, policy-oriented mitigation strategies presented through interactive, visual dashboards that showcase trade-offs, projected risk, and uncertainty ranges, empowering stakeholders to move beyond gut-feeling decisions in favour of policy decisions made over robust data-driven simulations, for example, the most cost-efficient location for a new water treatment facility, relying on predicted plume paths or prioritising scarce resources towards more urgent contamination hotspots, identified through geospatial AI, maximising the societal value across entire remediation portfolios. But going beyond the long-term view, AI directly disrupts on-the-ground cleanup itself with AI-Optimized Phytoremediation and Bioremediation Strategies, involving machine learning algorithms that process massive data sets ranging from soil chemistry, microbial genomics, plant physiology, and historical remediation effect, to provide extremely targeted biological treatment plans; recommendation engines to guide selection of the most effective chromium-hyperaccumulating plants and ideal microbial consortia based on the site's pH, organic matter content and chromium speciation, all while reinforcement learning algorithms control an automated system that regulates nutrient injection, aeration, and irrigation in real time to maximize microbial metabolic activity and plant uptake, effectively rendering the system a living, self-adjusting treatment system that is perpetually fine-tuned in response to AI's directions to make it as efficient as possible (Kumar et al., 2023b, 2023d; Sharma et al., 2023b, 2023c). In the most severe cases, Robotics and Automation in Remediation: AI-Guided Cleanup Operations involve autonomous systems operating in hazardous environments without human exposure: coordinated fleets of drones with hyperspectral imaging sensors survey the terrain from the air to construct an activity map of surface contamination; ground-based robots or submersible drones guided by computer vision and SLAM algorithms navigate toxic terrains, from toxic sludge ponds to contaminated groundwater, to target micro borehole injection points for nano-scale (e.g. zero-valent iron nanoparticles) remedial agents densely or gather core samples; while a central AI planner oversees these actions, dynamically re-routing them based on sensor feedback for global coverage and optimal application of remedial reagents much more safely and accurately with less human labour and no overhead of a human operator

(Devi et al., 2023a; Devi & Kumar, 2023a; Dey et al., 2023c; Hidangmayum et al., 2023; Sharma et al., 2023a; Upadhyay et al., 2023). Full-circle closure of the entire remediation life-cycle is provided by Real-Time Monitoring of Remediation Efficacy and System Adjustment, which forms a cyber-physical feedback loop between IoT sensors that continuously monitor post-intervention chromium concentrations and other key biogeochemical parameters, streaming this information back into AI models that perform comparative analysis with the predicted outcomes; any deviation from the anticipated remediation trajectory triggers an automated re-calibration of the models, which, critically, can then activate pre-programmed adjustments to the remediation systems themselves, whether modulating the flow rate of a pump-and-treat system, adjusting the dose of a chemical reductant in an automated injection well, or re-patterning the navigation course of a cleanup robot to engage a newly-identified recalcitrant hotspot, thus ensuring that the remediation is not a static enterprise but a dynamically self-adjusting regimen that iteratively learns to converge with maximal efficiency to the desired environmental cleanup objectives, thereby guaranteeing that the large-scale investments in remediation provide the highest return on environmental and human health preservation (Fig. 4).

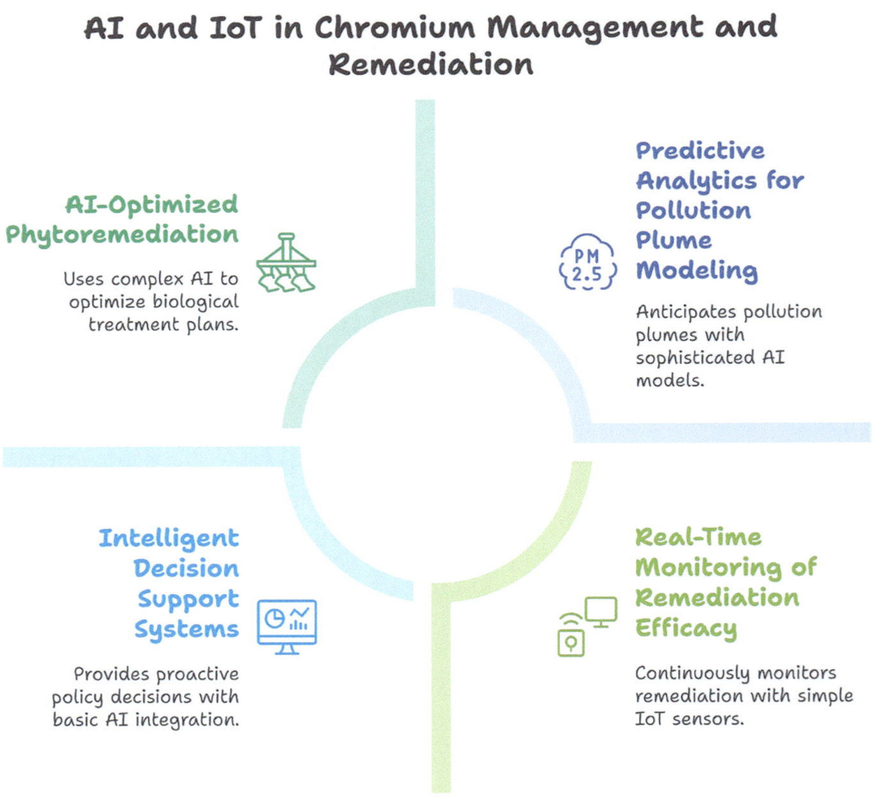

Fig. 4 AI and IoT in chromium management and remediation

5 Case Studies and Real-World Applications

This monitoring process, which has so far involved manual sampling at infrequent but predictable intervals, marks a paradigm shift from the traditional approach to a twenty-first-century smart and dynamic sentinel system. This system proactively manages chromium contamination by strategically deploying a multiplexing network of sensors at key sites. Utilizing ruggedized, waterproof probes with electrochemical and optical sensors, it places these sensors at critical subsurface depths in the soil profiles of contaminated areas—particularly where waste sludge was disposed of and chemicals stored. These sensors monitor total and hexavalent chromium (Cr (VI)) concentrations in the vadose zone. Modular low-power sensing equipment is suspended within targeted monitoring wells to provide high-resolution vertical profiles of the groundwater plume, including chromium speciation and key parameters such as pH, oxygen reduction potential (ORP), dissolved oxygen, and electrical conductivity, all of which influence chromium mobility and toxicity. Data from these sensors is transmitted via a hybrid LPWAN that combines LoRaWAN for aboveground soil sensors and NB-IoT for lower groundwater units, due to NB-IoT's superior ability to penetrate surface interference, ensuring real-time data reception even in submerged or remote locations of large, obstructed industrial sites (Devi et al., 2023a, 2023b; Dey et al., 2023c; Hidangmayum et al., 2023; Saini et al., 2023; Sharma et al., 2023a; Sharma & Kumar, 2023). This continuous stream of multi-parameter data feeds into a cloud-based AI platform. The first crucial step involves advanced data preprocessing and fusion from these diverse IoT sources, employing imputation algorithms to address gaps caused by sensor sleep cycles or temporary dropouts, and sophisticated anomaly detection models to identify and filter out false readings caused by biofouling, sediment disturbance, or electrical noise. This process validates the dataset and combines soil moisture data from time-domain reflectometry (TDR) sensors with geochemical data, producing a unified, contextual view of contaminant movement under various hydrological conditions, such as rainfall events or seasonal water table fluctuations. The core of the system's intelligence is a suite of machine learning models for chromium prediction and analysis. Supervised algorithms like Gradient Boosting Machines (XGBoost), trained on historical site investigation data, analyze real-time sensor inputs in parallel to estimate chromium concentrations at unsampled locations and depths, effectively creating a dynamic 3D model of the contamination plume (Kumar, 2023; Sharma & Kumar, 2023; Singh et al., 2023; Bharti et al., 2024; Silvanir et al., 2024). Unsupervised techniques, such as k-means clustering, autonomously identify contamination signatures that may correlate with specific site activities or reveal ongoing, previously undetected leaching from waste pits. Most critically, Long Short-Term Memory (LSTM) neural networks trained on time-series data accurately predict the movement and velocity of the plume, projecting weeks or months ahead to assess potential impacts on site boundaries or nearby sensitive receptors, such as drinking water wells. This predictive capability integrates with site-specific geological models and hydraulic conductivity data to run multiple simulations of future plume evolution under various climate scenarios. The system then functions as an early warning system, automatically alerting environmental managers via dashboards and SMS notifications if it forecasts exceeding regulatory limits at compliance boundaries—providing critical lead time for preemptive actions such as activating pump-and-treat systems or adjusting containment barriers. Over time, the system evolves into an intelligent decision support system (DSS) for policymakers and site engineers, combining AI-generated contamination forecasts with geospatial mapping tools and cost databases to evaluate and prioritize long-term remediation strategies. It displays trade-offs between options—such as monitored natural attenuation, in situ chemical reduction, or full-scale excavation—translating estimates of effectiveness, costs over decades, and risk reduction into visual formats that make complex stewardship tasks transparent and manageable. Importantly, if active remediation is ongoing, this approach allows real-time monitoring of remedy efficacy. For example, during in situ chemical reduction involving zero-valent iron injections, sensor data on ORP and Cr(VI) changes can be used by AI models to observe the influence radius and reaction time. The system can then automatically optimize injection parameters through a feedback loop, deploying maximum reagents where needed, ensuring comprehensive coverage, and monitoring for unintended side effects like the release of other metals due to drastic geochemical changes (Kumar et al., 2024; Moretti et al., 2024; Ghanbarlou et al., 2025; Islam & Ali, 2025; Khoshvaght & Khiadani, 2025; Robert & Kalpathy, 2025). Ultimately, implementing this advanced AI-IoT system on a former tannery site transforms it from a poorly understood, static liability into a dynamically managed asset. It reveals and predicts the complex processes of contaminant transport in real time, making them actionable. The significant investments already made for long-term stewardship can now be utilized with precision and efficiency, protecting public health and the environment while maintaining a transparent, verifiable record that supports regulatory compliance and informs future land-use decisions.

The implementation of a Real-Time E-F Monitoring system at an Electroplating Plant, based on an AI and IoT framework, marks a significant shift from traditional grab sampling and periodic lab analysis. This upgrade introduces a dynamic, intelligent EHS compliance management system that adapts to the fast pace of modern industry, transforming environmental monitoring from a reactive task into a proactive, real-time, process-integrated approach that balances operational efficiency with ecological health (Hakeem Anwer et al., 2024; Jayaraman et al., 2024; Islam & Ali, 2025; Mohd Adnan et al., 2025; Saha et al., 2025). The system is physically realized through the installation of an in-line, multi-layered sensor array placed directly in the plant's effluent collection sumps, primary equalization tanks, and, importantly, at the final discharge point before releasing to a municipal sewer or surface waters. These sensors, designed to withstand the challenging chemical environment of electroplating wastewaters, include solid-state ion-selective electrodes (ISEs), micro-fabricated electrochemical sensors, and anti-fouling coatings. They continuously monitor chromium (Cr(VI)), total chromium, and other regulated metals like cadmium, zinc, and nickel, using optical probes with UV-Vis spectrophotometry for colorimetric Cr(VI) detection—a common inorganic additive in electroplating—and conventional sensors for pH, temperature, flow rates, and total suspended solids. All sensors are housed in fully self-cleaning, pressurized enclosures with automated wipers and chemical cleaning ports to combat fouling from oily wastes, organics, and suspended metals, which traditionally plague computerized systems (Devi et al., 2023a, 2023c; Devi & Kumar, 2023b; Kumar et al., 2023a; Kumar & Kumar, 2023; Kumari et al., 2023). The sensor network transmits high-volume data through a reliable, low-latency Network Layer, typically wired Ethernet or optical fiber for robustness, possibly supplemented by a 5G private network for mobile sensors or drones used for tank inspections. This ensures precise, lossless transmission of critical data on pollutant levels and flow rates to a central Edge Computing node. This node performs immediate data preprocessing and fusion, handling real-time operations such as unit conversion, timestamp alignment, and basic outlier detection to remove erroneous readings caused by sensor faults. It produces clean, validated, and synchronized data for further analysis and interpretation. The core of the system resides in the Cloud/Edge Layer, where advanced Machine Learning Models conduct real-time analysis. Supervised models like XGBoost evaluate multivariate streams—correlating chromium, cyanide, pH, flow, and conductivity to generate estimates for parameters difficult to measure continuously, such as total dissolved solids, and to validate physical sensor data, adding redundancy and improving data integrity (Kumar et al., 2023a; Bharti et al., 2024; Silvanir et al., 2024; Ghanbarlou et al., 2025; Giwa et al., 2025). Unsupervised models, such as Isolation Forests, serve as anomaly detectors, learning the standard "biometric" signature of discharges under various production cycles and promptly alerting operators to anomalies, including sudden spikes of concentrated chromic acid or process failures, that trigger alarms and mobile alerts. Further, Deep Learning techniques, specifically Long Short-Term Memory (LSTM) networks, analyze months of production and effluent data to forecast short-term pollutant loads. This enables the system to predict violations minutes or hours before they occur, based on real-time influent quality changes, like ion-exchange resin regeneration releases. These predictions feed into a Pollution Prevention engine that guides operators to take preventive actions, such as redirecting high-strength waste to a holding tank or adjusting chemical dosages to prevent violations proactively. This comprehensive framework functions as a Decision Support System (DSS) for plant managers, integrating real-time compliance data with production schedules from the plant's MES. It offers actionable insights, identifying problematic plating lines or bath chemistries and estimating treatment costs per product unit, supporting root-cause analysis and process improvements to reduce waste at the source. In its most advanced form, the system can evolve into a closed-loop control architecture for AI-Optimized Remediation, where AI algorithms not only forecast but also dynamically adjust treatment parameters—such as dosing pumps and valves—in real-time. They optimize chemical use, minimize sludge formation, and ensure effluent consistently meets permit limits without human intervention (Hakeem Anwer et al., 2024; Jayaraman et al., 2024; Moretti et al., 2024; Islam & Ali, 2025; Li et al., 2025; Mohd Adnan et al., 2025; Saha et al., 2025). The system continuously monitors its performance via real-time KPI tracking, generating automated compliance reports for regulatory agencies. This creates a robust, self-optimizing, and transparent environmental management system that redefines effluent compliance from a cost burden to a standard of operational excellence and corporate responsibility.

It is the flagship product that "uses AI and IoT to enable fine-grained WS chromium management." The vision for predictable environmental care and precise intervention would be unthinkable without this. A systemic shift from point-based to holistic ecosystem management would enable an understanding of the complex hydrologic, geologic, and sociological forces that influence the fate and transport of this widespread pollutant across different temporal and spatial scales. This overall approach involves deploying an integrated, highly multiplexed, spatially and temporally diverse network of sensors throughout the entire hydrologic continuum of the watershed, from headwaters to estuary (Safeer et al., 2022; Giwa et al., 2025; Khoshvaght & Khiadani, 2025; Mohd Adnan et al., 2025; Robert & Kalpathy, 2025; Yadav & Sood, 2025). These sensors are placed at strategic locations, such as confluences of major tributaries, downstream of significant point sources like industrial clusters and wastewater treatment plants, within key agricultural runoff zones, and over sensitive aquifer recharge areas. They utilize a rich array of advanced sensing platforms such as submersible electrochemical analyzers for on-site chromium speciation in surface waters, rugged soil probe arrays with ion-selective membranes to monitor leaching into the vadose zone, passive samplers integrated with IoT transmitters for groundwater seepage monitoring, alongside remote sensing data from hyperspectral satellites and aerial drones that detect spectral signatures indicating chromium stress in vegetation or mineralogical changes in soil. This creates an unprecedented, multidimensional data tapestry that covers the watershed's health. Managing the flood of data from thousands of dispersed sensing nodes across vast areas is streamlined by a hierarchically organized Network Layer. This architecture selects communication protocols based on both location and power availability clusters of nodes and stationary stations in urban areas, for example (Moretti et al., 2024; Islam & Ali, 2025; Khoshvaght & Khiadani, 2025; Li et al., 2025; Mohd Adnan et al., 2025). Data processing is supported by regional edge computing centers, which are responsible for initial analysis and fusion. These centers employ techniques such as spatial kriging for gap filling, collective matrix factorization for anomaly detection, and time synchronization across asynchronous data streams. This curated data feeds advanced machine learning models operating at cloud scale. Geospatial AI plays a crucial role, with Convolutional Neural Networks (CNNs) identifying contamination hotspots and mapping chromium dispersion over floodplains, while Graph Neural Networks (GNNs) model river network connectivity, considering flow, sediment transport, and chemical reactions to predict contaminant movement. Ensemble methods like Random Forest regression combine land use, soil type, elevation, and climate data to produce high-resolution risk maps that identify areas at higher risk of chromium contamination, stemming from natural sources like weathering of chromium-rich rocks or human activities such as industrial operations and agriculture with chromium-based tanning (Devi & Kumar, 2023a; Karam et al., 2023; Kundathil et al., 2023; Sinam et al., 2023; Singh et al., 2023; Bharti et al., 2024).

These predictive models enable proactive watershed management, forecasting hydrologic and geochemical responses under changing climate conditions, such as how extreme rainfall might mobilize chromium from sediments or how droughts could concentrate pollutants, giving early warnings to downstream water treatment and agricultural facilities weeks in advance. Reinforcement learning algorithms continuously optimize digital twin models of the watershed, allowing managers to test thousands of intervention strategies virtually before implementation. This approach is supported by an intelligent Decision Support System (DSS) that guides hydrologists and policymakers. It incorporates biophysical, economic, and social factors, applying multi-objective optimization to balance competing goals, such as safeguarding drinking water, supporting agriculture, maintaining industrial operations, and protecting ecosystems. The system generates Pareto optimal solutions for strategies such as targeted sediment dredging, installing constructed wetlands for runoff treatment, and subsidizing industries to adopt zero-liquid-discharge technologies (Jayaraman et al., 2024; Silvanir et al., 2024; Ghanbarlou et al., 2025; Mahapatra et al., 2025; Sharma et al., 2025). The sensor network constantly monitors chromium levels in downstream waters, stream flow, and sediment stability after dredging, as well as the efficacy of agricultural best management practices. It also enables adaptive management through real-time monitoring of remediation efforts, allowing AI models to refine their predictions and recommendations over time. Ultimately, this comprehensive AI-driven watershed management framework aims to turn the challenge of chromium pollution into a manageable issue within sustainable water resource strategies. It empowers society to make informed, evidence-based decisions that protect ecosystem services, ensure public health through safe drinking water, promote economic growth, and preserve our water legacy for future generations—demonstrating how technological integration fosters resilient environmental governance and helps ecosystems cope with the legacies of industrial pollution during the Anthropocene.

6 Challenges, Limitations, and Future Perspectives

Data Privacy, Security, and Scalability, where the large volume of fine-grained environmental data being collected raises critical questions about ownership of this data, especially when it is collected on private property or in a manner that affects land value, and about its protection from cyberattacks. These attacks can spoof sensor readings to conceal signs of pollution or generate false alarms, causing public panic or market disruption (Kumar et al. 2023a; Kumar & Kumar, 2023; Kumari et al., 2023; Silvanir et al., 2024; Mahapatra et al., 2025). Addressing these issues requires strong encryption protocols, such as homomorphic encryption, which enables operations on encrypted data without decryption, along with zero-trust architectures and blockchain-based integrity authentication. These measures ensure data integrity throughout its lifecycle. The massive scale of managing millions of sensors across diverse, evolving environments stresses cloud infrastructure and demands innovative edge computing paradigms, where data reduction and model inferences happen closer to the source to prevent network congestion. Beyond data handling, there are profound scientific challenges: Model interpretability (XAI) and uncertainty quantification remain major barriers to acceptance by regulators and environmental scientists, who need transparent explanations for AI-driven decisions. The "black box" nature of complex models, such as LSTMs or CNNs, makes it difficult to explain how contamination forecasts are generated or which features are most influential. This is addressed through Explainable AI (XAI) methods, such as SHAP and LIME, which break down model predictions into understandable components. Equally important is rigorous uncertainty quantification, which accounts for data issues such as sensor noise, calibration drift, and model assumptions, including biases in the training data. Bayesian neural networks and ensemble methods provide confidence intervals and probabilistic predictions, helping policymakers understand the risks and reliability of AI recommendations (Devi & Kumar, 2023b; Dey et al., 2023c; Hidangmayum et al., 2023; Kumar et al., 2023d; Saini et al., 2023; Sharma et al., 2023b). Furthermore, progress is limited by pressing socio-economic challenges, including the costs, infrastructure, and barriers to the adoption of technology. The initial investment required for comprehensive sensor networks, edge computing devices, and AI platforms is substantial, especially in developing countries or small regulatory bodies with limited budgets and digital infrastructure, characterized by weak cellular coverage, unreliable power grids, and a lack of trust in automated solutions. Environmental professionals often lack expertise in AI, making it challenging to interpret AI outputs. Addressing these issues requires not only technological innovations but also financial solutions, such as sensing-as-a-service, workforce training, and compelling ROI studies that demonstrate long-term savings through predictive maintenance and targeted remediation. The next stage involves ambitious future directions, including next-generation smart sensors and self-powered IoT systems that go beyond simple data collection, integrating onboard preprocessing and AI inference with ultra-low-power microcontrollers. Energy harvesting technologies, such as triboelectric nanogenerators powered by water flows or vibrations and bio-photovoltaic cells using soil microbes, can enable battery-free sensors that operate indefinitely, eliminating the need for maintenance in inaccessible areas. Advanced AI approaches, such as transfer learning and federated learning, will address data scarcity and privacy concerns. Transfer learning pre-trains models on data-rich environments or simulations, then fine-tunes them locally, reducing calibration time and costs. Federated learning enables multiple organizations, including industrial plants and municipalities, to collaborate on training robust global models without sharing sensitive raw data. Instead, they exchange encrypted model updates, thereby preserving privacy while leveraging collective intelligence. An up-and-coming concept is the development of Digital Twins for chromium-impacted environments, a dynamic, high-fidelity virtual replica of entire ecosystems linked to real-time IoT sensors, mechanistic hydrogeological modeling, and AI predictions. This digital twin can run realistic simulations of intervention strategies, long-term climate change impacts on contaminant mobility, and 'what if' scenarios for planning, serving as a sandbox for sustainable environmental management (Bharti et al., 2024; Jayaraman et al., 2024; Moulahoum & Ghorbanizamani, 2024; Silvanir et al., 2024; Khoshvaght & Khiadani, 2025; Li et al., 2025). Finally, integrating blockchain technology for transparent data logging and compliance verification will revolutionize regulatory oversight. Every sensor reading, model prediction, and remedial action will be permanently recorded on a distributed ledger—creating an auditable, tamper-resistant trail of custody. These records can be shared transparently with regulators, community groups, and industry partners, automating compliance reporting through smart contracts that execute automatically when water quality parameters exceed permit limits. This reduces administrative costs, enhances accountability, and builds public trust in environmental monitoring efforts. In conclusion, although the path forward is long and challenging, the ongoing convergence of materials science, computer engineering, and environmental science points toward an increasingly autonomous and intelligent future for chromium management. Self-powered sensors, explainable AI, and collaborative digital twins will provide a resilient, transparent, and ultimately sustainable foundation for protecting human and ecological health from this persistent toxic legacy.

7 Conclusion

The comprehensive overview of an AI and IoT-enabled domain focusing on chromium sensing and management concludes with a summary emphasizing the transformative potential and practical steps ahead. This is supported by key findings showing that integrating advanced electrochemical, optochemical, and nanomaterial-based sensors with a robust, multi-protocol IoT infrastructure and advanced AI analytics fundamentally enhances our ability to diagnose and address chromium contamination. This applies across various contexts, including industrial effluent monitoring and legacy site management, shifting from reactive, periodic sampling to proactive, continuous, and predictive intelligence. Such advancements enable us to track contaminants' scale and movement in space and time, detect emergent threats through anomaly detection, and forecast future hazards with high spatial and temporal resolution, creating a credible evidence base for impactful intervention. This represents both the promise and the risk discussed in this special issue: a network of connected devices constantly streaming raw environmental data, a landscape beyond human oversight, monitoring, and control. In this future, Smart, Autonomous, Heterogeneous, and Automated Environmental Monitoring and Control Systems (SAEG-EMS) evolve from theoretical concepts to practical applications, transitioning from human-triggered alerts to deploying necessary technologies to address crises. They replace labor-intensive data collection with seamless data integration, analysis, and optimized responses, such as AI-driven chemical dosing for wastewater or drone-enabled nanoparticle targeting for resource recovery, operating without human intervention, delays, or excessive labor. This approach provides field staff with what they need most: relevant data continuously supplied by interconnected systems, not just big data machines that enable rapid decision-making through trusted, resilient, and transparent systems. These systems adapt based on evidence and iterative digital ecosystem models, similar to digital twins used to redesign buildings, infrastructure, or complex products. This optimizes resource use, anticipates emerging crises, and counters the fleeting nature of environmental impacts, ensuring regulation remains effective. Such a system becomes a constant source of relevant data, truthfully reflecting community expectations, thus enabling tighter controls without unnecessary overregulation. To realize this vision, urgent action on Final Recommendations for Stakeholders is essential: policymakers and regulatory agencies should update environmental permitting and compliance frameworks to trust continuous, AI-driven data streams, invest in digital infrastructure, and establish clear privacy and security guidelines; industry and site managers must prioritize early investment in these intelligent systems as sustainable practices for harm reduction, operational efficiency, and corporate responsibility; the scientific community needs to overcome technical hurdles through innovation in self-powering sensors, explainable AI, and robust uncertainty quantification; and public and community groups must engage and educate to build trust and promote equitable deployment. Ultimately, harnessing AI-IoT convergence promises to advance environmental stewardship while fostering a more transparent, accountable, and sustainable relationship with our natural resources.

References

Bharti, A., Jain, U., & Chauhan, N. (2024). From lab to field: Nano-biosensors for real-time plant nutrient tracking. *Plant Nano Biology, 9*, 100079. https://doi.org/10.1016/j.plana.2024.100079

Bin Abu Sofian, A. D. A., Majid, S. R., Kang, K., Kim, J.-K., & Show, P. L. (2025). Upcycling and recycling of spent battery waste for a sustainable future: Progress and perspectives. *Progress in Materials Science, 153*, 101478. https://doi.org/10.1016/j.pmatsci.2025.101478

Devi, P., Dey, S. R., & Kumar, P. (2023c). Integration of water resources management in rural areas. In *Water resources Management for Rural Development: Challenges and mitigation* (pp. 139–152). Elsevier.

Devi, P., Dey, S. R., Saini, L., Kumar, P., Panigrahi, S., & Dwivedi, P. (2023b). Toward sustainable agriculture: Strategies involving Phytoprotectants against reactive oxygen species. In *Reactive oxygen species: Prospects in plant metabolism* (pp. 229–247). Department of Agronomy, School of Agriculture, Lovely Professional University, Phagwara, India.

Devi, P., & Kumar, P. (2023a). Role of oxidative enzymes in the alleviation of oxidative stress in plant system. In *Microbial oxidative enzymes: Biotechnological applications* (pp. 109–128). De Gruyter, Department of Agronomy, School of Agriculture, Lovely Professional University.

Devi, P., & Kumar, P. (2023b). Effect of microbial oxidative enzymes on soil health. In *Microbial oxidative enzymes: Biotechnological applications* (pp. 37–61). De Gruyter.

Devi, P., Kumar, P., Sharma, K., & Singh, J. (2023a). Chapter 10—Is there a relationship between myconanotechnology and sustainable development? In M. Kuddus, I. Z. Ahmad, & C. M. Hussain, (Series Eds.), *Progress in Biochemistry and Biotechnology: Myconanotechnology and application of nanoparticles in biology* (pp. 175–207). Academic Press

Dey, S. R., Devi, P., & Kumar, P. (2023c). Organic contaminants in aquatic environments: Sources and impact assessment. In *Metal organic frameworks for wastewater contaminant removal* (pp. 299–317). Wiley.

Dey, S. R., Kumar, P., & Singh, J. (2023b). Growth of mustard under open field conditions with salicylic acid and antimicrobial Thiourea. *Biopesticides International, 19*(2), 183–192. https://doi.org/10.59467/BI.2023.19.183

Dey, S. R., Sharma, M., & Kumar, P. (2023a). Effects and responses of chromium on plants. In *Environmental science and engineering* (pp. 385–427). Springer Science and Business Media Deutschland GmbH.

Ghanbarlou, S., Kahforoushan, D., Abdollahi, H., Zarrintaj, P., Alomar, A., Villanueva, C., & Davachi, S. M. (2025). Advances in quantum dot-based fluorescence sensors for environmental and biomedical detection. *Talanta, 294*, 128176. https://doi.org/10.1016/j.talanta.2025.128176

Giwa, A. S., Maurice, N. J., Zelong, W., Claire, M. J., Vakili, M., Mabi, A., Liu, B., Lv, F., & Memon, A. G. (2025). Advancing resource recovery from sewage sludge with IoT-based bioleaching and anaerobic digestion techniques. *Journal of Environmental Chemical Engineering, 13*(3), 116293. https://doi.org/10.1016/j.jece.2025.116293

Hakeem Anwer, A., Saadaoui, M., Mohamed, A. T., Ahmad, N., & Benamor, A. (2024). State-of-the-art advances and challenges in wearable gas sensors for emerging applications: Innovations and future prospects. *Chemical Engineering Journal, 502*, 157899. https://doi.org/10.1016/j.cej.2024.157899

Hidangmayum, A., Dwivedi, P., Kumar, P., & Upadhyay, S. K. (2023). Seed priming and foliar application of chitosan ameliorate drought stress responses in Mungbean genotypes through modulation of Morpho-physiological attributes and increased Antioxidative Defense mechanism. *Journal of Plant Growth Regulation, 42*(10), 6137–6154. https://doi.org/10.1007/s00344-022-10792-1

Islam, M. T., & Ali, A. (2025). E-waste management in Saudi Arabia: Policy challenges, environmental hazards, and sustainable solutions. *Journal of Hazardous Materials Advances, 18*, 100738. https://doi.org/10.1016/j.hazadv.2025.100738

Jayaraman, P., Nagarajan, K. K., Partheeban, P., & Krishnamurthy, V. (2024). Critical review on water quality analysis using IoT and machine learning models. *International Journal of Information Management, IJIM Data Insights, 4*(1), 100210. https://doi.org/10.1016/j.jjimei.2023.100210

Karam, N., Kumar, P., Anand, Y. R., & Choudhury, D. (2023). Management of Rice Blast-a Review of the strategies available. *Biopesticides International, 19*(2), 107–116. https://doi.org/10.59467/BI.2023.19.107

Khoshvaght, H., & Khiadani, M. (2025). Chapter 6—Wastewater mining: a new frontier for artificial intelligence in mining. In R. Amir, & A. Mohsen (Series Eds.), *Cognitive Data Science in Sustainable Computing: Artificial intelligence in future mining* (pp. 249–307). Academic Press.

Kumar, P. (2023). Degradation of xenobiotics by bacteria and fungi: An overview. *Biopesticides International, 19*(2), 117–132. https://doi.org/10.59467/BI.2023.19.117

Kumar, P., Choudhury, D., Sahoo, J., & Devi, M. I. (2023a). Exploring the potential for extended investigation: A preliminary examination of Liriomyza trifolii and its bibliometric Foundation for Analytical Pursuits. *Journal Of Food Chemistry And Nanotechnology, 9*(Special Issue 1:S359–S366) https://doi.org/10.17756/jfcn.2023-s1-045

Kumar, P., Latha, D. H., & Singh, J. (2023c). Rhizoremediation of organic emerging soil contaminants: Green technology. In *Clean technologies toward a sustainable future: Physicochemical, biochemical and biotechnological approaches* (pp. 211–229). IWA Publishing.

Kumar, P., Saini, L., & Sharma, M. (2023b). Assessment of the phytochemical constituents and metabolites in the medicinal plants and herbal medicine used in the treatment and Management of Respiratory Diseases. In *Reference series in Phytochemistry*. Springer Science and Business Media B.V.

Kumar, P., Singh, S., Gacem, A., Yadav, K. K., Bhutto, J. K., Alreshidi, M. A., Kumar, M., Kumar, A., Yadav, V. K., Soni, S., Kumar, R., Qasim, M. T., Tariq, M., & Alam, M. W. (2024). A review on e-waste contamination, toxicity, and sustainable clean-up approaches for its management. *Toxicology, 508*, 153904. https://doi.org/10.1016/j.tox.2024.153904

Kumar, R., Mauryam A., Raj, A. (2023d). Emerging technological solutions for the management of paper mill wastewater: Treatment, nutrient recovery and fourth industrial revolution (IR 4.0). *Journal of Water Process Engineering*, 53, 103715. https://doi.org/10.1016/j.jwpe.2023.103715.

Kumar, V. A., & Kumar, P. (2023). Revolutionizing agriculture with satellite Technology for Farmers: A review. *Biopesticides International, 19*(2), 97–106. https://doi.org/10.59467/BI.2023.19.97

Kumari, P., Kumar, P., & Singh, J. (2023). Toxic contamination of aquatic habitat and its remediation. In *Spatial Modeling of environmental pollution and ecological risk* (pp. 209–220). Elsevier.

Kundathil, C., Viswan, H., & Kumar, P. (2023). Crop simulation Modeling: A strategic tool in crop management. *Journal Of Food Chemistry And Nanotechnology, 9(Special issue 1):S342–S358.* https://doi.org/10.17756/jfcn.2023-s1-044

Li, Y., Chen, B., Yang, S., Jiao, Z., Zhang, M., Yang, Y., & Gao, Y. (2025). Advances in environmental pollutant detection techniques: Enhancing public health monitoring and risk assessment. *Environment International, 197*, 109365. https://doi.org/10.1016/j.envint.2025.109365

Mahapatra, D. M., Kumar, A., Kumar, R., Gupta, N. K., Ethiraj, B., & Singh, L. (2025). Artificial intelligence interventions in 2D MXenes-based photocatalytic applications. *Coordination Chemistry Reviews, 529*, 216460. https://doi.org/10.1016/j.ccr.2025.216460

Malode, S. J., Alshehri, M. A., & Shetti, N. P. (2025). Revolutionizing human healthcare with wearable sensors for monitoring human strain. *Colloids Surfaces B Biointerfaces, 246*, 114384. https://doi.org/10.1016/j.colsurfb.2024.114384

Mohd Adnan, M. A., Majnis, M. F., Wan Md Adnan, W. N., Abdullah, N. H., Baharom, A. S., Muhd Julkapli, N. (2025). Deep insights into the integration of artificial neural networks (ANNs) for predicting the photocatalytic activities of metal-based catalysts in water pollutant reduction. *Journal of Environmental Chemical Engineering 13*(3),116350. https://doi.org/10.1016/j.jece.2025.116350

Moretti, A., Ivan, H. L., & Skvaril, J. (2024). A review of the state-of-the-art wastewater quality characterization and measurement technologies. Is the shift to real-time monitoring nowadays feasible? *Journal of Water Process Engineering, 60*, 105061. https://doi.org/10.1016/j.jwpe.2024.105061

Moulahoum, H., & Ghorbanizamani, F. (2024). Navigating the development of silver nanoparticles based food analysis through the power of artificial intelligence. *Food Chemistry, 445*, 138800. https://doi.org/10.1016/j.foodchem.2024.138800

Robert, B., & Kalpathy, S. K. (2025). Future prospects, challenges and opportunities in detecting heavy metal ions in water. *Journal of Water Process Engineering, 75*, 107992. https://doi.org/10.1016/j.jwpe.2025.107992

Safeer, S., Pandey, R. P., Rehman, B., Safdar, T., Ahmad, I., Hasan, S. W., & Ullah, A. (2022). A review of artificial intelligence in water purification and wastewater treatment: Recent advancements. *Journal of Water Process Engineering, 49*, 102974. https://doi.org/10.1016/j.jwpe.2022.102974

Saha, A., Roy, K., Nazir, M., & Saha, D. (2025). An empirical assessment of the impact of industrial contamination on Paddy production. *Pedosphere.* https://doi.org/10.1016/j.pedsph.2025.01.004

Saini, L., Sharma, K., & Kumar, P. (2023). Water efficiency in agriculture as a vital approach toward water management in rural areas. In *Water resources Management for Rural Development: Challenges and mitigation* (pp. 127–138). Elsevier.

Sharma, K., Devi, P., Dey, S. R., & Kumar, P. (2023b). Mercury phytovolatilization: An overview of the mechanism and mitigation. In *Role of green chemistry in ecosystem restoration to achieve environmental sustainability* (pp. 325–331). Elsevier.

Sharma, K., Devi, P., & Kumar, P. (2023c). Metal-organic frameworks for wastewater contaminants removal. In *Metal organic frameworks for wastewater contaminant removal* (pp. 95–117). Wiley.

Sharma, K., Devi, P., Kumar, P., Dey, A., & Dwivedi, P. (2023a). Hazardous phytotoxic nature of reactive oxygen species in agriculture. In *Reactive oxygen species: Prospects in plant metabolism* (pp. 135–146). Department of Agronomy, School of Agriculture, Lovely Professional University, Phagwara, India.

Sharma, K., & Kumar, P. (2023). Microbe-mediated impact on cell membrane stability and leaf pigmentation in pearl millet under toxic mercury stress in soil. *Biopesticides International, 19*(2), 217–228. https://doi.org/10.59467/BI.2023.19.217

Sharma, M., Dey, S. R., & Kumar, P. (2023d). Antioxidant defence: A key mechanism of chromium tolerance. In *Environmental science and engineering* (pp. 91–116). Springer Science and Business Media Deutschland GmbH.

Sharma, M., Mahajan, P., Alsubaie, A. S., Khanna, V., Chahal, S., Thakur, A., Yadav, A., Arya, A., Singh, A., & Singh, G. (2025). Next-generation nanomaterials-based biosensors: Real-time biosensing devices for detecting emerging environmental pollutants. *Mater Today Sustain, 29*, 101068. https://doi.org/10.1016/j.mtsust.2024.101068

Silvanir, F. W. H., Chia, W. Y., Ende, S., Chia, S. R., & Chew, K. W. (2024). Nanomaterials in aquaculture disinfection, water quality monitoring and wastewater remediation. *Journal of Environmental Chemical Engineering, 12*(5), 113947. https://doi.org/10.1016/j.jece.2024.113947

Sinam, V., Kumar, P., & Singh, J. (2023). Bioremediation and ecorestoration strategies of aquatic environment. In *Spatial Modeling of environmental pollution and ecological risk* (pp. 483–499). Elsevier.

Singh, H., Kabilan, G., Choudhury, D., & Kumar, P. (2023). Management of Banana Anthracnose, post-harvest practices, and future aspects—A review. *Biopesticides International, 19*(1), 1–10. https://doi.org/10.59467/BI.2023.19.1

Upadhyay, S. K., Devi, P., Kumar, V., Pathak, H. K., Kumar, P., Rajput, V. D., & Dwivedi, P. (2023). Efficient removal of total arsenic (As3+/5+) from contaminated water by novel strategies mediated iron and plant extract activated waste flowers of marigold. *Chemosphere, 313*, 137551. https://doi.org/10.1016/j.chemosphere.2022.137551

Yadav, U. S., & Sood, K. (2025). Chapter 1—IT-driven sustainability: Pathways to economic and environmental progress. In S. Grima, K. Sood, E. Özen, & G. RDBT (Eds.), *Science of sustainable systems* (pp. 1–12). Elsevier.

Biological Treatment Solutions Using Bioreactors for Chromium Removal from Environment

Sanjana Bhagat

Abstract Industrial uses and commercial processes emit significant amounts of Cr into environment (soil, surface water, or atmosphere), leading to severe human health issues due to its toxicity. Biological Cr removal presents an alternative to conventional physicochemical approaches. This is regarded as a sustainable technology with reduced environmental effect. Microorganisms exhibiting resistance, like fungi, bacteria, alongside algae, have been primary focus of research about this characteristic. Microorganisms have developed various methods to mitigate chromium toxicity. These methods encompass bioaccumulation, biotransformation (oxidation/reduction), and/or biosorption, and are regarded as alternatives for heavy metal removal. Review objective is to summarize Cr(VI) bioremediation technology focused on practical applications at a wider scale. Salient findings of various studies concerning process feasibility and resilience of diverse systems (pilot scale and bioreactors) engineered for Cr-removal capability are emphasized.

Keywords Bioremediation · Bioreactor · Pollution · Chromium

1 Introduction

Population increase and various anthropogenic activities exacerbated environmental contamination. Industrial effluents have been generated by integrating organic alongside inorganic pollutants, along with the discharge of HM (heavy metals) like chromium, copper, lead, cadmium, and selenium. Wastewater comprising HM is frequently released into environment (air, water, and soil) without adequate treatment, leading to a significant global socio-environmental issue.

They are nondegradable toxic pollutants, thereby persistent in nature and accumulating in food chain, that over time attain deleterious levels in living organisms, leading to various diseases like irritation and/or cancer in digestive tract and lungs, mortality in animals, reduced growth rates in plants, and other health modifications.

S. Bhagat (✉)
Government Nagarjuna PG College of Science, Raipur, Chhattisgarh, India
e-mail: sanjanabhagat@gmail.com

N. Kumar (ed.), *Chromium Toxicity*, https://doi.org/10.1007/978-3-032-14323-5_13

Chromium implies geochemical element extensively found in minerals, soils, rocks, alongside freshwater. Metal exhibits multiple oxidation states, although "more stable forms in environment are hexavalent [Cr(VI)] & trivalent [Cr(III)]. Owing to its oxidizing properties, Cr(VI) (predominantly CrO_4^{2-} at neutral or alkaline pH) is recognized as a mutagenic and carcinogenic substance to living organisms, like humans". This HM is classified as a priority pollutant in numerous nations and by the USEPA ("United States Environmental Protection Agency") (Juvera-Espinosa et al., 2006; Ksheminska et al., 2003).

Research indicates that Cr(VI) toxicity arises from the ability of metal complexes to traverse cellular membranes, leading to intracellular reactive oxygen species (ROS) accumulation and subsequent alterations in cellular structures. Cr(VI) is believed to be hundreds times more hazardous and thousands times "more mutagenic than Cr(III). Cr(III) is vital for human metabolism, contributing to regulation of cholesterol, glucose, and triglyceride levels, cellular membrane integrity, alongside stability as well as synthesis of proteins and nucleic acids. It has recently been observed that Cr(III) can no longer be regarded as a dietary supplement", as rats on a low trivalent chromium diet exhibited no deleterious effects when compared to rats on a diet with adequate Cr(III) levels. At elevated "concentrations, Cr(III) can form complexes with organic compounds, disrupting metalloenzyme systems" and potentially leading to health issues, like lung cancer, birth deficiency alongside diminished reproductive health.

Cr is an exceedingly hazardous metal to plants, micro-organisms, "as well as animal cells. Owing to its extensive industrial application, Cr has emerged as a significant pollutant in various environmental contexts. Hexavalent metalform", Cr(VI), is regarded as more dangerous than the more benign, less transportable Cr(III) form. Investigation of interactions b/w microbes and Cr has elucidated processes enabling organisms to endure elevated concentrations of Cr(VI), moreover, for detoxifying and eliminating oxyanions. "Diverse species of bacteria and fungi" exhibit various modes of interaction with Cr, like bioaccumulation, biosorption, conversion of Cr(VI) to Cr(III), alongside chromate efflux. Certain systems are suggested as prospective biotechnological instruments for Cr pollution bioremediation through bioreactors or in situ treatments.

Biological treatment using bioreactors for the removal of chromium from the environment has become a promising, eco-friendly solution for managing chromium contamination. Chromium is a HM that can be toxic to both aquatic ecosystems and human health when present in high concentrations. Biological treatments generally use microorganisms such as bacteria, fungi, and algae, or biological materials like activated sludge, to remove or neutralize contaminants.

Such work summarizes moreover discusses bioremediation potential of Cr(VI) at various scales (pilot scale and bioreactors) to reduce Cr(VI) concentrations to acceptable levels, considering socio-environmental influence of Cr in industrial effluents and its toxic effects on animal and human health.

2 Microbial Mechanism of Cr(VI) Resistance

There are 2 major chromium forms:

1. **Hexavalent chromium (Cr(VI))**: Highly toxic and carcinogenic.
2. **Trivalent chromium (Cr(III))**: Less toxic but could still pose environmental risks in high concentrations.

Biological treatment focuses primarily on reducing Cr(VI) to Cr(III), which is less toxic and can be more easily removed from the system.

2.1 *Biosorption*

It is a passive, fast, alongside reversible physicochemical procedure involving biosorbent (biological material) and sorbate (metal species) which implies a passive procedure, independent of cellular activity performed by either active/inactive microbe. Dead biomass utilization presents several advantages compared to living cells: there is no requirement for nutrient supplementation, which is resistant to toxicity and unfavorable operational conditions, metal recovery is facilitated through treatments that enable biomass regeneration, moreover biomass could be sourced economically as one industrial byproduct. Living cells exhibit a diverse array of pathways for metal accumulation, including extracellular complex formation, transport, and precipitation. Such procedures haven't been exclusive moreover might comprise metabolic alongside physicochemical absorption mechanisms that could also contribute to procedure. Metal recovery in biological procedures could be concerning, particularly if they are precipitated/compartmentalized internally.

Significance of utilizing micro-organisms as biosorbents arises from their high surface-to-volume ratio along with low production cost of biomass. Diverse microorganisms, including microalgae, algae, cyanobacteria, have been utilized for HM elimination.

2.2 *Bioaccumulation*

Various studies synthesized findings obtained from fungi, bacteria, algae, along with other plant-derived biomass based on their ability "to actively assimilate metals from aqueous solutions into their cells. Metal uptake by microbes is contingent upon initial metal concentration alongside contact duration. It is a metabolism-dependent mechanism that happens exclusively in living cells and necessitates energy for Cr(VI) transport" over membrane into cells. Primary constraint is the suppression of cell development at elevated metal concentrations. Comprehending the mechanism via which certain microbes accumulate Cr(VI) is crucial for developing procedures for removal, concentration, and recovery from contaminated solutions. Ksheminska et al. (2008) proposed "that Cr(VI) compounds, similar to anions PO_4^{2-} and SO_4^{2-} enter cells via oxidative, non-selective state-sensitive anion channel via enhanced diffusion. Cellular membranes have been frequently impermeable to Cr(III), likely due to complex formation with limited solubility.

2.3 Biotransformation

Hexavalent Cr biotransformation is seen as a detoxifying system, as trivalent form has been less toxic as well as more stable than Cr(VI). Current investigations have demonstrated "extracellular chromate reductase activity presence in microbial culture supernatant". In bacteria like *Bacillus, Pseudomonas,* and Arthrobacter, enzymatic decrease of Cr(VI) may be "related to soluble cytosolic proteins or insoluble membrane-bound enzymes. Various chromate reductases (YieF, NemA, ChrR, and LpDH) had been detected in soluble (cytoplasmic fraction) or related to the membrane, catalyzing reactions under anaerobic, aerobic, or occasionally both" circumstances. Ksheminska et al. (2008) detected numerous yeasts as optimal organisms for bioremediation research. "*Pichia guilliermondii*" (Ksheminska et al., 2008), *Rhodotorula" mucilaginosa* (Chatterjee et al., 2012), "and *Cyberlindnera jadinii* and *Wickerhamomyces anomalus* (Fernandez et al., 2016) have been documented for their ability to biotransform Cr(VI) into Cr(III). Ramírez-Ramírez et al. stated NAD-dependent chromate reducing activity in soluble protein fraction of *Candida maltosa.* Cr(VI) detoxification could additionally occur indirectly via chemicals (like riboflavin and sulfate) secreted by yeast cells to extracellular medium. Accordingly, detoxification of Cr(VI) could additionally transpire indirectly through chemicals, like sulfate alongside riboflavin, produced by yeast cells into extracellular media. Cell-free extracts of two yeasts, *Pichia anomala* and *Pichia jadinii,* separated from factory textile-dye effluents exhibit Cr(VI) reduction. Primary explanation for yeast resistance to Cr is attributed more to their restricted ion absorption than biological conversion of Cr(VI) to Cr(III).Gu et al.(2015) identified one chromate reductase of *Aspergillus niger* situated in the soluble fraction, and Singh and Bishnoi (2015) demonstrated enzymatic Cr(VI) decrease in "Aspergillus flavus," which was in conjunction with the biosorption procedure. "In this context, it is crucial to emphasize non-enzymatic reduction facilitated by nucleotides, sugars, amino acids, organic acids, vitamins", glutathione, alongside anaerobic metabolic byproducts of iron- and sulfate-reducing bacteria (Fe(II) & H2S). Fungal metabolites (organic acid citrate) alongside oxalate also diminish Cr(VI) via photocatalytic action of Fe(III) or via Mn in light's absence.

3 Biological Treatment Mechanisms in Bioreactors

3.1 *Microbial Reduction (Bioreduction)*

Bacteria (e.g., *Pseudomonas*, *Shewanella*, *Bacillus*, and *Geobacter*) have been identified as key players in biological decrease of Cr(VI) to Cr(III). Reduction process is often carried out by microbes utilizing Cr(VI) as electron acceptor during anaerobic respiration. Once Cr(VI) is decreased to Cr(III), it precipitates out of solution, making it easier to separate from water or soil.

3.2 *Bioaccumulation*

Certain microorganisms, particularly fungi (e.g., *Aspergillus*, *Penicillium*) and bacteria, can bioaccumulate Cr(VI) or Cr(III) within their cells. The accumulated chromium can then be removed by harvesting the microbial biomass.

3.3 *Bioadsorption*

Certain microorganisms or materials, such as activated sludge or biosorbents (e.g., chitosan, biosorbent algae, bacterial biofilms), can bind chromium ions to their surface. The removal of chromium occurs through a process called adsorption, where the chromium ions stick to the surface of the biosorbents.

3.4 *Phytoremediation*

Although not strictly a bioreactor process, certain plants are capable of absorbing, accumulating, or transforming chromium from the environment. These plants can be integrated into bioreactor systems for chromium removal.

4 Types of Bioreactors Used for Chromium Removal

The utilization of microbes in Cr(VI) removal process is being assessed across various systems employing fed-batch, batch, or continuous methodologies. Suitable bioreactor selection must correspond to many operational characteristics necessary for Cr(VI) removal, including mass transfer, hydrodynamics, alongside growth conditions. Such a technique is adaptable due to growth potential alongside microbial

Table 1 Removal of chromium using bioreactor approaches

Microorganism	Bioreactors	Initial concentration of Cr in (mg /L)	Mechanism and removal efficiency	References
Bacillus sp.	Immobilized enzyme	300/Cr(VI) in soil	Reduction 100%	Kathiravan et al. (2010)
Trichoderma viride	Stirred tank bioreactor	300/Cr(VI) in Synthetic solution	Reduction 60%	Morales-Barrera and Cristiani-Urbina (2006)
Aspergillus niger	Airlift bioreactor	Tanning wastewater, 1300/Cr(III)	Adsorption 88%	Sepehr et al. (2012a)
Consortium (*A. junii, E. coli, B. subtilis*)	Packed bed reactor	Contaminated water 100/Cr(VI)	Adsorption 24%	Samuel et al. (2013)

adaptation associated with modern biotechnology. Bioreactor system is generally more efficient, predictable, manageable, moreover easier to handle inside a limited area than in situ systems or those generated in solid phase. Removal of chromium using bioreactor approaches is tabulated in Table 1.

The standard classifications of systems employed in this context include:

4.1 Stirred Tank Reactors

Reactor design incorporates a stirrer and is capable of operating in both batch and continuous modes; in certain instances, it allows for tank feeding until reaction volume is attained. Benefits of stirred tank reactors pertain to straightforwardness and convenience of replicating tests; yet, the operating expenses are higher because of energy consumed for agitation. Such tool is often utilized for smaller industrial effluent quantities moreover where biosorption has been seasonal.

Cr(VI) solution to be treated is continually supplied. They facilitate examining fluctuations in Cr(VI) concentration, modify hydraulic retention duration based on microbial growth rate, and operate sequentially with various reactors. Drawbacks pertain to diminished contamination alongside cell viability. Chromium biosorption experiments utilizing this operational mode are crucial for assessing technical viability of an actual procedure concerning effluent to be treated and subsequent Cr(VI) elimination.

Furthermore, efficacy of numerous bioreactors in reducing Cr(VI) under diverse operating circumstances was assessed. "Arthrobacter rhombi, a Cr(VI) decreasing strain separated from Cr-contaminated soil, had been employed in one continuous reactor utilizing three distinct systems for Cr(VI) biotransformation: (i) aerobic attached, (ii) aerobic suspended, and (iii) anoxic attached growth". Anoxic attached alongside Aerobic suspended growth systems conducted poorer than "aerobic attached growth system obtaining a 98% Cr(VI) reduction efficiency employing aerobic attached growth" system.

4.2 Fixed-Bed Reactors

In these systems, microorganisms are immobilized on a solid support (such as activated carbon or biofilms) within the bioreactor. These systems can be operated in both continuous-flow or batch modes and are suitable for treating water with low to medium chromium concentrations.

Cr(VI) biosorption tendency underscores necessity to extend operational lifespan of "biosorbent and assess Cr(VI) desorption alongside recovery cycles. Biosorbent is positioned in a fixed bed" within a column that facilitates metal-contaminated solution passage. Packed bed columns (PBCs) have been utilized the most for conducting biosorption research. Biomass "can be immobilized through encapsulation, entrapment, alongside bonding onto textile fibers, polymeric materials," moreover inorganic compounds. Primary advantages comprise utilization of big particles for biosorbent immobilization alongside ease of installation as well as "operation (Rosca et al., 2015) multiple variations of this system were created".

4.3 Fluidized-Bed Bioreactors

A fluidized-bed bioreactor uses a finely divided support material (such as sand) that is fluidized by the incoming liquid stream. The support material carries the microbial biomass, and these systems are typically more efficient in high-flow, continuous treatment applications. Fluidized-bed bioreactors have been dependent over microbial biofilm establishment over solid particles which facilitate microbial growth. "Counterflow mechanism within a column sustains particles in continuous motion fluidized condition" moreover facilitates their continual migration over the full volume of column. As flow rate continues to rise, bed transitions into one fluidized state. For such reactors, one biosorbent is positioned inside one column alongside one solution having "metals is provided in counter-flow system which facilitates biomass retention" within reactor, enabling operation at reduced hydraulic retention times.

Chirwa and Wang (2001) documented "biological Cr(VI) reduction in a laboratory-scale continuous-flow biofilm reactor" with no fresh biomass intermittent addition into system. They had been pioneers in demonstrating efficacy of fixed-film bioreactor for Cr(VI) reduction. In such "reactors, electron donors" had been included into wastewater as required (Elangovan & Philip, 2009).

4.4 Airlift Reactors

Airlift alongside bubble columns have been bioreactors agitated pneumatically, which exert minimal shear stress. They have been optimal choice when desired microbe is a fungi, as they mitigate issues related to multicellular filament growth alongside rheological broth properties. Design minimizes pellet formation moreover facilitates development of dispersion mycelium and hyphae, hence enhancing surface contact area. Airlift bioreactors have been frequently utilized in bio procedures in where gas–liquid interaction is crucial moreover have been regarded as viable as well as sustainable substitutes for stirred tank reactors (Choi et al., 2007; Cozma & Gavrilescu, 2010). Aeration function is to meet the demand for dissolved oxygen while facilitating pneumatic mixing and associated mass transfer.

4.5 Activated Sludge Reactors

These have been most prevalently utilized bioreactors in wastewater treatment. They consist of a suspended microbial population that degrades pollutants, including chromium, by decreasing Cr(VI) to Cr(III). They are most efficient when combined with anaerobic conditions, where Cr(VI) could be decreased by specific bacteria.

4.6 Membrane Bioreactors

These combine biological treatment and membrane filtration into a single unit. They are ideal for removing low concentrations of chromium from large volumes of water. The membrane allows for the separation of microbial biomass from the treated water, making the system more efficient for long-term operation.

4.7 Constructed Wetlands and Biofilters

These systems are often used for large-scale applications, such as treating industrial effluents or mining wastewater. In these systems, plants, microbes, and organic materials interact to remove contaminants, including chromium Lytras et al. (2017).

5 Factors Influencing Chromium Removal in Bioreactors

1. **pH**:

The pH of the environment is a key factor for Cr(VI) to Cr(III) reduction. Most microbial reduction processes occur effectively in neutral to slightly alkaline conditions (pH 6.5–8.0). Acidic conditions (low pH) can inhibit microbial activity and limit chromium reduction.

2. **Temperature**:

The activity of chromium-reducing microorganisms is influenced by temperature, with optimal removal often occurring between 20 and 40 °C.

3. **Organic Carbon Availability**:

Microbial reduction of Cr(VI) is often coupled with the metabolism of organic carbon. A sufficient carbon source is necessary for the microbes to thrive and carry out the reduction.

4. **Presence of Other Metals**:

The presence of other metals or chemicals may either inhibit or promote chromium reduction, depending on the system's design and microbial communities.

5. **Retention Time**:

Longer retention times generally allow for more efficient reduction of chromium. However, this may need to be balanced with operational constraints and costs.

6 Advantages of Biological Treatment Using Bioreactors

- **Eco-friendly**: Biological treatment is sustainable and less expensive than traditional methods such as chemical precipitation or ion-exchange.
- **Efficiency**: It can remove chromium to concentrations below environmental standards, especially when used in conjunction with other treatment methods.
- **Low Energy Requirements**: Biological systems typically require less energy compared to other treatment processes.
- **Versatility**: Bioreactors can be adapted to treat a wide variety of chromium-containing wastewater streams (e.g., from tanneries, electroplating, and mining operations).

7 Challenges

- **Slow Process**: Biological reduction of chromium can take longer than physical or chemical methods.
- **Microbial Adaptation**: Some microorganisms need time to adapt to the presence of chromium, especially at high concentrations.

- **Waste Biomass Management**: The biomass that accumulates during the process must be properly managed and disposed of.
- **System Design**: Optimizing bioreactor design to handle varying concentrations of chromium and adjusting operational parameters to maintain effective removal can be complex.

8 Conclusion

Bioreactor-based biological treatment is a promising solution for chromium removal from the environment, offering a more sustainable alternative to traditional chemical methods. However, its efficiency can depend on several factors, including the type of bioreactor used, environmental conditions, and the presence of other pollutants. Advances in microbial biotechnology, reactor design, and operational optimization will likely enhance the effectiveness of these systems for large-scale industrial applications.

Diverse industrial alongside anthropogenic activities contaminate environment; Cr(VI) & Cr(III) exhibit distinct behaviors, with great solubility and mobility of Cr(VI) enhancing its prevalence. Cr(III) implies non-toxic moreover comparatively insoluble in aqueous environments. Bioremediation provides a technique for HM elimination by methods employed by microorganisms, including bioaccumulation, biosorption, and biotransformation. The utilization of an environmentally sustainable, multifunctional, and cost-effective instrument is essential for the extraction of this HM from soil, water, and sediments. Biosorption method has demonstrated success in relation to technical advancements. It is evident that biological Cr(VI) detoxification is evaluated in several systems. The process can be conducted in batch, fed-batch, or continuous modes, depending on varying operational parameters, to optimize hydrodynamic, mass transfer, and growth environments. By enhancing genetic knowledge and establishing dedicated experimental sites in transferring bioremediation, such prospects present opportunities for significant progress; nevertheless, extensible investigation must be conducted in this field to address challenges related to large-scale remediation procedures.

References

Chatterjee, S., Chatterjee, N. C., & Dutta, S. (2012). Bioreduction of chromium (VI) to chromium (III) by a novel yeast strain Rhodotorula mucilaginosa (MTCC 9315). *African Journal of Biotechnology, 11*(83), 14920–14929.

Choi, H., Stathatos, E., & Dionysiou, D. D. (2007). Photocatalytic TiO2 films and membranes for the development of efficient wastewater treatment and reuse systems. *Desalination, 202*(1–3), 199–206.

Cozma, P., & Gavrilescu, M. (2010). Airlift reactors: Hydrodynamics, mass transfer and applications in environmental remediation. *Environmental Engineering and Management Journal, 9*(5), 681–702.
Elangovan, R., & Philip, L. (2009). Performance evaluation of various bioreactors for the removal of Cr (VI) and organic matter from industrial effluent. *Biochemical Engineering Journal, 44*(2–3), 174–186.
Fernández, P. M., Figueroa, L. I., & Fariña, J. I. (2010). Critical influence of culture medium and Cr (III) quantification protocols on the interpretation of Cr (VI) bioremediation by environmental fungal isolates. *Water, Air, and Soil Pollution, 206*(1), 283–293.
Fernandez, P. M., Cruz, E. L., Vi~narta, S. C., & Figueroa, L. I. C. (2016). Optimization of culture conditions for growth associated with Cr (VI) removal by Wickerhamomyces anomalus M10. *The Bulletin of Environmental Contamination and Toxicology, 98*(3), 400–4006.
Gu, Y., Xu, W., Liu, Y., Zeng, G., Huang, J., Tan, X., & Wang, D. (2015). Mechanism of Cr (VI) reduction by Aspergillus niger: Enzymatic characteristic, oxidative stress response, and reduction product. *Environmental Science and Pollution Research, 22*(8), 6271–6279.
Juvera-Espinosa, J., Morales-Barrera, L., & Cristiani-Urbina, E. (2006). Isolation and characterization of a yeast strain capable of removing Cr (VI). *Enzyme and Microbial Technology, 40*(1), 114–121.
Kathiravan, M. N., Rani, R. K., Karthick, R., & Muthukumar, K. (2010). Mass transfer studies on the reduction of Cr (VI) using calcium alginate immobilized Bacillus sp. in packed bed reactor. *Bioresource Technology, 101*(3), 853–858.
Ksheminska, H., Jaglarz, A., Fedorovych, D., Babyak, L., Yanovych, D., Kaszycki, P., & Koloczek, H. (2003). Bioremediation of chromium by the yeast Pichia guilliermondii: Toxicity and accumulation of Cr (III) and Cr (VI) and the influence of riboflavin on Cr tolerance. *Microbiological Research, 158*(1), 59–67.
Ksheminska, H., Fedorovych, D., Honchar, T., Ivash, M., & Gonchar, M. (2008). Yeast tolerance to chromium depends on extracellular chromate reduction and Cr (III) chelation. *Food Technology and Biotechnology, 46*(4), 419–426.
Lytras, G., Lytras, C., Argyropoulou, D., Dimopoulos, N., Malavetas, G., & Lyberatos, G. (2017). A novel two-phase bioreactor for microbial hexavalent chromium removal from wastewater. *Journal of Hazardous Materials, 336*, 41–51.
Morales-Barrera, L., & Cristiani-Urbina, E. (2006). Removal of hexavalent chromium by Trichoderma viride in an airlift bioreactor. *Enzyme and Microbial Technology, 40*(1), 107–113.
Nkhalambayausi-Chirwa, E. M., & Wang, Y. T. (2001). Simultaneous chromium (VI) reduction and phenol degradation in a fixed-film coculture bioreactor: Reactor performance. *Water Research, 35*(8), 1921–1932.
Rosca, M., Hlihor, R. M., Cozma, P., Com_anit¸_a, E. D., Simion, I. M., & Gavrilescu, M. (2015). Potential of biosorption and bioaccumulation processes for heavy metals removal in bioreactors. In *E-Health and Bioengineering Conference (EHB)*. https://doi.org/10.1109/EHB.2015.7391487
Samuel, J., Pulimi, M., Paul, M. L., Maurya, A., Chandrasekaran, N., & Mukherjee, A. (2013). Batch and continuous flow studies of adsorptive removal of Cr (VI) by adapted bacterial consortia immobilized in alginate beads. *Bioresource Technology, 128*, 423–430.
Sepehr, M. N., Nasseri, S., Zarrabi, M., Samarghandi, M. R., & Amrane, A. (2012). Removal of Cr (III) from tanning effluent by Aspergillus niger in airlift bioreactor. *Separation and Purification Technology, 96*, 256–262.
Singh, R., & Bishnoi, N. R. (2015). Biotransformation dynamics of chromium (VI) detoxification using Aspergillus flavus system. *Ecological Engineering, 75*, 103–109.

Chromium Removal from Aqueous Solutions Using Bioadsorbents

Krishnendu Sen, Soumen Barat, Madhurima Dutta, Avipsa Bhanja, Sudip Bhattacharya, Subhas Patar, Kalyanbrata Pal, and Kuntal Ghosh

Abstract Hexavalent chromium (Cr(VI)) is an extremely toxic and mobile pollutant produce from tannery, electroplating and textile effluents, and sustainable remediation strategies are required to comply with strict environmental standards (e.g. WHO: 0.05 mg/L in drinking water). This chapter elaborates in detail about bioadsorption as an eco-friendly cost-effective method of Cr(VI) removal from aqueous solutions using renewable bioadsorbents consisting of agricultural byproducts (e.g. rice husk, walnut shells), microbial biomass (bacterial, fungal, algal), chitosan derivatives and biochar derived from biomass. The chemistry of Cr(VI) speciation ($HCrO_4^-$ at pH 1–6, CrO_4^{2-} at pH > 7) and toxicity are the basis for the design of bioadsorption processes comprising ion exchange, electrostatic attraction, surface complexation and reductive transformation to Cr(III) supplemented with bioaccumulation in living cells. Advanced bioadsorbent modifications—chemical activation (e.g. citric acid, NaOH), nanocomposites (e.g. magnetite-chitosan), ternary nanoadsorbents (Fe_2O_3-MnO_2-SnO_2), green synthesis have improved the adsorption capacity reaching 100–315 mg/g, which is higher than many synthetic adsorbents. Factors affecting the efficiency such as pH (optimal 2–3), initial Cr(VI) concentration, bioadsorbent dosage, contact time (60–250 min until equilibrium), temperature and effect of competing ions (SO_4^{2-}, NO_3^-), are modelled using Langmuir, Freundlich, Temkin and Redlich-Peterson isotherms, using pseudo-second-order kinetics and intra-particle diffusion identify chemisorption domination. Real world applications in wastewaters from tanneries and electroplating evident that removal of Cr(VI) of 90–99%, with continuous systems and regeneration (80–95% recovery over 5–10 cycles) refining practicality, as shown by specific trials with tannery effluent. Synchrotron-based techniques (e.g. XANES, EXAFS) and FTIR/XPS analyses help to understand the Cr binding and reduction processes, whereas the techno-economic analyses reinforce the scale-up feasibility. Challenges such as low native capacities (< 50 mg/g), pH

K. Sen (✉) · M. Dutta · A. Bhanja · S. Bhattacharya · S. Patar · K. Pal · K. Ghosh
Department of Paramedical and Allied Health Science, Midnapore City College, Midnapore, West Bengal, India
e-mail: krishnendusen10@gmail.com

S. Barat
Department of Molecular Microbiology, Institute of Health Sciences, Presidency University, Kolkata, West Bengal, India

N. Kumar (ed.), *Chromium Toxicity*, https://doi.org/10.1007/978-3-032-14323-5_14

sensitivity and spent bioadsorbent disposal—are being overcome by such emerging trends as hybrid technologies (bioadsorption-photocatalysis), genetic engineering of a microbial consortia, microbiome engineering for synergistic effects, artificial intelligence for optimisation processes and circular economy approaches valorising spent biomass into biofertilisers. The chapter concludes by promoting the need for pilot-scale validation and policy-driven involvement of the community to realise the transformative potential of bioadsorption for Cr(VI) remediation.

Keywords Bioadsorption · Hexavalent chromium · Nanocomposites · Chitosan · Circular economy · Synchrotron techniques

1 Introduction

Hexavalent chromium (Cr(VI)) is known to be a highly toxic, cancer-causing pollutant emitted primarily from the industrial activities like leather tanning, electroplating and textiles manufacturing (Sharma et al., 2022). Its high solubility and mobility in water make it a very serious threat to the environment and to human health, with the World Health Organisation's strict permissible limit set at a level of 0.05 mg/L in drinking water (Bandh & Mushtaq, 2025). Conventional treatments such as chemical reduction, precipitation, ion exchange and membrane filtration often have disadvantages such as high cost, incomplete removal and secondary pollutant generation (Ayach et al., 2024).

Bioadsorption has become an alternative excepting metallic adsorption that is green, inexpensive and efficient, employing naturally occurring or biomass-based products, to adsorb and eliminate Cr(VI) of aqueous solutions (Das et al., 2024). Agricultural waste (rice husk, walnut shells) or microbial biomass (bacteria, fungi, algae), chitosan and biochar are all examples of common bioadsorbent, and each of them contains significant concentrations of functional groups that can bind and reduce Cr(VI) and transform it into less toxic Cr(III) (Singh & Vijayan, 2024). Adsorption capacities have been improved through chemical activation with acids or bases, the manufacture of nanocomposites (e.g. magnetite-chitosan) or green preparation techniques up to 300 mg/g, and bioadsorbents are competitive with synthetic adsorbents (Bakhtiari et al., 2024; Putra et al., 2025). Schematic workflow of Cr(VI) bioadsorption studies is depicted in Fig. 1. Instrumental approaches to evaluate Cr(VI) removal by biosorbents is presented in Table 1.

Optimal removal of Cr(VI) is generally obtained in acidic conditions (pH 2–3), where Cr(VI) forms (e.g. $HCrO_{4-}$) strongly adsorb to the protonated bioadsorbent sites (Liu et al., 2025). Critical parameters affecting the adsorption efficiency are the initial concentration of Cr(VI), dosage of bioadsorbent, contact time (usually from 60 to 250 min), temperature and competing ion such as sulphate, nitrate (Xia et al., 2019). Isotherm models like Langmuir and Freundlich generally exhibit a good fit for the adsorption data and from the kinetics, pseudo-second-order kinetics is suggested with chemisorption being the dominant process (Gürsoy et al., 2025).

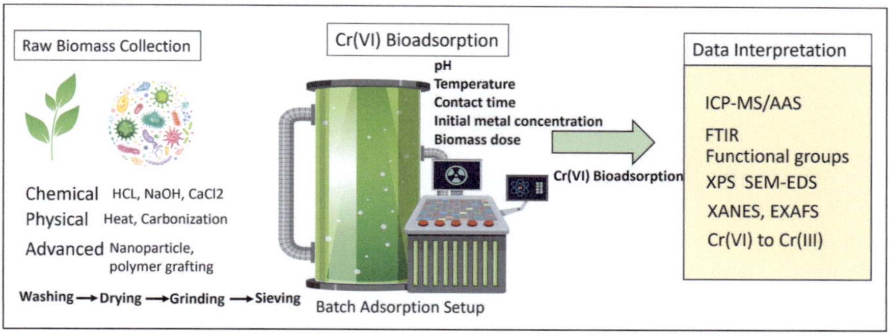

Fig. 1 Schematic workflow of Cr(VI) bioadsorption studies

Recently, hybrid technologies have been popular to improve Cr(VI) removal by combining bioadsorption and photocatalysis or other oxidation processes to overcome the limitations associated with bioadsorption (Zeng et al., 2025). There is also a focus on regeneration and reuse of bioadsorbents for their economical viability and in keeping with circular economy principles (Amesho et al., 2024). Real applications in wastewater effluent of the tanneries and electroplants have shown the removal of Cr(VI) (90–99%) using bioadsorbents, and the regeneration processes retain 80–95% efficiency after multiple reuse cycles (Kumari et al., 2025; Zhigalenok et al., 2025). This chapter summarises the recent developments in the design, modification and use of bioadsorbents for the removal of Cr(VI), emphasising the mechanisms, factors that affect the removal, challenges and current trends for practical removal in the environment.

2 Chemistry and Toxicity of Chromium in Aqueous Solutions

2.1 *Speciation of Chromium*

Chromium is found in a number of oxidation states, but in the water environment, the most common forms of chromium are trivalent chromium [Cr(III)] and hexavalent chromium [Cr(VI)]. The chemical behaviour of these species is strongly dependent on pH and redox potential. Cr(III) is usually in the form of cation Cr^{3+}, $Cr(OH)^{2+}$, or as the precipitate $Cr(OH)_3$, which is sparingly soluble in water. In contrast, Cr(VI) forms highly soluble oxyanions such as chromate (CrO_4^{2-}), dichromate ($Cr_2O_7^{2-}$) and hydrogen chromate ($HCrO_4^-$). Under acidic conditions $HCrO_4^-$ is predominant whereas at neutral to alkaline pH, CrO_4^{2-} is predominant (Nishad et al., 2025). Experimental investigations emphasise the complexity of chromium speciation in natural and contaminated systems. For example, soils collected from electroplating waste sites in China had total chromium concentrations of 7.2 –7735 mg/kg with

Table 1 Instrumental approaches to evaluate Cr(VI) removal by biosorbents

SI. no	Techniques	Purpose	Outcome	References
1	ICP–MS/AAS	Quantify residual chromium in solution	Accurate concentration, limits of detection (Cr total or Cr(III)/ Cr(VI) when hyphenated with separation)	El Hajam et al., 2023
2	UV–Vis (diphenylcarbazide method/ spectrophotometry)	Routine Cr(VI) concentration checks in batch tests	Fast Cr(VI) concentration from calibration curve (λ often ~ 540 nm)	Naseem et al., 2022
3	FTIR (Fourier transform IR)	Identify functional groups on adsorbent (before/ after)	Reveal–OH,–COOH,–NH_2, C = O changes and band shifts indicating binding	Naseem et al., 2022
4	SEM—EDS (SEM/ EDX)	Surface morphology and elemental mapping	Images of surface texture; EDS shows presence/distribution of Cr on sorbent	Aké et al., 2024
5	TEM	High-res morphology/ nanoparticles observation	Particle size, core–shell structure, nanodeposit imaging	Permporn et al., 2023
6	XPS (X-ray photoelectron)	Determine surface oxidation states	Distinguishes Cr(VI) vs Cr(III) on the adsorbent surface (binding energies)	Liu et al., 2023
7	XRD	Identify crystalline phases/new oxide phases after adsorption	Detects crystalline metal oxides, Fe_3O_4, ZnO, or new Cr-containing phases	Naseem et al., 2022; Liu et al., 2023
8	BET (N_2 adsorption)	Surface area and porosity (link to capacity)	Specific surface area (m^2/g), pore volume and size distribution	Liu et al., 2023
9	TGA/DTA (thermal analysis)	Thermal stability and composition	Mass loss steps, decomposition temperatures; evidence of adsorbed species	Liu et al., 2023
10	LA-ICP-MS (or micro-mapping)	Spatial mapping of metal on solid samples	Localised elemental distribution and semi-quantitative maps	Liu et al., 2023
11	Zeta potential/ pHPZC	Surface charge and point of zero charge	Explains electrostatic attraction/ repulsion vs pH (mechanistic)	Aké et al., 2024
12	XANES/EXAFS (synchrotron)	In-depth oxidation state and local structure	Direct speciation (Cr K-edge): confirms Cr(VI) → Cr(III) transformation	Permporn et al., 2023

Cr(VI) contributing to an average of 138.7 mg/kg (Sun et al., 2022). Leaching tests showed that the Cr(VI) fraction was up to 40% water soluble, indicating the potential for groundwater contamination (Sultanbayeva et al., 2025). X-ray photoelectron spectroscopy showed that Cr(VI) was present as calcium chromate ($CaCrO_4$) and Cr(III) as $Cr(OH)_3$ and Cr_2O_3 (Biesinger et al., 2004). Similar results have also been observed in water treatment investigations, in which magnetic amino-modified silica-coated glycidyl methacrylate nanocomposites not only adsorbed Cr(VI) species from solution, but also catalysed their reduction to Cr(III); maximum adsorption capacities up to 64 mg/g with pH 2 were reported (Suručić et al., 2023).

Natural aquatic environments are a further example of the sensitivity of chromium speciation to physicochemical conditions. At pH values less than 5, Cr(VI) is likely to be reduced to Cr(III) in the presence of organic matter and Fe(II), whereas at alkaline pH values greater than 8 in combination with oxidising redox potentials greater than + 500 mV, Cr(VI) is stable and mobile (Richard & Bourg, 1991). These dynamics underpin the fact that Cr(VI) is found predominantly in industrial effluents and alkaline waters, and Cr(III) is predominant in reducing environments with a high organic matter content.

2.2 *Toxicity and Environmental Impact*

The toxicological competencies of Cr(III) and Cr(VI) are extremely different. Cr(III) is known to be one of the micronutrients for glucose and lipid metabolites, but in trace levels. High concentrations of Cr(III) in soluble or complexed forms may still be toxic to biological systems (Peralta-Videa et al., 2022). In contrast, Cr(VI) is highly toxic, and has been listed as a Group 1 human carcinogen by the International Agency for Research on Cancer. Its toxicity is based on its capacity to cross the cellular membranes through the sulphate transport channels, and its intracellular reduction to Cr(V), Cr(IV) and Cr(III). This process leads to the production of reactive oxygen species and causes oxidative stress, DNA damage, mitochondrial dysfunction and finally carcinogenesis (Zhitkovich et al., 2021). The environmental implications of Cr(VI) contamination are equally worrisome. Laboratory and field studies do show substantial bioaccumulation and trophic transfer. For example, Singh et al. (2013) showed that Cr(VI) exposure of spinach plants at 100 uM concentration decreased chlorophyll concentration by 30%, decreased photosynthetic efficiency and induced increased lipid peroxidation. While maximum accumulation happened in roots, significant translocation to shoots was measurable and suppressed growth followed. Such plant-level impacts are not isolated, as chromium is easily absorbed by the food chain, finally impacting animals and humans. Animal studies have confirmed the existence of oxidative stress in hepatic and renal tissues after Cr(VI) exposure, consistent with the human health effects seen in the health effects in mining and industrial regions. A case in point is that of Sukinda Valley in India, which contains one of the largest reserves of chromite in the world, where it has been shown that chronic exposure is very severe. Communities who live near mining sites report high cases of skin ulceration, respiratory issues and cancers, which have been directly linked to Cr(VI) contamination in soil and water (Mohanty et al., 2023). Groundwater analysis near electroplating and tanning industries has shown that the concentration of Cr(VI) can be as high as 4.6 mg/L, and is almost two orders of magnitude higher than the limit recommended by the World Health Organisation for Drinking Water (0.05 mg/L) (Gezahegn et al., 2021). These findings certainly highlight the urgent need for remediation technologies that can reduce Cr(VI) exposure in the environmental and occupational exposure scenarios.

2.3 *Sources of Chromium Pollution*

Chromium contamination of aquatic systems results from a combination of natural and anthropogenic activities, but the latter is qualitatively and quantitatively the predominant source of contamination. Natural processes like weathering of ultramafic rocks, volcanic eruptions and soil erosion are major contributors of Cr(III) with negligible Cr(VI) under reductive conditions. Only in oxidising environments does Cr(III) change to Cr(VI), and such occurrences are fairly rare in unpolluted ecosystems (Apte et al., 2006). Anthropogenic activities, on the other hand, release enormous amounts of chromium, frequently under the form of Cr(VI). Electroplating wastewater typically has a Cr(VI) concentration of 10–200 mg/L. Tannery wastewater contains mixtures of Cr(III) (20–100 mg/L) and Cr(VI) (5–10 mg/L). Textile dyeing processes are also a major contributor with concentrations of 15–60 mg/L reported (De Silva et al., 2020; Kalsoom & Batool, 2020). In mining areas, like the Sukinda Valley, more than 30 million tonnes of waste rock and tailing with high concentration of Cr(VI) is dumped each year which causes contamination of the surrounding rivers where its concentration between 0.1 and .5 mg/L Cr(VI) have been recorded (Mohanty et al., 2023). In addition to water and soil pollution, atmospheric pollution is an often-overlooked pathway of chromium pollution. Combustion of coal and chromate production facilities emit airborne Cr(VI) that is adsorbed onto fine particulate matter. A study by Cha et al. (2025) found Cr(VI) concentrations higher in $PM_{2.5}$ of 6.5 to 25.7 ng/m^3 in various cities of China. Given the route of exposure (direct inhalation) these concentrations represent serious risks to respiratory health, especially in urban populations. All these various sources have led to a global burden of chromium pollution, with industrial discharges playing the most important role in human and ecological diversity.

3 Principles of Bioadsorption for Chromium Removal

3.1 Definition and Mechanisms

Bioadsorption is defined as a passive fixation of metal ions onto both living and dead biological surfaces through interaction of physicochemical between cell surfaces, functional groups and exterior cell polymers. Bioadsorption is an independent process of metabolism, unlike bioaccumulation that needs metabolism to work, and can enable a broad range of biomaterials to be used to remover metals in aqueous systems. Recent research provides a few prominent mechanisms which explain chromium sequestration.

One of the most common reported mechanisms is ion exchange, where Cr(VI) oxyanions get replaced with naturally existing anions like the sulphate or phosphate on the surface of bioadsorbent. As an illustration, readily exchangeable $HCrO_4^-$ and CrO_4^{2-} assay using a mixture of agricultural residues including peanut husks demonstrated that these two groups can bear an exchange without 90% efficiency at pH 2 (Liu et al., 2024). Electrostatic interaction is also quite essential, particularly in acidic conditions where protonation of surface groups can generate positive sites which attract negatively charged Cr(IV) species.

Surface complexation, where Cr species reacts covalently or coordinate with functional groups, including amino functional groups, hydroxyl functional groups and phosphate functional groups has been confirmed by FTIR and XPS studies. Biomass of fungus including *Aspergillus niger* has been demonstrated to bind Cr(VI) by carboxyl and amine functionalities and subsequent reduction to Cr (III), and subsequently becomes more firmly attached to the cell tool (Ahluwalia et al., 2010). This reduction is specifically necessary since the Cr(VI) can be reduced enzymatically or chemically into the less toxic Cr(III) during the adsorption process. As an example, simultaneous adsorption and reduction was observed in biosorbents onto *Chlorella vulgaris* microalgae and of up to 65% of the Cr(VI) converted to Cr(III) in contact with a biosorbent over an 48 h period (Chen et al., 2016).

There is the bioaccumulation mechanism of active uptake in living cells where chromium ions enter cell membranes and then accumulate intracellularly. It is a slower, energy-dependent process than passive biosorption but in engineered microbial systems can result in a higher uptake capacity. The critical intracellular reduction of Cr(VI) behaviour of genetically modified *Escherichia coli* strains expressing chromate reductase demonstrates strong unused possibilities of metabolic engineering in the future (Bose et al., 2023). Performance of modified biosorbents and composites in Cr(VI) adsorption studies is tablulated in Table 2.

Table 2 Performance of modified biosorbents and composites in Cr(VI) adsorption studies

SI no	Biosorbent/composite	Modification/ treatment	Test conditions	Outcomes	References
1	Fe_3O_4/vine-shoots activated carbon nanocomposite	Fe_3O_4 decorated on activated carbon	Optimum pH 3; 25 °C; high removal efficiency 90%; Freundlich fit	Very high capacity; good real-water performance	Bagherzadeh et al., 2023
2	*Craticula subminuscula* (freshwater microalga)	Raw algal biomass (biobased adsorbent)	Optimum pH 1.09; dose 10.9 $mg \cdot L^{-1}$; 25 °C; 129.5 min	Very large q_m; endothermic and spontaneous; 95% removal at optimum	Sbihi et al., 2024a, 2024b
3	Acid-modified banana peel (ABP)	H_2SO_4 acid treatment	pH 2–3 optimum; 20–50 °C tested; adsorption rapid (pseudo-second order)	High q_m after acid activation; partial reduction of Cr(VI) to Cr(III)	Huang et al., 2024
4	Chitosan-coated magnetic carbon (MC@CS)	Chitosan coated on magnetic carbon nanoparticles	pH 3; standard batch tests; good cyclic regeneration (≥70% after 10 cycles)	High removal at acidic pH; magnetic separation and reusability	Li et al., 2023
5	nZVI@chicken-manure biochar (nZVI@CMBC)	Nano-zero valent iron dispersed on biochar	Acidic conditions (optimum pH 2.2); measurements under N_2; multi-step kinetics	Strong removal due to combined reduction + adsorption; good synergy	Fang et al., 2022
6	Nano-zero-valent iron loaded corn-straw biochar (nZVI/BC)	nZVI supported on corn-straw biochar	Max removal at pH 2; pseudo-second-order kinetics; multi-mechanism (adsorption + reduction)	Composite outperforms nZVI or BC alone; up to 99% removal at optimal dose	Wei et al., 2022
7	Chitosan-modified *Mimosa pigra* biochar (CMPBC)	Chitosan grafting on *Mimosa pigra* biochar	Batch tests; pH acidic favourable; regeneration cycles reported	Chitosan modification significantly improves q and reusability vs raw biochar	Perera et al., 2023

3.2 *The Benefits of Bioadsorption*

Bioadsorption has quite a number of benefits over traditional physicochemical procedures like chemical precipitation, ion exchange resins and membrane filtration. It is cost efficient because a lot of bioadsorbents can be prepared as a result of some of the byproducts found in large quantities at agricultural or industrial activities which include sawdust, fruit peels or even microbial biomasses. Moreover, biosorbents are usually highly affined and selective towards chromium species in acidic situation. Bioadsorption does not need much input of energy as compared to high energy in use by chemical procedures in addition to producing minimal secondary pollution. A number of works have also shown bioadsorbent sorption capacities of 40–120 mg/g, which matches and even surpasses the sorption values of commercial activated carbons (Kolodynska et al., 2017; Hammad et al., 2025).

3.3 *Limitations*

In spite of these benefits, there are a number of limitations affecting the use of bioadsorption in industrial wastewater treatment. Biosorbent efficacy is highly pH-dependent, and the bonding activity is generally optimal at acidic pH (1–3), which is not representative of overall pH field of most industrial effluents. Sorption performance may be greatly compromised by competition with other co-existing anions which include sulphate, nitrate and phosphate anions. In addition, the repeated adsorption–desorption performances of raw biomass involve structural instability, which reduces reusability. The regeneration process necessitates a lot of harsh chemical agents, which undermine biosorbent integrity. Scaling up difficulties are not less, with the majority of works being carried out on a very small scale, in batch facilities, but a continuous flow at industrial level is required (Kumar et al., 2024).

4 Types of Bioadsorbents for Chromium Removal

4.1 *Agricultural and Industrial Byproducts*

Agricultural and industrial wastes are extensively investigated as contributing to a low-cost bioadsorbent in chromium removal due to their abundance and renewable quality as well as abundant in lignocellulosic substances capable of producing cellulose, hemicellulose and lignin. This is because their functional groups (hydroxyl, carboxyl, phenolic and amino groups) have the potential to bind chrome ions easily.

Using the citric acid in the chemical modification of banana peels, Langmuir isotherm modelling further established that the highest capacity of the compounds towards adsorption of Cr(VI) was 87.4 mg/g at pH 2 at maximum (Das et al., 2023).

Equally, the sugarcane bagasse biochar treated with nitric acid exhibited a higher sorption capacity of 112.6 mg/g and good reduction of Cr(VI) to Cr(III) which suggested dual adsorption-reduction bonds (Elhamdy et al., 2024). The byproducts of industries like fly ash and red mud have also been investigated but their lower surface area tends to be ineffective unless treated or modified.

4.2 Microbial Biomass

4.2.1 Bacterial Biomass

Recent experimental studies have reported many times combinations of hexavalent chromium in aqueous solution treated with bacterial biomasses in detoxified forms, where removal is influenced by species, pretreatment/immobilisation, contact time and pH and starting metal concentration in a highly dependent manner. One soil isolate of *Bacillus amyloliquefaciens* was indicated to realise rapid latch-up (nearing 79.9% removal in 60 min) with a Langmuir maximum capacity (q_m) of 48.44 mg.g-1 of Cr(VI) and excellent recoverability proved through desorption/regeration tests (Ramachandran et al., 2022). In another paper, *Pseudomonas alcaliphila* NEWG-2 1 abeled the growth attained approximately 96.6% elimination of 200 mgL^{-1} Cr(VI) in 48 h, and biosorption was confirmed by SEM and EDST, correspondingly, where the presence of functional surface groups was verified with FTIR (El-Naggar et al., 2020). Further techno-applied studies that have used immobilised bacteria or biopolymer/alginate bead formulations (e.g. of *Pseudomonas* spp. and native *Pseudomonas koreensis* biofilters) have reliably demonstrated high removal efficiencies, indicating that immobilisation increases mechanical stability and reusability, although the apparent (q_m) values might be slightly low in comparison to chemically modification sorbents (Diaz-Jimenez et al., 2023; El-Naggar et al., 2020). Mechanistic analyses of these recent studies (FTIR, SEM–EDX/EDS, desorption tests) recurringly emphasise the significance of cell-wall functional groups (carboxyl, hydroxyl, amino, phenolic and phosphoryl functional groups), electrostatic attraction at acidic conditions, and in some cases partial enzymatic or abiotic reduction Cr(VI) to Cr(III), and thereby reported capacities can only be meaningfully compared if experimental factors like species/biomass condition, pH, temperature, contact time and initial concentration are taken (Diaz-Jimenez et al., 2023; El-Naggar et al., 2020; Ramachandran et al., 2022).

4.2.2 Fungal Biomass

Biomasses of fungi have demonstrated significant ability to capture hexavalent chromium (Cr(VI)) in aqueous environment under acidic environments, but uptake differs significantly by organom, treatment and state of biomass. As a reference, Aspergillus niger spores reached the highest level of uptake of about 97.1 mg.g^{-1}

at pH 2.0, 40 40 °C with a starting Cr(VI) concentration of 300 $mg.L^{-1}$ (Ren et al., 2018). Elsewhere, an acid-treated *Trichoderma gamsii* biomass exhibited approximately 44.8 $mg.g^{-1}$ uptake of electroplating effluent with initial Cr(VI) concentration of approximately 500 $mg.L^{-1}$, and equilibrium was reached after approximately 240 min (Kavita & Keharia, 2012). Under optimum pH and conditions, Tunali et al. reported that *Neurospora crassa* had approximately 15.85 $mg.g^{-1}$ capacity in acid pre-treated form (Tunali et al., 2005). Reportedly waste mushroom biomass, e.g. *Calocybe indica* fruiting-body waste, treated chemically, have also been reported, with individual values of wes being valued at ~ 205.8 $mg.g^{-1}$ of Cr(VI) uptake by treated biomass and at least 72.85 $mg.g^{-1}$ of Cr(VI) uptake by untreated biomass under favourable conditions (Beig & Shah, 2022). These differences represent the variations in fungal species, pretreatment of biomass, pH of biosorption, temperature and metal concentration.

4.2.3 Algal Biomass

It has been reported repeatedly (most recently) in peer-reviewed sources (2020–2024) that algal biomasses (including algae (seaweeds), microalgae and algal-borne materials, e.g. alginate-residue, algal biochars and activated/carbonised), algal biomasses affordably and frequently act as effective, low-cost biosorbents capable of removing chromium (both Cr(VI) and Cr(III)) in aqueous solutions, but greener capacities and removal *Craticula subminuscula* freshwater microalga Canada Freshwater microalga *Craticula subminuscula* demonstrated tremendous high performance under optimised acidic circumstances with experimental Langmuir q_m values of approximately 277–296 $mg.g^{-1}$ (experimental q_m of approximately 277.6 $mg.g^{-1}$; Langmuir q_m of approximately 295.9 $mg.g^{-1}$) and validated a reasonably good Cr(VI) removal of approximately 95% at pH 1.1 (Sbihi et al., 2024a, 2024b). Hybrid systems of microalgae/activated-sludge bacteria (e.g. *Chlorella vulgaris*) were able to achieve very high Cr(VI) and total-Cr removal efficiencies (maximum Cr(VI) removals of about 91–93% under reactor conditions), with FTIR showing successful purification in carboxyl, carbonyl, amino and hydroxyl groups lowering removal efficiencies) and recovery rates (ahead of desorption cycles) of about 78%. As a crude biosorbent, coastal brown seaweed residues (*Sargassum* spp.) into activated carbons or chemically modified versions have been successfully used in both batch and fixed-bed systems and with significantly increased adsorption towards Cr(VI) (Alvarez-Galván et al., 2022; da Costa et al., 2022). The algal biomasses—in particular optimised microalgae and algal residues/activated carbons—can be scaled to tens to a number of hundreds of mg Cr/g biosorbent under laboratory conditions, and cannot be dismissed as promising so far unless pilot-scale testing, regeneration/desorption and speciation (Cr(VI) vs Cr(III)) studies are incorporated in development activities.

4.3 Chitosan and Derivatives

Chitosan and its chlorinated products are the frequently examined and useful biosorbent in the removal of chromium (Cr(III) and Cr(VI)). The richness of coloured and hydroxyl functionalities makes chitosan's bind chromium in exchange of ions, electrostatically attracted and complexed and chemically modified (grafting, crosslinking, quarterisation, magnetic loading, carbon-dot or MOF incorporation) which significantly enhances capacity, selectivity and reusability. Simple chitosan hydrogels in simple chitosan fibre shapes regularly indicate high performance in Cr(VI) uptake at acidic conditions (e.g. composite chitosan-cellulose-carbon-dot hydrogels with q_m 217.8 mg.g^{-1} and tetraethylenepentamine-crosslinked chitosan oligosaccharide hydrogels with $q_m > 150$ mg.g^{-1}), with performance clearly assumed through porous 3-D networks and introduced chelating locations. Anionic Cr(VI) can be concentrated by surfactant-supported and quaternised chitosan material components with good regeneration at low pH with one CS@CTAB formulation schemes approaching 490 mg.g^{-1} fully in the case of Cr(VI) although coupled regeneration is also effective (Akl et al., 2023). Nano-structured chitosan-based composites can be easily separateable, and adsorption may be combined with partial reductive detoxification through magnetic- and nano-structured chitosan variants and MOF-chitosan composites have been shown to enhance mechanical stability and viable recyclability with high capacities and demonstrated recently through review and experimentation (da Silva Alves et al., 2021; Wang et al., 2023). In most reports, selectivity often is further adjusted with new chitosan derivatives (e.g. carboxymethyl, xanthate-modified, Schiff-base and EDTA-grafted chitosans), although reported capacities vary with pH, initial concentration of Cr, contact time and definition of q_m value; so, it cannot be fairly compared with other new chitosan derivatives side-by-side unless all use identical experimental conditions (Muddin et al., 2024). Moreover, the research establishes that the chitosan-based sorbents are multipurpose, can be tuned and achieve extremely high levels of chromium uptake with optimisation as trade-offs of synthetic complexity, the regeneration process and cost, requirement that must be tackled in scale-up.

4.4 Biochar and Activated Carbons from Biomass

Under laboratory conditions, the hexavalent chromium (Cr(VI)) could be adsorbed using biochar and activated carbon materials (wastes (bagasse, rice straw, walnut shell, poultry manure, etc.)) or modified with iron/zero-valent iron and polymer Coatings. Altered biochars (such as been loaded with iron or subjected to amide/acid-treatment) tend to have significantly greater uptake and faster kinetics than pristine biochar. In this case, magnetic bagasse biochar (BMBC) reported a Langmuir Qmax of Cr(VI) was of about 29.1 mg.g^{-1} at 25 25 °C and was found to be reusable after several cycles (Liang et al., 2020). Organic studies of the biochar made of corn-stalk suggest that the adsorption of Cr(VI) is controlled by the electrostatic binding of the fraction and surface complexation, as well as filling pore and partial reduction to Cr(III) by organic contents (organic and inorganic) present in the sample (Guo et al., 2020). In general, fruit-waste and rice-straw biochars are produced chemically modified, yielding adsorption capacities and removals usually in wide range depending on the activation and functionalisation and meets pseudo-second-order kinetics and Langmuir/Freundlich isotherms in typical studies (Chen et al., 2023; Daffalla et al., 2023). The agro-waste (walnut shell, eucalyptus, etc.) has been shown to have a high removal efficiency (void of optimisation 1 shorter 90%): activated carbons frequently demonstrate high removal efficiencies (>90% under optimised conditions) and, in supplement with computational studies, provide good binding of Cr(VI) to oxygen and nitrogen-containing prototiles (Yazid et al., 2024). Composites that combine biochar with nanoscale zero-valent iron (nZVI@biochar) or with conducting polymers (biochar@PANI) achieve enhanced removal by coupling adsorption with reductive transformation of Cr(VI) to Cr(III), improving apparent capacities and practical performance in real waters (Fang et al., 2022; Yang et al., 2024).

5 Parameters that Affect Efficiency of Bioadsorption

Chromium removal is situations in which the capacity of bioadsorbents is significantly influenced by the parameters of operation and the environment. These parameters change both the chemical interactions of the chromium species and the functional groups of the bioadsorbent surface along with the equilibrium and the rate of adsorption.

5.1 pH

One of the most important parameters in chromium biosorption is the pH of the aqueous solution since it influences both the chromium speciation and ionisation of chemical groups on the bioadsorbent surface. Acidic conditions optimise adsorption in the case of hexavalent chromium because the conditions promote the prevalence of $HCrO_4^-$ species and active group protonation. According to Sbihi et al., (2024a, 2024b), *Craticula subminuscula* biomass reached the highest biosorption capacity of 277.57 mg/g in alkaloid conditions at pH 1.09, which is an indication of high electrostatic attraction (Sbihi et al., 2024a, 2024b). In the case of trivalent chromium, it is preferable to take it off under slightly acidic to neutral conditions. *Fucus vesiculosus* recorded the most uptake with 14.12 mg/g in comparison to other pH 5, this can be attributed to the interactions between the Cr^{3+} ions and the carboxyl as well as the hydroxyl groups on the algal surface (Asaad, 2024).

5.2 Preliminary Chromium Concentration

Primary chromium concentration determines the driving force of the mass transfers and the degree of the active sites saturation. A study by Zhang et al. (2024) explored the biomass of *Bacillus amyloliquefaciens* and was able to find it to remove 79.9% of Cr(VI) in a 50 mg/L solution in 60 min with a Langmuir maximum of 48.44 mg/g (Khan et al., 2025). Likewise, using research indicated that dead and living *Arthrobacter viscosus* biomass had concentration-dependent Cr(VI) uptake with sorption capacities increasing with an increase in the initial concentration of chromium, during both batch and column experiments (Silva et al., 2009).

5.3 Bioadsorbent Dosage

Concentration of bioadsorbent directly influences the chromium removal by offering increased binding sites. But, increased doses can cause a decrease in adsorption per unit mass because of underutilisation of sites. An adjustment in dosage makes methods of *Theobroma cacao* pericarp (TCP) biosorbent perform better at increasing the Cr(VI) removal capacity and decreasing the per-mass equilibrium without any change in specific capacity (q_e); observed TCP doses were 0.5–3.0 g.L^{-1} (for C_0 = 100 mg.L^{-1}, C_0 = 100 mg.L^{-1}, pH = 2, t = 180 min), and reported an optimal qmax = 48.5 mg·g^{-1} (at pH 2, dose = 0.5 g·L^{-1}, C_0 = 100 mg·L^{-1}) (Fernández-Pezúa et al., 2024).

5.4 Contact Time

Equilibrium reaction rate depends on contact time. Under ideal conditions (pH 6, 40 °C, 0.5 g/L biomass, contact time 60 min) *Bacillus subtilis* can remove Cr(VI) effectively on synthetic wastewater; the kinetics and the isotherm datasets can be described well by pseudo-second order and Langmuir models. Conversely, Sbihi et al., (2024a, 2024b) found that *Craticula subminuscula* took about 129 min to reach its maximum uptake, which indicated disparities in the pore structure of bioadsorbents, and diffusion resistance (Sbihi et al., 2024a, 2024b).

5.5 *Temperature*

Temperature affects biosorption via changing solubility, diffusivity and energy of adsorption reactions. Temperature affects biosorption via changing solubility, diffusivity and energy of adsorption reactions. The Langmuir uptake of Cr(VI) in *Saccharomyces cerevisiae* was found to equal to about 32.6 $mg.g^{-1}$ at 25 °C, and the biosorption rate diminished as temperature increased, implying that biosorption was exothermic (Ozer & Ozer, 2003). However, in some algal bioadsorbents, little or no adsorption with temperature change is observed; this points to preponderance of physical adsorption processes.

5.6 *Competing Ions*

Adsorption of chromium is usually hampered by the presence of other co-existing ions of wastewater because of intermolecular forces. Biomass-based adsorbents are highly sensitive to competing ions, which determine the chromium (Cr) sorption capacity. Biosorption was inhibited by the presence of zinc (Zn^{2+}) and cadmium (Cd^{2+}), with a stronger competitive action of Zn^{2+} than Cd^{2+} being observed with orange waste biomass (Pérez-Marín et al., 2024). In a similar manner, Macena et al. (2025) found that the presence of competing ions, including calcium (Cd^{2+}), magnesium (Mg^{2+}) and sodium (Na^{+}), reduced the biomass-based biosorbent capacity to remove Cr in aqueous outliers, noting that further studies are needed on the bioactivity and ionic composition of aqueous solutions when comparing the potential of biomass-based biosorbents in removing Cr (Macena et al., 2025).

5.7 *Surface Modifications*

Chemical modification and improvement of the optimisation processes can effectively boost chromium (Cr(VI)) biosorption with the assistance of biomass-based adsorbents. *Sargassum* sp. immobilised on alginate in a packed-bed column obtained the best Cr(VI) removal at the initial concentration of 25 $mg.L^{-1}$, Bed Height and Service Time modelled with service time using ANN-GA, and kinetic analysis modelled the adsorption coefficient with Bed Depth Service Time and Yoon-Nelson (Prabhu et al., 2020). In a similar fashion, chemically modified arecanut leaf sheath (CALS) achieved the most significant adsorption capacity of 109.89 mg/g at pH 2.0, according to pseudo-second-order kinetics and the Langmuir isotherm, and affording easy regeneration and reuse, which implies the potential success of chemical modification in improving Cr(VI) uptake (Pant et al., 2022). Collectively, these works indicate that the selection of biomass and both chemical or physical alteration of the biomass and the process of optimising operational conditions are essential in maximising Cr(VI) biosorption performance.

6 Modelling and Kinetics of Bioadsorption

Mathematical modelling and kinetic studies give understanding on the mechanisms of chromium uptake by bioadsorbents, which are used to predict the performance at various step of operation. Through the application of isotherm models, investigators are able to measure adsorption capacity and affinity, and kinetic models elucidate rate-limiting steps and may identify possible mechanisms, e.g. by film diffusion, intra-particle diffusion and chemisorption. The thermodynamic studies also indicate the spontaneity and energy variation of the biosorption processes and hence help in designing the effective treatment systems.

6.1 Adsorption Isotherms

Adsorption isotherms characterise how much chromium is adsorbed at equilibrium corresponding to the concentration of the chromium in the solution. The Langmuir model which relies on the notion of monolayer adsorption on a homogeneous surface has been extensively used in studies of biosorption. According to Sbihi et al., (2024a, 2024b), necessitating pattern *Craticula subminuscula* biomass Cr(VI) fitted the Langmuir model, necessitating a maximum capacity of 289.01 mg/g, demonstrating the good and uniform binding of chromium ions to functional sites on the algal surface (Sbihi et al., 2024a, 2024b). Conversely, the multilayer adsorption on heterogeneous surface as later described by Freundlich usually offers a superior fit to natural biomass. Asaad (2024) found that the biosorption of Cr(III) onto *Fucus vesiculosus* was most suitable to the Freundlich model, implying the participation of heterogeneous and energetically different adsorption space (Asaad, 2024). Other models have also been used to describe variations to the Langmuir and Freundlich assumptions including the Temkin and Redlich-Peterson model. The Temkin model showed that Cr(VI) adsorption by *Theobroma cacao* pericarp biomass could be described using Temkin model plus the adsorption energy distribution was affected by the adsorbent-adsorbate interactions (Fernández-Pezúa et al., 2024).

6.2 Kinetic Models

The kinetic behaviour of chromium bioadsorption by *Staphylococcus aureus* CC1956 showed that the pseudo-second-order kinetic model that refers to chemisorption by electronic exchange or sharing between functional groups of bacterial cell surface and the chromium ions explain the kinetic behaviour of adsorption process best. The significant correlation coefficients (R^2 values close to one) were taken to establish the validity of this model compared to pseudo-first-order kinetics and that the rate of biosorption is controlled by chemical interactions and not mere diffusion. Chemisorption mechanisms are also dominant and the role of hydroxyl, phosphoryl, carboxyl and amine groups in the binding, determined by the FTIR analysis, also testifies to this. Besides, the presence of chromium on the cell surface was confirmed by SEM–EDX, supporting the pseudo-second-order description of kinetics (Harboul et al., 2025). In a similar fashion, the biosorption of Cr(VI) with *Bacillus subtilis* was also found to exhibit pseudo-second-order kinetics in its equilibrium, with an equilibrium time of 60 min (Alsamhary, 2025). Pseudo-first-order, pseudo-second-order and Vermeulen models were used to determine the kinetics of adsorption of Cr(VI) onto the combined adsorption of tolerant bacteria and nanoparticles of zero-valent iron (nZVI). The findings showed that both, the pseudo-first-order and pseudo-second-order model did not effectively capture the adsorption process, but the Vermeulen model gave the best explanation of the kinetic data, which is of an intra-particle diffusion-controlled mechanism. The diffusion coefficient obtained in the Vermeulen model was 3.14×10^{-5} cm^2/s which confirmed the diffusion of Cr(VI) to the adsorbent pores is a gradual step in the reaction before surface binding. Furthermore, the equilibrium was established in 90 min, and the adsorbed bacteria-nZVI bi-adsorbent had the highest removal efficiency of up to 93% and high adsorption capacity relative to single bacteria or the nanoparticle (Garg et al., 2024). Elovich model has been successfully used in chemisorption of heterogeneous surfaces, which have high energies whereas intra-particle diffusion models tend to follow multi-stage adsorption, i.e. an initial adsorption on the surface after which the diffusion is slow into pores. Sbihi et al., (2024a, 2024b) established that the intra-particle diffusion played a major role in the biosorption of Cr(VI) using *Craticura subminuscula*, meaning that diffusion resistance was a partial controller of the kinetics (Sbihi et al., 2024a, 2024b).

6.3 Thermodynamic Insights

Thermodynamic measures of chromium biosorption give important evidence regarding the possibility, spontaneity and energy of the process. In the study by Musah et al. (2022), acid-pre-treated *Chlorella vulgaris* and *Spirulina platensis* were used as a biosorbent of Cr(VI), and it was found that the process was not spontaneous at low temperatures (positive $\Delta G°$, 2.59–1.75 kJ/mol), whereas the positive $\Delta H°$ (41.98 kJ/mol) and $\Delta S°$ (135.28 J/molK) indicated that it was a non-Conversely, Göçenoğlu Sarikaya (2019) reported that when using *Agaricus campestris* to biosorption Cr(VI), the process was found to be spontaneous and exothermic with a negative $\Delta G°$ (−1.635 kJ/mol at 45 °C) and $\Delta H°$ (−9.582 kJ/mol) values. The negative $\Delta S°$ (−24.992 J/mol·K) further showed that the system randomness was making the sorption decreasing, and more binding interactions are more orderly and stronger.

Collectively, these results indicate that the thermodynamics of chromium biosorption can differ significantly across the type of biosorbent and treatment, with the algal biomass endothermically and entropically adsorbing and the fungal biomass adsorbing enthalpically.

7 Advances in Bioadsorbent Design

Chromium bioadsorption has developed considerably since using crude and untreated biomass. Recent studies involve strategic engineering of advanced bioadsorbents to overcome the inherent weaknesses of low adsorption capacity, limited stability as well as selectivity. The main attributes to all these advancements are through chemical functionalisation, multiplexing of nanomaterials and the creation of multi-component advanced composite with the point of view of improving performance in order to be put into practise.

7.1 Chemical Modifications

The chromium adsorption performance is highly improved by chemical modification of biomass and increases its structural, surface and functional characteristics. The study by Zhao et al. (2020) showed that TiO_2-loaded pine biomass (Pine/TiO_2) possessed a Cr(VI) adsorption capacity of 12.8 mg/g, an order of magnitude higher than that of Cr(III), because the $TiOH_2^+$ functional groups, which enabled electrostatic binding and reduction of Cr(VI) to Cr(III) were introduced (Zhao et al., 2020). Correspondingly, Langmuir isotherm and pseudo-second-order kinetics of adsorption Ayele et al. (2022) found that phosphoric acid (H_3PO_4) and potassium hydroxide (KOH) functionalisation of Teff straw biosorbents considerably improved Cr(VI) efficiencies to 92.5% and 95.2%, respectively, under optimised conditions, and the adsorption follows the Langmuir isotherm and pseudo-second-order kinetics, which underpins Ekanayike et al. (2023) also demonstrated that citric acid-modified biochar made of *Mimosa pigra* had better Cr(VI) selectivity (78%) than raw biochar, with the process being endothermic and spontaneous with adsorption driven by the carboxyl and hydroxyl groups that were introduced by esterification (Ekanayake et al., 2023). Another strategy Ren et al. (2021) chemically adsorbed *Aspergillus niger* spores with cetyltrimethylammonium bromide (CTAB), and the result was that the adsorption capacity has significantly increased to 177 mg/g. The change added positive surface charges and secondary amine groups, thereby allowing electrostatic binding and reducing Cr(VI) to Cr(III), then coordination with functional groups (carboxyl and protonated amines) (Ren et al., 2021).

7.2 *Nanocomposites and Magnetic Bioadsorbents*

The combination of bioadsorbents and nanomaterials has brought about a new generation of high-performance hybrids which takes advanced advantage of the synergistic characteristics of the biological and synthetic components. One of the significant changes in this sphere is the creation of magnetic bioadsorbents. Magnetic bioadsorbents are a superior approach to Cr(VI) removal since they have a high adsorption capacity, are easily separated and reused. As an example, $Fe_3O_4^-$ loaded corn straw biochar (Fe + CSBC) attained a monolayer adsorption capacity of 138.8 $mg.g^{-1}$, and Cr(VI) adsorption efficiencies of 90.3% in groundwater and 72.6% in wastewater, with the recyclability of the compound in four reuse cycles (Chu et al., 2023). More recently, Cui et al. (2023) prepared a bifunctionalised magnetic biochar (BMBC) by functionalising rice-straw biochar with cystamine and glutaraldehyde. Under favourable conditions the BMBC removed 100% of Cr(IV) with the highest adsorption capacity of 66.10 $mg.g^{-1}$ at 50 °C and PH 2 and the combined action of reduction (Cr(VI) to Cr(III)) reduction, electrostatic attraction and complexation and retained 61.12% of efficiency after five cycles. In addition to magnetisation, nanocarbon-based hybridisations have also worked well: Cr(VI) removal through graphene oxide which includes- microalgae nanohybrids obtained efficiencies of 99.6% towards Cr(VI), which is a result of the combination of biomass reduction sites and the dense surface area of GO (Mulovhedzi et al., 2025). Functionalisation, magnetisation and nanohybrid techniques significantly enhance the efficiency of adsorption, selectivity and reusability of the biomass-based sorbent to chromium-contaminated wastewater.

7.3 *Ternary Nanoadsorbents*

Recently, ternary oxide nanoadsorbents are proven to be the effective materials in removing Cr(VI), due to its synergetic physicochemical characteristics, large surface area and high redox potential. Using an easy pathway, a new Fe_2O_3–MnO_2–SnO_2 nanocomposite exhibited the maximum Cr(VI) adsorption capacity of 69.2 mg/g^{-1} under the optimal conditions (pH 2, 298 K, 90 min). The pseudo-second-order kinetic model and Langmuir isotherm best described the adsorption process, as well as indicating a chemisorption-controlled monolayer process. Spontaneity and entropy-driven adsorption was demonstrated by thermodynamic analysis, and reusability tests allowed identifying that the material remained effective (>80%) over the course of four cycles. Notably, nanocomposite was able to remove Cr(VI) to 91.89% in tannery wastewater and therefore it has realistic application in the treatment of industrial effluent (Uddin et al., 2023). In the second strategy, ternary layered CuFeCr ternary hydroxides (L-CuFeCr) were prepared through a combination of a microwave hydrothermal reaction, in which nanosheet-layered structures of spheres were made with specific surface area of 100.3 $m^2{\cdot}g^{-1}$. The optimal adsorption capacity of the adsorbent was 22.24 mg·g^{-1} at pH 2, and the kinetics of adsorption were consistent with the pseudo-second-order equation and equilibrium parameters were consistent with the Langmuir isotherm, which characterises a monolayer chemisorption mechanism. The synergistic interaction in the LDH structure of Cu, Fe and Cr strengthened redox activity and ion exchange capacity, as the L-CuFeCr was a potential solution to Cr(VI) removal in aqueous systems (Zhang et al., 2021). Ternary nanoadsorbents, and in particular those that are co-loaded with transition metal oxides or layered double hydroxides, are best in adsorption capacity, redox characteristics and reusability relative to materials composed of just one or two components, and are therefore promising agents to use in chromium-contaminated wastewater treatment.

7.4 Green Synthesis

Along with the enhancement of performance, the great and intensive accentuation of enhancing the sustainability of the very process of synthesis is observed. Green synthesis involves the application of non-toxic processes that are normally used to prepare nanocomposites based on biological extract of plants, fungi or bacteria. These extracts include reducing and capping agents (e.g. polyphenols, flavonoids) which are able to transform metal salts to nanoparticles without cause of toxic chemicals. The strategy is in tandem with principles of green chemistry in terms of waste reduction and energy use. As illustration, an extract of green tea leaves was used to make a magnetic bioadsorbent which removed iron salts and coated Fe_3O_4 nano particles on a biochar base directly. In this sustainable synthesis, adsorbent is able to remove Cr(VI) in aqueous solution and manifest high reusabilities, along with the synthesis path which did not utilise conventional toxic agents (Anjum et al., 2023). This green tendency of synthesis guarantees that the lifecycle of bioadsorbent, its production to application, is environmentally friendly.

8 Challenges and Limitations

Although the development of bioadsorbents has made great progress and there is a good encouraging laboratory scale performance, there are a few real challenges in the translation of this development to large scale application of the technology in industry. These restrictions, which cut across the basic material properties, engineering and economic constraints, are among the major obstacles that should be countered in the forthcoming research and development studies.

8.1 Low Adsorption Capacity

Even though modified and composite bioadsorbents usually perform significantly better than non-modified versions, most of them are still prone to limitations in the cases of high-Cr(VI) industrial effluents. To give some examples, industrial electroplating wastewater may contain concentrations of Cr(VI) in the range of tens to hundreds and higher, where bioadsorbents frequently have adsorption capacities in tens to low hundreds of mg/g, and their removal efficiencies are often worse than would happen to synthetic Cr solutions—because of competing ions, organic fractions and blockage of the pores. When studying Fe@CSBC biochar, the removal percentage dropped down to approximately 72.6%compared to 90.3% in groundwater and to approximately 50.6% in real wastewater of comparable conditions and reuse cycles performance diminished slightly (Chu and Nguyen, 2023). These issues imply that very large quantities of adsorbent might be necessary to treat very high-concentration Cr(VI) wastewaters, and this raises both costs and operating load.

8.2 Environmental Sensitivity

The performance of bioadsorption is highly sensitive to environmental parameters especially pH and this poses an operational problem when it comes to treating wastewaters that have variable nature. Functional groups (e.g. amines, carboxyls, hydroxyls) are subject to considerable pH-dependent changes in protonation state, which can impact adsorption capacity and mechanism. The efficacy of bioadsorption is highly contingent on conditions within an environment, particularly the pH. Research case studies such as the *Vietnamese corncob biochar* (CCBC) indicate that the removal of Cr(VI) is greatest at pH 2.0 with declining removal efficiency with increases in pH. (Nguyen et al., 2024). Temperature is also a factor: at higher temperatures Cr(VI) uptake is frequently increased, which is also compatible with endothermic adsorption, but the precise effects depend on the adsorbent.

8.3 Scale-Up Issues

The engineering difficulties with moving a batch laboratory experiment to continuous fixed-bed or pilot-scale systems arise. Such studies as the one conducted by Cruz-Olivares et al. (2022) modelling the study with the use of the Bed Depth Service Time (BDST) model show that the parameters of the packed-bed design, in particular, bed height, flow rate and inlet concentration, should be optimised to properly delay the breakthrough and use the adsorption capacity efficiently. The experiments with Quercus crassipes acorn shell in fixed beds demonstrate that the removal capacity and the breakthrough curves are highly dependent on flow rate, bed height, pH and influent concentration (Aranda-García & Cristiani-Urbina, 2020). Although these studies define the significance of operational parameters in scale-up, structural resistance amid numerous adsorption–desorption cycles and comprehensive economic comparisons are less well-developed in the literature.

8.4 Disposal of Spent Bioadsorbents

Chromium-based spent adsorbents are a major concern in the environment that is still not well managed. Although regeneration and reuse are the most preferable, the majority of bioadsorbents suffer a significant capacity loss after 3–7 cycles and, at a certain point, may no longer be used and must be discarded. The question of the fate of the adsorbents contaminated with chromium is also still a pressing environmental challenge. Although regeneration can be enabled, the capacity decreases over successive cycles; in one study with modified biochar of spent bricks, Cr(VI) removal was reduced to around 93.7% by the third cycle (Daffalla, 2023). Pyrolysis of tannery sludge at high temperatures facilitates the stabilisation of chromium, whereby the compounds containing the Cr-bearing compounds are converted to more crystalline and inert mineral forms (including $(FeMg)Cr_2O_5$) and entrap them in the carbon skeleton, thus significantly diminishing the ratio of environmentally mobile Cr(VI) (Li et al., 2024). Tannery sludge is also converted into biochar, which enhances the incorporation of chromium into more resistant mineral and carbon structures, which minimises the risk of Cr(VI) (Li et al., 2024). Designing of efficient and safe disposal processes or valorisation of spent adsorbents is a key study requirement.

8.5 Lack of Mechanistic Understanding

Although kinetics and isotherms of adsorption have been widely studied, the underlying biomolecular specificities concerning the binding process are still elucidated poorly on a large number of bioadsorbent-chromium samples. The particular coordination chemistry between chromium species and functional groups, the influence of topography and pore architecture, and the interaction dynamics of concurrent adsorption-reduction processes are the aspects that still need to be better clarified with the help of advanced characterisation methods (Sinha et al., 2022; Tripathi et al., 2024). Such a knowledge gap hinders the logical engineering towards better adsorbents and prevents the predictive modelling accurateness of actual wastewater practises. Among them in particular, there is a lack of studies to collectively characterise adsorption processes using a combination of different analytical methods (XPS, EXAFS, FTIR, XRD) with environmentally relevant conditions.

9 Future Perspectives

The development of the chromium removal technology based on bioadsorbents is to cross the bridge between basic research and the practical one with the help of strategic combination of existing method with system-based approaches. Some of these possible areas of improvement exist amid existing constraints and would have the potential to enable radical advancements in remediation functions alongside managing the issue of economic and environmental sustainability.

9.1 Hybrid Technologies

The combined use of bioadsorption and other complementary treatment methods is the least distant route to better chromium remediation. Combined adsorption-membrane filtration also demonstrates specific potential, wherein bioadsorbents receive a wastewater treatment to minimise chromium levels before subjecting it to end polishing with nanofiltration or reverse osmosis. This method reduces substantially membrane contamination, as well as prolongs the working life. Recent developments in photocatalytic-bioadsorbent hybrids exhibit synergy, help of semiconductor materials (e.g. TiO_2, ZnO) over bioadsorbents to adsorbate and reduce Cr(VI) to Cr(III) simultaneously by visible light, which is said to be indeed synergistic. Photocatalysts of such systems generate regeneration and preserve high removal efficiencies with repeated cycles without any chemical regenerants (Du et al., 2024).

9.2 Genetic Engineering

Genetic engineering and synthetic biology have presented game-changing strategies to improve bioadsorbent functions on a molecular scale. Genetic engineering has facilitated the improvement of chromium tolerance, reduction and adsorption in microorganisms. As an illustration, genetically engineered *Escherichia coli* that expressed human metallothionein (MT2A) removed approximately 89% of 25 mg/L Cr(VI) and reduced approximately 76% under said conditions (*E. coli* MT2A) (Akkurt et al., 2022). In another study, the chromate reductase gene (ChrR) of *Pseudomonas putida* was transferred into *E. coli* DH5 of cells, producing transformants ready to reduce Cr(VI) significantly in tannery effluent conditions than was possible with wild-types (Sundararaj et al., 2023). M52 strains of Recombinant *Sporosarcina saromensis* that contain orf2987, as well as identified in boosting and reducing Cr(VI), have been confirmed by protein expression and surface analyses (An et al., 2022).

9.3 Artificial Intelligence and Machine Learning

The idea of computer-aided optimisation of bioadsorption systems based on multivariate experimental data is reconstitutively being performed through machine learning (ML) or neural network models. To illustrate, an ANN model made it possible to make accurate predictions of Cr(IV) elimination by algal biochar on textile effluents even in the case of different pH, dose of adsorbent and duration of exposure (Khan et al., 2023). Metal–organic framework (MOF) adsorbents were also screened using ML: models were trained based on structural characteristics of the MOF and experimental data of Cr(VI) adsorption were used to identify promising adsorbents with an appropriate pore size, surface area and cavity sizes (Chang & Lee, 2024; Jiang et al., 2024). Equally, an ANN model was trained to simulate Cr(VI) extraction by nanoparticle-loaded biochar (MnO_2 and Fe_2O_3 on cellulose biochar) under different conditions (pH, temperature, etc.), which enhanced predictive power and minimised experimental workload (Belcaid et al., 2024).

9.4 Circular Economy Approaches

A shift towards the models of circular economy is one of the most essential changes in the modelling and implementation of bioadsorbent. Research aims at developing value chains in which waste biomass in one industry can be used as raw material in making adsorbents and spent chromium-contaminated adsorbents are reused as an additive in construction material or as catalysts in industry. Life cycle assessment procedures are being incorporated into adsorbent designing procedures to make sure that its environmental effects would be net-positive. Advanced regeneration modes, such as electrochemical and photocatalytic processes, allow a tendency to reuse a number of times without significant production of secondary wastes. Such strategies make remediation neither a cost centre but a possible revenue stream with resource recovery as well as waste valorisation (Kokkinos & Zouboulis, 2020).

9.5 Policy and Community Engagement

The practise of education system is known as policy and community engagement, which is mentioned in the policy abbreviations. The goal of the successful use of bioadsorption technologies will be the simultaneous development of the policy frameworks and the methods of stakeholder engagement. Commercial adoption and regulatory approval will be facilitated by development of standardised testing protocols and certification systems of bioadsorbent materials. The market penetration can be expedited with policy instruments such as green procurement mandates, tax incentives on sustainable technologies and long-term producer responsibility regulations.

Models of community-based participatory research will ensure that the developed solution considers the local contextual variables, and the culture especially in areas that are highly impacted by chromium pollution. Clear information on performance indicators and environmental goodwill fosters the requisite confidence between the regulators, industries and the communities (Kokkinos & Zouboulis, 2020; Madela and Skuza, 2021). The intersecting capabilities of these multidisciplinary strategies, in terms of materials science, biotechnology, computational modelling, economic systems and social science bringing forth new opportunities to devise effective, sustainable and socially responsible remedies to chromium contamination. In a bid to maximise bioadsorption technologies, future studies should give priority to combined strategies that consider technical performance and implementation issues.

10 Conclusion

The detailed analysis of chromium removal with the help of bioadsorbents discloses not only impressive improvement but also a lot of difficulties on the way to sustainable solutions of wastewater treatment. Even more recent studies have shown significant enhancement of adsorption capacities using advanced material engineering strategies, such as the design of chemical functionalization, nanocomposites and hybrid systems. These developments have been used to stretch the limits of bio-based materials performance levels and a number of the modified bioadsorbents can now perform on a par with conventional synthetic adsorbents in terms of performance in particular applications. The development of simple biomass into high ternary composites and genetically modified materials is a paradigm shift in thought process and a transition driven more by active characteristics rather than passive adsorption to produce the capability of selective recognition and transformation of chromium species and their recovery. The use of magnetism elements has solved such important issues that constitute solid–liquid separation, and production of adsorbents has become more environmentally conscientious by utilising green syntheses. These advancements have continued to shrink the performance difference between bioadsorbents and traditional technologies of treatment at the expense of not losing benefits of renewable sourcing and lower environmental impact. Nevertheless, there have been continued impediments to the transfer of laboratory success to industrial use pertaining to scale, economic feasibility and stability of the system under real wastewater regimes. The high sensitivity of bioadsorption processes to environmental factors, counter ions and organic components necessitates complex process management strategies, which can make the implementation complex. The recycle of used adsorbents is also not a well-discussed issue that should be better considered in the next research rounds. The future will probably rely on the achievements of interdisciplinary solutions, integrating materials science with biotechnology, computer modelling and circular economy principles. Working together with machine learning and artificial intelligence should provide new avenues to optimise the design of adsorbents and their use, and genetic engineering could facilitate radically new approaches to chromium cleaning using biology. Finally, policy frameworks, economic models and stakeholder engagement practises will have to be developed in tandem with bioadsorption technologies to achieve the successful implementation of bioadsorption technologies and to establish enabling environments of sustainable innovation.

References

Ahluwalia, S. S., & Goyal, D. (2010). Removal of Cr (VI) from aqueous solution by fungal biomass. *Engineering in Life Sciences, 10*(5), 480–485.

Akkurt, Ş, Oğuz, M., & Alkan Uçkun, A. (2022). Bioreduction and bioremoval of hexavalent chromium by genetically engineered strains (*Escherichia coli* MT2A and *Escherichia coli* MT3). *World Journal of Microbiology and Biotechnology, 38*, 45. https://doi.org/10.1007/s11274-022-03235-2

Akl M. A., et al. (2023). Surfactant-supported chitosan for efficient removal of Cr(VI) and anionic food-stuff dyes from aquatic solutions. *Scientific Reports, 13*, 15786. https://doi.org/10.1038/s41598-023-43034-9

Alsamhary, K. E. (2025). Optimizing the process conditions for the biosorption of chromium (VI) by *Bacillus subtilis* in artificial wastewater. *Electronic Journal of Biotechnology, 76*, 57–65. https://doi.org/10.1016/j.ejbt.2025.03.005

Alvarez-Galván Y., Minofar B., Futera Z., Francoeur M., Jean-Marius C., Brehm N., Yacou C., Jauregui-Haza U. J., Gaspard S. (2022) Adsorption of hexavalent chromium using activated carbon produced from Sargassum ssp.: Comparison between lab experiments and molecular dynamics simulations. *Molecules, 27*, 6040. https://doi.org/10.3390/molecules27186040

Amesho, K.T., Shopati, A.K., Shangdiar, S., Kadhila, T., Iikela, S., Sithole, N.T., Edoun, E.I. and Chowdhury, S. (2024). Integrating circular economy and sustainability strategy in water remediation applications. In *Circular Economy* (pp. 1–26). CRC Press.

An, Q., Zhang, M., Guo, D., Wang, G., Xu, H., Fan, C., Li, J., Zhang, W., Li, Y., Chen, X., & You, W. (2022). Cr (VI) removal by recombinant *Escherichia coli* harboring the main functional genes of *Sporosarcina saromensis* M52. *Frontiers in Microbiology, 13*, Article 820657. https://doi.org/10.3389/fmicb.2022.820657

Anjum, A., Mazari, S.A., Hashmi, Z., Jatoi, A.S., Abro, R., Bhutto, A.W., Mubarak, N.M., Dehghani, M.H., Karri, R.R., Mahvi, A.H. and Nasseri, S. (2023) A review of novel green adsorbents as a sustainable alternative for the remediation of chromium (VI) from water environments. *Heliyon, 9*(5).

Apte, A. D., Tare, V., & Bose, P. (2006). Extent of oxidation of Cr (III) to Cr (VI) under various conditions pertaining to natural environment. *Journal of Hazardous Materials, 128*(2–3), 164–174.

Aranda-García, E., & Cristiani-Urbina, E. (2020). Hexavalent chromium removal and total chromium biosorption from aqueous solution by *Quercus crassipes* acorn shell in a continuous up-flow fixed-bed column: Influencing parameters, kinetics, and mechanism. *PLoS ONE, 15*(1), Article e0227953. https://doi.org/10.1371/journal.pone.0227953

Asaad, A. A. (2024). Sorption of chromium from aqueous solutions using Fucus vesiculosus algae biosorbent. *BMC Chemistry, 18*(1), 145.

Ayach, J., El Malti, W., Duma, L., Lalevée, J., Al Ajami, M., Hamad, H., & Hijazi, A. (2024). Comparing conventional and advanced approaches for heavy metal removal in wastewater treatment: An in-depth review emphasizing filter-based strategies. *Polymers, 16*(14), 1959.

Ayele, A. L., Tizazu, B. Z., & Wassie, A. B. (2022). Chemical modification of Teff straw biomass for adsorptive removal of Cr(VI) from aqueous solution: Characterization, optimization, kinetics, and thermodynamic aspects. *Journal of Chemistry, 2022*, 1–12. https://doi.org/10.1155/2022/5820207

Bagherzadeh, M., Aslibeiki, B., & Arsalani, N. (2023). Preparation of Fe_3O_4/vine-shoots derived activated carbon nanocomposite for improved removal of Cr(VI) from aqueous solutions. *Scientific Reports, 13*, 3960. https://doi.org/10.1038/s41598-023-31015-x

Bakhtiari, S., Salari, M., Shahrashoub, M., Zeidabadinejad, A., Sharma, G., & Sillanpää, M. (2024). A comprehensive review on green and eco-friendly nano-adsorbents for the removal of heavy metal ions: Synthesis, adsorption mechanisms, and applications. *Current Pollution Reports, 10*(1), 1–39.

Bandh, S. A., & Mushtaq, B. (2025). Drinking water treatment process. In *Wastewater Treatment Technology: Technical Progressions and Advances* (pp. 165–227). Cham: Springer Nature Switzerland.

Beig, S.-U.-R., & Shah, S. A. (2022). Biosorption of Cr(VI) by acid-modified based-waste fungal biomass from *Calocybe indica* fruiting bodies production. *International Journal of Phytoremediation, 25*(10), 1269–1288.

Belcaid, A., Beakou, B. H., Bouhsina, S., & Anouar, A. (2024). Biosorption of cobalt and chromium from wastewater using manganese dioxide and iron oxide nanoparticles loaded on cellulose-based biochar: Modeling and optimization with machine learning (artificial neural network). *International Journal of Biological Macromolecules, 282*, Article 136855. https://doi.org/10.1016/j.ijbiomac.2024.136855

Biesinger, M. C., Brown, C., Mycroft, J. R., Davidson, R. D., & McIntyre, N. S. (2004). X-ray photoelectron spectroscopy studies of chromium compounds. *Surface and Interface Analysis: An International Journal Devoted to the Development and Application of Techniques for the Analysis of Surfaces, Interfaces and Thin Films, 36*(12), 1550–1563.

Bose, R., Kulkarni, S., Lakkakula, J., & Wagh, N. S. (2023). Genetic engineering for chromium removal. In *Chromium in Plants and Environment* (pp. 139–166). Cham: Springer Nature Switzerland.

Cha, Z., Zhang, X., Zhang, K., Zhou, G., Gao, J., Sun, S., Gao, Y., & Liu, H. (2025). Atmospheric heavy metal pollution characteristics and health risk assessment across various type of cities in China. *Toxics, 13*(3), 220.

Chang, J., & Lee, J.-Y. (2024). Machine learning-based prediction of the adsorption characteristics of biochar from waste wood by chemical activation. *Materials, 17*(21), 5359. https://doi.org/10.3390/ma17215359

Chen, Y., Yang, J., & Abbas, A. (2023). Enhanced chromium (VI) adsorption onto waste pomegranate-peel-derived biochar for wastewater treatment: Performance and mechanism. *Toxics, 11*(5), 440. https://doi.org/10.3390/toxics11050440

Chen, Z., Song, S., & Wen, Y. (2016). Reduction of Cr (VI) into Cr (III) by organelles of Chlorella vulgaris in aqueous solution: An organelle-level attempt. *Science of the Total Environment, 572*, 361–368.

Chu, T. T. H., Nguyen, M. V., et al. (2023). Improved Cr(VI) adsorption performance in wastewater and groundwater by synthesized magnetic adsorbent derived from Fe_3O_4 loaded corn straw biochar. *Environmental Research, 216*(Pt 4), Article 114764. https://doi.org/10.1016/j.envres.2022.114764

Cruz-Olivares, J., Pérez-Alonso, C., Martínez-Barrera, G., Roa-Morales, G., López-Téllez, G., & del Campo-López, E. M. (2022). Modeling and scaling up of the Cr (VI) adsorption process by using mexicalcite natural mineral in a packed bed column. *Results in Engineering, 16*, Article 100687.

Cui, X., Wang, J., Zhao, Q., Li, C., Huang, J., Hu, X., Li, J., Li, M., et al. (2023). Application of a novel bifunctionalized magnetic biochar to remove Cr(VI) from wastewater: Performance and mechanism. *Separations, 10*(6), 358. https://doi.org/10.3390/separations10060358

da Costa, T. B., da Silva, T. L., Costa, C. S. D., da Silva, M. G. C., & Vieira, M. G. A. (2022). Chromium adsorption using Sargassum filipendula algae waste from alginate extraction: Batch and fixed-bed column studies. *Chemical Engineering Journal Advances, 11*, Article 100341. https://doi.org/10.1016/j.ceja.2022.100341

da Silva Alves, D. C., Healy, B., Pinto, L. A. D. A., Cadaval, T. R. S. A., Jr., & Breslin, C. B. (2021). Recent developments in chitosan-based adsorbents for the removal of pollutants from aqueous environments. *Molecules, 26*(3), 594. https://doi.org/10.3390/molecules26030594

Da Silva, G. S., Dos Santos, F. A., Roth, G. C. L. C., & Frankenberg, C. L. C. (2020). Electroplating for chromium removal from tannery wastewater. *International Journal of Environmental Science and Technology, 17*(2), 607–614.

Daffalla, S. (2023). Adsorption of Chromium (VI) from aqueous solution using palm leaf-derived biochar: Kinetic and isothermal studies. *Separations, 10*(4), 260. https://doi.org/10.3390/separations10040260

Das, L., Das, P., & Bhowal, A. (2024). Recent advances in potential application of biomass waste and biomass-derived materials for bioremediation of pollutants from wastewater. *Circular Economy*, 127–139.

Das, R., Saha, S., & Roymahapatra, G. (2023). Removal of hexavalent chromium ion (Cr^{6+}) from industrial effluents using low cost bioadsorbents: A review. *Education in Chemical Science and Technology, 12*, 53–60.

Diaz-Jimenez, L., Garcia-Torres, S., & Carlos-Hernandez, S. (2023). High adsorption of hazardous Cr(VI) from Water using a biofilter composed of native *Pseudomonas koreensis* on alginate beads. *International Journal of Environmental Research and Public Health, 20*(2), 1385. https://doi.org/10.3390/ijerph20021385

Du, Y., Ye, X., Hui, Z., Jiao, D., Xie, Y., Chen, S., & Ding, J. (2024). Synergistic effect of adsorption-photocatalytic reduction of Cr(VI) in wastewater with biochar/TiO_2 composite under simulated sunlight illumination. *Physical Chemistry Chemical Physics, 26*, 15891–15901. https://doi.org/10.1039/D4CP01226G

Ekanayake, A., Rajapaksha, A. U., Ahmad, M., & Vithanage, M. (2023). Enhanced adsorption of hexavalent chromium from aqueous solution by citric acid-modified biochar from invasive plant biomass. *Water, Air, and Soil Pollution, 234*, 461. https://doi.org/10.1007/s11270-023-06456-9

Elhamdy, W. A. (2024). Sustainable phosphorous porous biochar generated from sugarcane bagasse for effective adsorption coupled reduction of hexavalent chromium: Evaluating the method's greenness. *Inorganic Chemistry Communications, 169*, Article 112931.

El-Naggar, N. E. A., El-Khateeb, A. Y., Ghoniem, A. A., El-Hersh, M. S., & Saber, W. I. A. (2020). Innovative low-cost biosorption process of Cr^{6+} by *Pseudomonas alcaliphila* NEWG-2. *Scientific Reports, 10*, 14043. https://doi.org/10.1038/s41598-020-70473-5

Fang, S., Huang, X., Xie, S., Du, J., Zhu, J., Wang, K., Zhuang, Q., & Huang, X. (2022). Removal of chromium (VI) by a magnetic nanoscale zerovalent iron–assisted chicken manure-derived biochar: Adsorption behavior and synergetic mechanism. *Frontiers in Bioengineering and Biotechnology, 10*, Article 935525. https://doi.org/10.3389/fbioe.2022.935525

Fernández-Pezua, M., Lavado-Meza, C., De la Cruz-Cerrón, L., Gamarra-Gómez, F., Sacari-Sacari, E., Lavado-Puente, C., & Dávalos-Prado, J. Z. (2024). Biosorption of Cr(VI) by *Theobroma cacao* pericarp. *Environmental Science and Pollution Research, 31*, 59700–59711. https://doi.org/10.1007/s11356-024-34971-7

Garg, S., Singh, S., Khan, N. A., Samuel, J., Ramamurthy, P. C., & Singh, J. (2024). Equilibrium and kinetic modeling of Cr(VI) removal by novel tolerant bacteria species along with zero-valent iron nanoparticles. *Scientific Reports, 14*, 8611. https://doi.org/10.1038/s41598-024-57835-z

Gezahegn, A. M., Feyessa, F. F., Tekeste, E. A., & Beyene, E. M. (2021). Chromium laden soil, water, and vegetables nearby tanning industries: Speciation and spatial distribution. *Journal of Chemistry, 2021*(1), 5531349.

Göçenoğlu Sarıkaya, A. (2019). Kinetic and thermodynamic studies of the biosorption of Cr(VI) in aqueous solutions by *Agaricus campestris*. *Environmental Technology, 42*(1), 72–80. https://doi.org/10.1080/09593330.2019.1620867

Guo, X., Liu, A., Lu, J., Niu, X., Jiang, M., Ma, Y., Liu, X., & Li, M. (2020). Adsorption mechanism of hexavalent chromium on biochar: Kinetic, thermodynamic, and characterization studies. *ACS Omega, 5*(42), 27323–27331. https://doi.org/10.1021/acsomega.0c03652

Gürsoy, S., Zeytinci, N. K., Zaman, B. T., Bakırdere, S., & Öztürk Er, E. (2025). Study of linear and nonlinear isotherm and kinetic parameters of hexavalent chromium adsorption onto reduced graphene oxide coated iron oxide. *Scientific Reports, 15*(1), 25206.

Hammad, W. A., Hawash, S. A., Abdel-Latif, M. S., Kuku, M., & Amr, M. H. A. (2025). High efficiency adsorption of hexavalent chromium using bioderived activated carbon kinetics, isotherms, and thermodynamics. *Scientific Reports, 15*(1), 25871.

Harboul, K., Amakdouf, H., Hammani, K., & El-Karkouri, A. (2025). Efficient removal of Cr(VI) ions by *Staphylococcus aureus* CC1956: Biosorption isotherm and kinetic modeling. *BioMetals*. https://doi.org/10.1007/s10534-025-00734-1

Huang, Z., Campbell, R., & Mangwandi, C. (2024). Kinetics and thermodynamics study on removal of Cr(VI) using acid-modified banana peel adsorbents. *Molecules, 29*, 990. https://doi.org/10.3390/molecules29050990

Jiang, M., Fu, W., Wang, Y., Xu, D., & Wang, S. (2024). Machine-learning-driven discovery of metal–organic framework adsorbents for hexavalent chromium removal from aqueous environments. *Journal of Colloid and Interface Science, 662*, 836–845. https://doi.org/10.1016/j.jcis.2024.02.084

Kalsoom, A., & Batool, R. (2020). Biological and nonbiological approaches for treatment of Cr (VI) in Tannery Effluent. In *Emerging Eco-friendly Green Technologies for Wastewater Treatment* (pp. 147–170). Singapore: Springer Singapore.

Kavita, B., & Keharia, H. (2012). Biosorption potential of *Trichoderma gamsii* biomass for removal of Cr(VI) from electroplating industrial effluent. *International Journal of Chemical Engineering*, 2012.

Khan, A. A., Naqvi, S. R., Ali, I., Arshad, M., AlMohamadi, H., & Sikandar, U. (2023). Algal-derived biochar as an efficient adsorbent for removal of Cr (VI) in textile industry wastewater: Non-linear isotherm, kinetics and ANN studies. *Chemosphere, 316*, Article 137826. https://doi.org/10.1016/j.chemosphere.2023.137826

Khan, M. T. A., Al-Battashi, H., Al-Hinai, M., Almdawi, M., Pracejus, B., Elshafey, E. S. I., & Abed, R. M. (2025). Isolation of Aerobic heterotrophic bacteria from a microbial mat with the ability to grow on and remove hexavalent chromium through biosorption and bioreduction. *Applied Biochemistry and Biotechnology, 197*(1), 94–112.

Kokkinos, E., & Zouboulis, A. I. (2020). The chromium recovery and reuse from tanneries: A case study according to the principles of circular economy. In *Leather and Footwear Sustainability: Manufacturing, Supply Chain, and Product Level Issues* (pp. 123–157). Singapore: Springer Singapore.

Kołodyńska, D., Krukowska, J. A., & Thomas, P. (2017). Comparison of sorption and desorption studies of heavy metal ions from biochar and commercial active carbon. *Chemical Engineering Journal, 307*, 353–363.

Kumar, P., Prajapati, K. B., Mahajan, A. K., Pant, D., Meena, N. K., Kumar, A., & Kumar, P. (2024). Evaluating feasibility of biosorption technique for heavy metals removal: Limitations and future perspective. *International Journal of Environmental Analytical Chemistry, 104*(18), 6401–6425.

Kumari, A., Kamaraj, N., Selvaraj, R., & Nanoth, R. (2025). Emerging trends and future outlook on chromium removal in the lab, pilot scale, and industrial wastewater system: An updated review exploring 10 years of research. *Environmental Monitoring and Assessment, 197*(5), 547.

Li, Z., Yu, D., Wang, X., Liu, X., Xu, Z., & Wang, Y. (2024). A novel strategy of tannery sludge disposal–converting into biochar and reusing for Cr (VI) removal from tannery wastewater. *journal of environmental sciences, 138*, 637–649.

Li, Z., et al. (2023). Chitosan-modified magnetic carbon nanomaterials for chromium removal. *Environmental Science and Pollution Research*. https://doi.org/10.1007/s11356-023-25302-3

Liang, M., Chen, M., Song, Q., Wang, Z., Yin, D., & Wang, D. et al. (2020) Removal of aqueous Cr(VI) by magnetic biochar derived from bagasse. *Scientific Reports*, 10(2020). https://doi.org/10.1038/s41598-020-78142-3

Liu, B., Xin, Y. N., Zou, J., Khoso, F. M., Liu, Y. P., Jiang, X. Y., Peng, S., & Yu, J. G. (2023). Removal of chromium species by adsorption: Fundamental principles, newly developed adsorbents and future perspectives. *Molecules, 28*(2), 639.

Liu, Y., Liang, Y., Cui, W., Zhai, H., & Ji, M. (2024). Efficient removal of Cr (VI) from wastewater by magnetic biochar derived from peanut hull. *Water, Air, and Soil Pollution, 235*(2), 100.

Macena, M., Pereira, H., Cruz-Lopes, L., Grosche, L., & Esteves, B. (2025). Competitive adsorption of metal ions by lignocellulosic materials: A review of applications, mechanisms and influencing factors. *Separations, 12*(3), 70. https://doi.org/10.3390/separations12030070

Madeła, M., & Skuza, M. (2021). Towards a circular economy: Analysis of the use of biowaste as biosorbent for the removal of heavy metals. *Energies, 14*(17), 5427.

Mohanty, S., Benya, A., Hota, S., Kumar, M. S., & Singh, S. (2023). Eco-toxicity of hexavalent chromium and its adverse impact on environment and human health in Sukinda Valley of India: A review on pollution and prevention strategies. *Environmental Chemistry and Ecotoxicology, 5*, 46–54.

Muddin, N. A. I., Badsha, M. M., Arafath, M. A., Merican, Z. M. A., & Hossain, M. S. (2024). Magnetic chitosan nanoparticles as a potential bio-sorbent for the removal of Cr (VI) from wastewater: Synthesis, environmental impact and challenges. *Desalination and Water Treatment, 319*, Article 100449. https://doi.org/10.1016/j.dwt.2024.100449

Mulovhedzi, R., Yokwana, K., Ajiboye, T. O., Ogunlaja, A. S., Mhlanga, S. D., et al. (2025). A novel structured graphene oxide@microalgae nanohybrids as adsorbents for removal of Cr(VI) ions in aqueous solutions. *Adsorption, 31*, 63. https://doi.org/10.1007/s10450-025-00614-7

Musah, B. I., Wan, P., Xu, Y., Liang, C., & Peng, L. (2022). Biosorption of chromium (VI) and iron (II) by acid-based modified *Chlorella vulgaris* and *Spirulina platensis*: Isotherms and thermodynamics. *International Journal of Environmental Science and Technology, 19*, 11087–11102. https://doi.org/10.1007/s13762-021-03873-3

Nguyen, D.-K., Ly-Tran, Q.-B., Dinh, V.-P., Duong, B.-N., Nguyen, T.-P.-T., & Tuyen, P. N. K. (2024). Adsorption mechanism of aqueous Cr(VI) by Vietnamese corncob biochar: A spectroscopic study. *RSC Advances, 14*, 39205–39218. https://doi.org/10.1039/D4RA07455F

Nishad, V., Kumar, S., & Sastry, S. V. A. R. (2025). Kinetics, isotherms and thermodynamics studies of Cr (VI) removal using zero-valent iron nanoparticles synthesized from Aegle marmelos (Bael) plant extract. *Nanotechnology for Environmental Engineering, 10*(3), 66.

Ozer, A., & Ozer, D. (2003). Comparative study of the biosorption of Pb(II), Ni(II) and Cr(VI) ions onto *Saccharomyces cerevisiae*: Determination of biosorption heats. *Journal of Hazardous Materials, 100*(1–3), 219–229. https://doi.org/10.1016/S0304-3894(03)00109-2

Pant, B. D., Neupane, D., Paudel, D. R., Gautam, S. K., Pokhrel, M. R., & Poudel, B. R. (2022). Efficient biosorption of hexavalent chromium from water by modified arecanut leaf sheath. *Heliyon, 8*(4), Article e09283. https://doi.org/10.1016/j.heliyon.2022.e09283

Pérez-Marín, A. B., Ortuño, J. F., Aguilar, M. I., Lloréns, M., & Meseguer, V. F. (2024). Competitive effect of zinc and cadmium on the biosorption of chromium by orange waste. *Processes, 12*(1), 148. https://doi.org/10.3390/pr12010148

Prabhu, A. A., Chityala, S., Jayachandran, D., Deshavath, N. N., & Veeranki, V. D. (2020). A two step optimization approach for maximizing biosorption of hexavalent chromium ions (Cr(VI)) using alginate immobilized *Sargassum* sp. in a packed bed column. *Separation Science and Technology, 55*(1), 90–106. https://doi.org/10.1080/01496395.2019.1708933

Putra, N. R., Zaini, M. A. A., Kusuma, H. S., Darmokoesoemo, H., & Faizal, A. N. M. (2025). Advances in chromium removal using biomass-derived activated carbon: A comprehensive review and bibliometric analysis. *Environmental Progress and Sustainable Energy, 2025*, Article e14598.

Ramachandran, G., Chackaravarthi, G., Rajivgandhi, G. N., Quero, F., Maruthupandy, M., Alharbi, N. S., Kadaikunnan, S., Khaled, J. M., & Li, W.-J. (2022). Biosorption and adsorption isotherm of chromium (VI) ions in aqueous solution using soil bacteria *Bacillus amyloliquefaciens*. *Environmental Research, 212*, Article 113310. https://doi.org/10.1016/j.envres.2022.113310

Ren, B., Jin, Y., Zhao, L., Cui, C., & Song, X. (2021). Enhanced Cr(VI) adsorption using chemically modified dormant *Aspergillus niger* spores: Process and mechanisms. *Journal of Environmental Chemical Engineering, 9*(5), Article 106955. https://doi.org/10.1016/j.jece.2021.106955

Ren, B., Zhang, Q., Zhang, X., Zhao, L., & Li, H. (2018). Biosorption of Cr(VI) from aqueous solution using dormant spores of *Aspergillus niger*. *RSC Advances, 8*, 38157–38165.

Richard, F. C., & Bourg, A. C. M. (1991). Aqueous geochemistry of chromium: A review. *Water Research, 25*(7), 807–816. https://doi.org/10.1016/0043-1354(91)90160-R

Sbihi, K., Elhamji, S., Lghoul, S., Aziz, K., El Maallem, A., Mabrouki, J., El-Sheekh, M., & Aziz, F. (2024b). Biosorption of hexavalent chromium by freshwater microalgae *Craticula subminuscula* from aqueous solutions. *Sustainability, 16*(2), 918. https://doi.org/10.3390/su16020918

Sbihi, K., Elhamji, S., Lghoul, S., Aziz, K., El-Maallem, A., Mabrouki, J., El-Sheekh, M., & Aziz, F. (2024a). Biosorption of Hexavalent Chromium by Freshwater Microalgae Craticula subminuscula from Aqueous Solutions. *Sustainability, 16*, 918. https://doi.org/10.3390/su1602 0918

Sharma, P., Singh, S. P., Parakh, S. K., & Tong, Y. W. (2022). Health hazards of hexavalent chromium (Cr (VI)) and its microbial reduction. *Bioengineered, 13*(3), 4923–4938.

Silva, B., Figueiredo, H., Neves, I. C., & Tavares, T. (2009). Biosorption of chromium(VI) by dead and living Arthrobacter viscosus biomass. *Chemical Engineering Journal, 149*(1–3), 139–145. https://doi.org/10.1016/j.cej.2008.10.010

Singh, H. P., Mahajan, P., Kaur, S., Batish, D. R., & Kohli, R. K. (2013). Chromium toxicity and tolerance in plants. *Environmental Chemistry Letters, 11*(3), 229–254. https://doi.org/10.1007/s10311-013-0407-5

Singh, A., & Vijayan, J. G. (2024). Biosorbents for wastewater treatment. In *Application of Nanotechnology for Resource Recovery from Wastewater* (pp. 254–294). CRC Press.

Sinha, R., Kumar, R., Sharma, P., Kant, N., Shang, J., & Aminabhavi, T. M. (2022). Removal of hexavalent chromium via biochar-based adsorbents: State-of-the-art, challenges, and future perspectives. *Journal of Environmental Management, 317*, Article 115356. https://doi.org/10.1016/j.jenvman.2022.115356

Sultanbayeva, G., Kaiynbayeva, R., Chernyakova, R., Temel, H., Jussipbekov, U., & Tassibekov, K. (2025). Sustainable chromium remediation: sorption of chromium from leaching solutions of refined. *Sustainability, 17*(6), 2726.

Sun, J., Luo, Y., Ye, J., Li, C., & Shi, J. (2022). Chromium distribution, leachability and speciation in a chrome plating site. *Processes, 10*(1), 142.

Sundarraj, S., Sudarmani, D. N. P., Samuel, P., & Sevarkodiyone, S. P. (2023). Bioremediation of hexavalent chromium by transformation of *Escherichia coli* DH5α with chromate reductase (ChrR) genes of *Pseudomonas putida* isolated from tannery effluent. *Journal of applied microbiology*, *134*(1), lxac019. https://doi.org/10.1093/jambio/lxac019

Suručić, L., Janjić, G., Marković, B., Tadić, T., Vuković, Z., Nastasović, A., & Onjia, A. (2023). Speciation of hexavalent chromium in aqueous solutions using a magnetic silica-coated amino-modified glycidyl methacrylate polymer nanocomposite. *Materials, 16*(6), 2233.

Tripathi, M., Pathak, S., Singh, R., Singh, P., Singh, P. K., Shukla, A. K., Maurya, S., Kaur, S., & Thakur, B. (2024). A comprehensive review of lab-scale studies on removing hexavalent chromium from aqueous solutions by using unmodified and modified waste biomass as adsorbents. *Toxics, 12*(9), 657. https://doi.org/10.3390/toxics12090657

Tunali, S., Kiran, I., & Akar, T. (2005). Chromium(VI) biosorption characteristics of *Neurospora crassa* fungal biomass. *Minerals Engineering, 18*(5–6), 681–689.

Uddin, M. N., Saha, G. C., Hasanath, M. A., Badsha, M. A. H., Chowdhury, M. H., Islam, A. R. M. T., et al. (2023). Hexavalent chromium removal from aqueous medium by ternary nanoadsorbent: A study of kinetics, equilibrium, and thermodynamic mechanism. *PLoS ONE, 18*(12), Article e0290234. https://doi.org/10.1371/journal.pone.0290234

Wang, K., Zhang, F., Xu, K., Che, Y., Qi, M., & Song, C. (2023). Modified magnetic chitosan materials for heavy metal adsorption: A review. *RSC Advances, 13*, 6713–6736. https://doi.org/10.1039/D2RA07112F

Wei, Y., Chu, R., Zhang, Q., et al. (2022). Nano zero-valent iron loaded corn-straw biochar for efficient removal of hexavalent chromium. *RSC Advances, 12*, 26953. https://doi.org/10.1039/D2RA04650D

Xia, S., Song, Z., Jeyakumar, P., Shaheen, S. M., Rinklebe, J., Ok, Y. S., Bolan, N., & Wang, H. (2019). A critical review on bioremediation technologies for Cr (VI)-contaminated soils and wastewater. *Critical Reviews in Environmental Science and Technology, 49*(12), 1027–1078.

Yang, Z., Wang, J., Zhao, N., Pang, R., Zhao, C., Deng, Y., Yang, D., Jiang, H., Wu, Z., & Qiu, R. (2024). A novel biochar-based 3D composite for ultrafast and selective Cr (VI) removal in electroplating wastewater. *Biochar, 6*(1), 46. https://doi.org/10.1007/s42773-024-00338-x

Yazid, H., Bouzid, T., El Mouchtari, E. M., Bahsis, L., El Himri, M., Rafqah, S., & El Haddad, M. (2024). Insights into the adsorption of Cr(VI) on activated carbon prepared from walnut shells: Combining response surface methodology with computational calculation. *Clean Technologies, 6*(1), 199–220. https://doi.org/10.3390/cleantechnol6010012

Zeng, W., Luo, Y., Wang, C., Liu, Z., Xue, M., & Xie, X. (2025). A review on recent advances of adsorption and photoreduction of Cr (VI). *Environmental Pollutants and Bioavailability, 37*(1), 2493057.

Zhang, Y., Jing, C., Zheng, J., Yu, H., Chen, Q., Guo, L., Pan, D., Naik, N., Shao, Q., & Guo, Z. (2021). Microwave hydrothermal fabrication of CuFeCr ternary layered double hydroxides with excellent Cr(VI) adsorption. *Colloids and Surfaces a: Physicochemical and Engineering Aspects, 628*, Article 127279. https://doi.org/10.1016/j.colsurfa.2021.127279

Zhao, J., Boada, R., Cibin, G., & Palet, C. (2020). Enhancement of selective adsorption of Cr species via modification of pine biomass. *Science of the Total Environment, 719*, Article 143816. https://doi.org/10.1016/j.scitotenv.2020.143816

Zhigalenok, Y., Tazhibayeva, A., Kokhmetova, S., Starodubtseva, A., Kan, T., Isbergenova, D., & Malchik, F. (2025). Hexavalent chromium at the crossroads of science, environment and public health. *RSC Advances, 15*(27), 21439–21464.

MIX
Papier aus verantwortungsvollen Quellen
Paper from responsible sources
FSC® C105338

If you have any concerns about our products,
you can contact us on
ProductSafety@springernature.com

In case Publisher is established outside the EU,
the EU authorized representative is:
Springer Nature Customer Service Center GmbH
Europaplatz 3, 69115 Heidelberg, Germany

Printed by Libri Plureos GmbH
in Hamburg, Germany